STRANGE SPACES

Strange Spaces
Explorations into Mediated Obscurity

Edited by
ANDRÉ JANSSON
Karlstad University, Sweden

and

AMANDA LAGERKVIST
Uppsala University, Sweden

LONDON AND NEW YORK

First published 2009 by Ashgate Publishing

Published 2016 by Routledge
2 Park Square, Milton Park, Abingdon, Oxon OX14 4RN
711 Third Avenue, New York, NY 10017, USA

Routledge is an imprint of the Taylor & Francis Group, an informa business

British Library Cataloguing in Publication Data
Strange spaces : explorations into mediated obscurity.
   1. Space and time in mass media. 2. Spatial behavior.
   3. Environmental psychology. 4. Geographical perception.
   I. Jansson, André. II. Lagerkvist, Amanda, 1970-
   302.2'3'01-dc22

Library of Congress Cataloging-in-Publication Data
Jansson, André.
   Strange spaces : explorations into mediated obscurity / by André Jansson and Amanda Lagerkvist.
      p. cm.
   Includes bibliographical references and index.
   ISBN 978-0-7546-7461-0 (hardback) 1. Space. 2. Space--Psychological aspects. 3. Space--Social aspects. 4. Space perception. 5. Space and time in mass media. I. Lagerkvist, Amanda, 1970- II. Title.

   HM654.J36 2009
   304.2'3--dc22

                                                                          2009005400

ISBN 9780754674610 (hbk)

Contents

List of Figures

Notes on Contributors

Phil Carney has trained in medicine and psychiatry and has also taken degrees in medical sociology and criminology. In the process of conversion to a book, his doctoral thesis, *The Punitive Gaze*, uses photographic, critical and post-structuralist theory to examine case studies in which the photograph is both the scene and means of punishment. In contributions to the growing field of cultural criminology he has written on photography and the spectacle, and on critical aspects of cultural criminology. A lecturer at the University of Kent, he teaches in the fields of crime, media and culture, the sociology of crime and deviance, and also coordinates the international Common Study Programme in Critical Criminology.

Staffan Ericson is a PhD in Media and Communications, and an Associate Professor at Södertörn University, Sweden. His dissertation from 2004 compares the dramatic forms of modernist playwright August Strindberg and television dramatist Dennis Potter and how the dream-play form is integrated in Raymond Williams' and Walter Benjamin's theories on mass culture and modernism. He is currently the co-leader of the project 'Media Houses: Architecture, Media and the (Re-)centralisation of Power', which is studying the buildings of major media institutions in a historical context (the results of which are to be presented in a forthcoming anthology, published by Peter Lang, New York).

Ylva Habel, PhD, is a Research fellow at the Department of Cinema Studies, Stockholm University, Sweden. She wrote her dissertation, 'Modern Media, Modern Audiences: Mass Media and Social Engineering in the 1930s Swedish Welfare State', in 2002. Her forthcoming book focuses upon postcolonial and intermedial aspects of the Swedish reception of the artists Josephine Baker and Paul Robeson.

Cynthia Imogen Hammond teaches architectural history in the Department of Art History, Concordia University, Montreal. Her doctoral thesis on the role of women in the architectural history of world heritage site, Bath, England (2002), won the Canadian Governor General's gold medal for excellence. Her research continues to engage with questions of urban images, public memory and architectural heritage in diverse sites, such as Pécs, Hungary and Montreal, Canada. Hammond has published several monographs on the impact of non-architects on the built environment, including influential British hospital reformer, Florence Nightingale (1820–1910), and American housing activist, educator and urban planner, Catherine Bauer Wurster (1905–1964). In addition to her research and teaching, Hammond maintains a studio practice and has recently co-founded a design firm, *pouf! art + architecture*. Hammond's current research project, 'Renegade Ornament: Museum Cities and the Artful Urban Gesture', explores practices of heritage and civic regeneration as they relate to temporary, site-specific and often anonymous urban interventions.

Rikke Hansen is a researcher based at Tate Britain. Her educational background is in art and philosophy. She has taught Critical Studies and Photographic Studies at Norwich School of Art and Design, and Contemporary Critical Studies in the Department of Art at Goldsmiths, University of London. She is an art critic and a regular contributor to the UK-based journal *Art Monthly*. Her current research centres on the interface between animal studies and twentieth-century aesthetics.

Steven Jacobs, PhD, is an art historian specialized in the photographic and cinematic representations of architecture, cities, and landscapes. He has published in journals such as *Andere Sinema*, *Exposure*, *History of Photography*, *Journal of Architecture*, *De Witte Raaf* and in numerous books and catalogues. As a member of the Ghent Urban Studies Team (GUST), he co-edited and co-authored *The Urban Condition: Space, Community, and Self in the Contemporary Metropolis* (1999) and *Post Ex Sub Dis: Urban Fragmentations and Constructions* (2002). In 2007, he published *The Wrong House: The Architecture of Alfred Hitchcock* (010 Publishers, Rotterdam). He has taught at several universities and art schools in Belgium and the Netherlands. He currently teaches film history at Saint-Lukas College of Art Brussels, the Academy of Fine Arts in Ghent and the University of Antwerp.

André Jansson is a Professor of Media and Communication Studies at Karlstad University, Sweden. He is the editor of *Geographies of Communication: The Spatial Turn in Media Studies* (with Jesper Falkheimer, 2006) and has recently published articles in journals such as *Journal of Visual Culture*, *European Journal of Cultural Studies* and *Space and Culture*. He currently leads two research projects: 'Rural Networking/Networking the Rural', studying mediatization and cultural identity in the Swedish countryside, and 'Secure Spaces', studying the intersections of media, consumption and surveillance in social life.

Amanda Lagerkvist has a PhD in Media and Communication Studies and is Research Fellow in North American Studies at Uppsala University, Sweden. She works within the fields of media studies, urban theory and transnational American Studies. Centering on sensory-emotive and experiential dimensions of communication geographies, she has in a number of articles addressed the relationships between travelling and mediation and the spatio-temporalities of media cities, in particular the co-workings of media, memory and futurity in New Shanghai. Her ongoing research project engages American expatriate spaces and mobilities in the city. She has published in the *International Journal of Cultural Studies*, *The Senses and Society*, *European Journal of Communication* and *Space and Culture* (forthcoming).

Orvar Löfgren is Professor of European Ethnology at the University of Lund. His main field is the cultural analysis of everyday life and he is currently leading a project on 'the cultural dynamics of the inconspicuous' as well as participating in a multidisciplinary study of 'the management of overflow'. He has worked with studies of domestic consumption, tourism and travel, as well as studies of national identities and transnational processes. His latest books include *Off the Edge: Experiments in Cultural Analysis* (together with Richard Wilk, 2006) and *Doing Nothing: An Ethnography of Waiting, Daydreaming and other Non-events* (together with Billy Ehn, in press).

Vincent Miller is a Lecturer in Sociology at the University of Kent. His main research interests are in the areas of space, representation and community (offline and online). He has published a number of articles on representation, community and vagueness within a variety of urban contexts and cultures, and has looked at the roles of search engines and 'phatic' communication within online digital culture.

Lisa Parks, PhD. is Associate Professor and Chair of the Film and Media Studies Department at University of California-Santa Barbara. She is the author of *Cultures in Orbit: Satellites and the Televisual* and co-editor of *Planet TV: A Global Television Reader*. She has also published essays in numerous anthologies and journals and is currently working on two new books: *Coverage: Media Space and Security after 911* (forthcoming, Routledge) and *Mixed Signals: Media Infrastructures and Cultural Geographies*. In addition, she is co-editing *Down to Earth: Satellite Technologies, Industries and Cultures* with James Schwoch (forthcoming, Rutgers University Press).

David L. Pike teaches Literature and Film at American University. His most recent books are *Subterranean Cities: The World beneath Paris and London, 1800–1945* and *Metropolis on the Styx: The Underworlds of Modern Urban Culture, 1800–2001*, both from Cornell University Press. He is co-editor of the *Longman Anthology of World Literature,* and has published widely on nineteenth- and twentieth-century urban literature, culture, and film. He is currently writing a history of Canadian cinema since 1980, to be published by Wallflower Press, and a study of subterranean settings in film.

Michael Sappol is curator-historian at the National Library of Medicine (National Institutes of Health), Bethesda, MD, USA, and the author of *A Traffic of Dead Bodies: Anatomy and Embodied Social Identity in 19th-Century*

America (Princeton University Press, 2002) and *Dream Anatomy* (GPO, 2006). He received a PhD in history from Columbia University in 1997. In 2008 he curated *An Iconography of Contagion*, an exhibition on twentieth-century health posters at the National Academy of Sciences, and was a fellow-in-residence at the Clark Art Institute. His current work focuses on twentieth-century modernist medical illustration and displays, nineteenth-century medical narrative, and the history of medical film.

Johanne Sloan is an Associate Professor in the Department of Art History, Concordia University, in Montreal, Canada. She is currently completing further research on postcards, including the essay 'Postcards and the chromophilic visual culture of Expo 67', to be published in the book, *Expo 67: Not Just a Souvenir*, edited by Rhona Richman Kenneally and Johanne Sloan (University of Toronto Press, 2009).

Will Straw is Professor within the Department of Art History and Communications Studies at McGill University in Montreal. He is the author of *Cyanide and Sin: Visualizing Crime in 50s America* and over 70 articles on music, film, cities and popular culture. A collection of his articles on popular music, *Popular Music: Scenes and Sensibilities*, is forthcoming from Duke University Press. Dr Straw's current research focuses on tabloid newspapers in New York in the early twentieth century.

Chris Wilbert teaches Geography and Tourism at Anglia Ruskin University, (Chelmsford and Cambridge) England. He is co-editor of *Technonatures: Environments, Technologies, Spaces and Places in the Twentieth First Century* (Wilfred Laurier University Press, 2009); a special issue of *Science as Culture* on 'Technonatural Time-Spaces' (2006); and a Reader of Colin Ward's writings (for AK Press, 2009) with Damian White. His work has also focused on geographies of human-animal interactions, avian flu in SE Asia, and environmental politics. He currently researches cultural geographies of tourism to do with animals and food, as well as film and media.

Eva Åhrén, PhD, is a historian dealing with the cultural history of the body: visual and museological aspects of anatomy and pathology; representations of the body and the uses of media in medicine; dealings with the dead body. She is currently working on the history of the anatomical museums of the Karolinska Institute in Stockholm. Her book *Death, Modernity, and the Body: Sweden 1870–1940* is forthcoming at Rochester University Press, 2009. Åhrén is a Research Fellow at the Department of the History of Science and Ideas at Uppsala University and Research Associate at the Smithsonian Institution, Washington, DC.

Chapter 1
What is Strange about Strange Spaces?

André Jansson and Amanda Lagerkvist

Introduction

When crossing the Malaccan strait between Malaysian Penang and Indonesian Sumatra travellers board a ferry – or perhaps a facsimile of a ferry – that used to commute between the two Swedish islands of Öland and Gotland in the Baltic Sea in the 1970s and 80s. Now ostensibly transposed to this part of Southeast Asia, the boat is a *strange space*, refurbished with worn-out and ill-suited airplane seats, reeking of cigarette smoke. By force of being in the wrong place, these seats carry heavy connotations of disaster: in (and even *under*) water, they seem to tell a story about a recent or ominous emergency. The boat travels in waters repeatedly hit by pirate attacks in one of the most important cargo routes for shipping trade in the world. Such circumstances bring additional thrill or fear to this space, invoking a cinematic sense of emotional estrangement. News reports about capsized and submerged ferries in this area add to the strangeness of these ferry lines on a symbolic and representational level; at least for certain foreign travellers. For Swedish tourists spending their past summers on Öland or Gotland, additional obscurity is afforded to the experience since the relocated boats evoke the sunny memories of other holiday geographies, far away and far back in time.

This ferry constitutes an example of what Michel Foucault suggestively termed a *heterotopia*: a space in which spatial and temporal discrepancies converge and where different sites that are in themselves incompatible or irreducible to one-another are juxtaposed (Foucault 1967/1998). The boat, Foucault argues, is the 'heterotopia par excellence'. As it moves it is nowhere and somewhere at the same time. It is also a finite space which is a reserve for infinite imaginations: 'The boat is a floating piece of space, a place without a place, that exists by itself, that is closed in on itself and at the same time is given over to the infinity of the sea'. Foucault holds that boats fuel cultural fantasies and that '[i]n civilizations without boats dreams dry up, espionage takes the place of adventure, and the police take the place of pirates' (1967/1998, 244). The arcane and semiotically overloaded space of the ferry is not only a symbolic 'watershed', a transitional vessel that moves back and forth between one national and cultural terrain and another, but also a space of displacement on several levels, bringing about an immensely obscure spatial experience.

   This book approaches those bewildering and sometimes unspeakably bizarre spaces where disruption or disarray leave social subjects *estranged* and out of place. It engages the *emotional* and *mediated* geographies of uncertainty and in-betweenness; of cognitive displacement, loss, fear or exhilaration. It expands on *why* space is sometimes estranging and for *whom* is it strange. What kinds of perceptual, material and mediated transformations render space strange and obscure? What does it mean to be in estrangement?

   In the *Oxford Dictionary of the English Language* estrangement has nine entries: to remove something from its familiar place; to make someone a stranger to a condition of place; to withhold from a person's perception or knowledge; to render alien; to alienate in feeling or affection; to make unlike oneself; to render strange or unfamiliar in appearance; to be astonished. Such definitions have in common the sense that strange spaces are produced by and produce moments when we are faced with *a transformed state of affairs*: either by a dazzling light, a sudden flash, or by gradual unfolding of moments foreboding something unknown or new. Overlapping with affections evoked by otherness, such as the 'exile', the 'obscene', the 'deviant' or the 'queer', strange spaces also call for a separate discussion ranging from unplanned spaces to vanishing spaces in states of ruin or decay; from spaces of decadence or disorder to spaces glowing with celebrification and wonder. Strange spaces are here conceived of in terms of *change,* involving processes when the consciousness registers a form of loss or difference 'as the habitual suddenly or by degrees is transformed into the site of exile, discomfort, and sometimes novelty, astonishment and awe' (Smith 1996, 4).

   This is to say that in this book we want to approach strangeness not merely in terms of the 'unfamiliar' or 'other', but rather as a psycho-cultural spectrum of spatial opacity, marked not only by interpretive conflict or surprise, but also by an embodied sense of paradox, bewilderment and moral unease. Still, given these demarcations, there is of course a great body of relevant literature covering aspects of both strangeness and spatial production; literature waiting to be reinterpreted and combined in new ways. Similarly, empirical explorations of the mediated aspects of strange spaces, or the strangeness of media spaces, are dispersed and most often non-explicit. The purpose of this chapter, then, is to offer a realm of theoretical associations and linkages in order to initiate a profound discussion of 'what is strange about strange spaces'; a question that each of the following 14 chapters will provide empirically grounded, yet partial answers to. We begin with a broad exploration of views, concepts and methodologies within social and cultural theory, under the headings 'Space is Strange' and 'Everyday Elsewheres'. A discussion of the role of 'communication', 'mediation' and 'representation' follows, aiming to position the book within the emerging field of *communication geography*. Finally, we present the thematic structure of the book.

Space is Strange

Following Henri Lefebvre, this volume places the *opaque* at the centre of spatial exploration (1974/1991). Inaugurating an analysis of the socially produced interplay between spatial practices, representations of space and lived experience, Lefebvre showed that while representations of space and representational spaces seemingly naturalized space, they simultaneously made them made them contradictory, even making them evaporate. Lefebvre argued that 'the illusion of immanence' had made us believe that space was utterly and straightforwardly *knowable* while in fact it is encoded by multiple and differing meanings. And although representations of space express a lot, Lefebvre maintains that they hide, lose and set aside a great deal more – that is, those unfathomable, vague and strange dimensions of spatiality.

If opacity informs the fundamental human experience of space, for Julia Kristeva, estrangement and exile lie at the heart of subjectivity. Subjective existence is unhomely and the most intense forms of estrangement are produced by poetic language (Kristeva 1991). The poetic word takes subjects beyond themselves when language compensates for an original loss of the mother. The landscapes of literature are inhabited by a foreignness which makes all of us travelling there imaginatively into exiles: in order to take pleasure in reading, one needs to let go of the habitual, to become strangers to ourselves. What is strange about strange spaces, however, lies beyond imaginative travels and subject formation by means of *reading experiences* and arises out of a shift in perception and a change of the state of affairs, involving the whole body, the senses and the emotions. But their strangeness also, as this volume seeks to explore, have a great deal to do with the current transitions in the media landscape and the momentous and significant mediatization of society during the twentieth century. Coming to terms with strange spaces thus forces our attention on to a particular form of spatial obscurity: mediated obscurity. In our media age, incongruities sometimes occur because of a kind of overburdening of space by means of representation. Thirdspaces (as will be further elaborated below) are the real-and-imagined spaces where the mythological and symbolic landscapes of mediation converge with physical spaces resulting in ambivalences, ambiguities and conflations between the material and mediated (Soja 1996). Obscurity is thus increasingly, we argue, a mediated and mediatized trait of our media age and *Strange Spaces* pays particular attention to the relationships between space and media and to the specificities of mediated strangeness. Moving from physical to textual/visual/ mediated spaces into representational spaces or mediatized outer space, or further into the unutterable thirdspaces, nowheres and underworlds, this book endeavours to unearth both the architectonics and agencies of strange spaces.

Keeping in play these uncertainties and opacities of space, the volume seeks to expand on Foucault's enigmatic notion of the *heterotopia*, a concept which may allow for strange and unforeseen explorations. The heterotopia is a counter-site that simultaneously expresses core ideals and values of a respective society. A heterotopia is an 'effectively enacted utopia in which the real sites, all the other real sites that can be found within the culture, are simultaneously represented, contested and inverted' (1967/1998: 239). These counter-sites, existing in every society, *are there*. They are both openly acknowledged as existing, yet not freely accessible; they are institutionalized places, places which are outside of all other places, but nevertheless linked to them (such as the cemetery, the Chinese garden, the ward or the museum). Heterotopias come in two forms: heterotopias of crisis (hidden or forbidden places of people in crisis or of disorder: adolescents or menstruating women) and of deviance (rest homes and prisons for example).

What is *strange* about the heterotopias? Some specific heterotopian traits may crucially inform our discussion of strange spaces. Heterotopias are subject to historical changes and they may function in different ways at different times; they juxtapose irreconcilable sites within themselves and they come about *when people are displaced* through a break with the ordinary routines of temporality. What is strange about them is further the twofold way that they either create a space of illusion which discloses that the real world is even more illusory or the way that they create an *other space*; which makes us aware of the incompleteness of our messy and jumbled spaces (Foucault 1967/1998, 243).

Heterotopias are known physical spaces with certain predictable functions. Strange spaces may however appear to us in ways we never expected. They are not even in our wildest dreams. We neither had knowledge of such strange spaces, nor could we ever have imagined their existence. They shake our spirits. Still the strangeness of space sets in motion a search for apprehending something else 'beyond each plane surface, beyond each opaque form' (Lefebvre 1974/1991, 183). This is why strange spaces, that leave us speechless and bemused, are neither impenetrable nor completely unknowable: 'Everything knows itself, but not everything says itself, publicizes itself. Do not confuse silence with secrets! That which is forbidden from being said, be it external or intimate, produces an obscure, but not a secret, zone' (Lefebvre 1992/2004, 17). Taking our cue from Lefebvre, we are in search of whys, hows and whereabouts of estranging spaces – of their social production. Michel Foucault, who was most of all concerned with the prohibitions of/in space and with the margins of spatial organization (the prison, the cemetery, the psychiatric clinic) shed light on *deviance* in a way that contributes to our inquiry into strange spaces: deviance may sometimes be both the product and the process of estrangement. Resembling this is Lefebvre's notion of the power invested in space and the power exerted on subjects *by* space: 'Activity in space is restricted by that space, space "decides" what activity may occur, but even this "decision" has limits placed upon it. Space lays down the law because it implies a certain order and hence also a certain disorder (just as what may be seen defines what is *obscene*) [...]' (1974/1991, 143). This implies that the social production of space always incorporates the opposites and the unspoken 'obscenities', as constitutive of social and spatial arrangements.

A dominant allegory for the strange space of the 'postmodern city' is that of *collage*. Within this imaginary it is posited as a largely illegible 'text', a shifting complex entity, an unfathomable, heterogeneous web of practices and relations. Depending upon where you stand, Michel de Certeau argues, you appropriate the city differently: either in terms of an aerial overview and an omniscient gaze upon an urban spectacle in the distance – those mediated image ideals that express the grand narratives of 'concept space' or from the street level, encompassing the messy routines, spatial practices and seemingly unreadable urban cultures of the *quotidien* (de Certeau 1984). Calling into question the very idea about coherence, the collage city is an ambiguous and contrary space where different voices collide and where order and disorder, sanitation and defiance, planning, and uncertainty, image ideals and urban cultures are irresolvably entangled. If we embrace this position of the city as a differentiated space belonging to several ontologies – to poetry and geometry, emotion and struggle, image and materiality – it cannot be understood as a synthetic totality. The city is a multi-sited spatial formation produced across diverse discursive regimes, representational modes and material everyday practices, perhaps in some respects strange by its very form (cf. Balshaw and Kennedy 2000). This volume seeks to bring out the polyvalency of strangeness yet without resorting to a comfortable postmodern position where for example city space is posited as complex, chaotic and inscrutable *a priori*. We intend to move beyond the face value of the obscure and to expound the *sociality* of strange spaces and attend to them as produced within society – as laden with power geometries (Massey 1994, 2005). Important questions to be raised are: for whom is the naturalized, invisible spatial texture we take for granted strange? For whom is the strangest of places perfectly lucid, familiar, even 'at home'?

As spatial experiences are contextually bound, these meanings shift over time, but they also depend upon the intersectional power asymmetries that produce spaces lived and experienced differently by different people (see e.g. Rose 1993). Sexual or racial discrimination for example – in its varied forms and across multiple intersections – is apart from humiliating also quite estranging, resulting in bizarre orders, obscure spaces and systems of oppression (for example the slavery system, the brothel, the harem, the suburban home of 1950s America). What was termed 'consciousness-raising' by second-wave feminism, the ability within emancipatory movements of whirling (spatial) norms – of estranging what seems normal or natural in a specific societal, historical and cultural context – will convey a counter-narrative about strange spaces. Queer theory for example raised the awareness of heteronormativity, a term suited for disclosing and turning upside-down the normality of straight spaces (Butler 1990; Bell and Binnie 1994; Skeggs et al. 2004).

But what is perhaps less projectable is that being positioned in estranging, yet *normative* ways, may also engender 'strange responses' or even a form of unanticipated empowerment emanating from the obscene (or non-seen) off-spaces of the oppressed (de Lauretis 1987). An obscure order – patriarchy – turns woman into, for example, an impossible position to occupy. The strange space of the female body is massively burdened by representation and hence it is left, as Teresa de Lauretis says, vacant. Such vacancies potentially generate new and unexpected positions. de Lauretis's critical feminist project evoked a subject for feminism through theorizing those undefinable (and yet potentially liberatory) elsewheres that we are curiously roaming in throughout this book:

> For that 'elsewhere' is not some mythic distant past or some utopian future history. It is the elsewhere of discourse here and now, the blind spots or the space-off, of its representations. I think of it as spaces in the margins of hegemonic discourses, social spaces carved in the interstices of institutions and in the chinks and cracks of the power-knowledge apparati. (ibid.: 25)

Teresa de Lauretis describes the potential of empowerment from within: of movements from spaces represented by/in discourse and by/in the gender system to the 'the space not represented yet implied (unseen) in them' (ibid., 26). This is described as a movement 'between the (represented) discursive space of the positions made available by hegemonic discourses and the space-off, the elsewhere of those discourses'. Such discursive and social other spaces, de Lauretis argues, exists *within* hegemonic discourses, albeit on their margins (ibid.). Strange emancipatory space is constituted, according to de Lauretis, of what the representation leaves out, or, more pointedly, what it makes unrepresentable. The movement between represented space and off-space is not a dialectic or an integration but is made up of tensions of contradiction, multiplicity and heteronomy.

Our attempt to get a grasp on strange spaces may also be conceived as the retaining of such irreducible contradictions and as a project which potentially relocates subjects in ways that may generate *new spaces* for critical exchange and radical response. Silence and prohibition, deviance and marginality may thus be conceived of as *one* dimension of strangeness, one facet that does not exhaust the matter. We wish to bring spatial obscurities – which arise either out of outlawing and outright discrimination, or out of surprises, shifts, transitions and complexities in our everyday lives – onto a plane where they become observable, noticeable and analysable, always in awareness that as much as strange spaces need to be teased out in relation to their social cultural and political context, their opacities and obscurities are never reducible to them. In addition, their strangeness cannot be reduced into one formula or form of explanation. From such a vantage point we may for example acknowledge that modernist spaces of ordered perfection, reason, abstraction and spectacle (Lefebvre 1974/1991; Debord 1967/1994) may be equally estranging as their messy odds and ends, worn out leftovers, unplanned in-betweens and ageing residues sometimes are. And, perhaps counter-intuitively, leisure spaces of phantasmagoria in (postmodern) cities – theme parks, movie theatres, laser domes, online game worlds etc. – are not always the given sites of emotive disarray or dislocation, but rather lucid in their spatial organization, rigorously following the 'pleasure principle'. In consequence, as the contributions will display, strange spaces need to be addressed through less conventional methods and approaches. The following section will detect ways of understanding strangeness in the 'taken for granted' realm of the everyday.

Everyday Elsewheres

If it is true that strangeness follows from transition, as we have argued above, it implies that what is strange about strange spaces tends to escape the explanatory fixity of perceptive and theoretical categorizations (cf. Sheringham 2006, 223). An understanding of a certain space as strange involves a sense of being *here and now*, while at the same time *elsewhere*, losing oneself between what *normally is* and *what seems to happen*, not primarily to space, but to *normality as such*. Phenomena that seem 'normal' or unproblematic are no longer ready to be grasped and enacted, although they are clearly *there*. What we *do understand*, or feel, under such conditions, is that we cannot rely upon our own senses and judgments, but are mentally displaced. We cannot locate or fix what is strange, since its estranging force seems to reside *elsewhere* – beyond the space at hand; thus we must suspect that it resides within our own minds. Strangeness as so conceived thus implies a distortion of the interpretive schemes and embodied understandings we apply for making sense of everyday life.

What we encounter here seems to be a world of daydreaming and hallucinations – and to a certain extent it is. But it is also, as we will see, daydreaming in reverse – an ethnomethodological world of mundane estrangement. 'In point of fact', Gaston Bachelard contends in *The Poetics of Space*, 'daydreaming, from the very first second, is an entirely constituted state. We do not see it start, and yet it always starts the same way, that is, it flees the object nearby and right away it is far off, elsewhere, in the space of *elsewhere*' (1958/1994, 183–4, italics in original). What Bachelard sets out to explore is the phenomenological constitution of 'intimate immensity', the sense of limitless mental movement fostered through daydreaming. While daydreaming is indeed experienced as a realm of fantasy and wonder (see also Schutz and Luckmann 1973) it is not, however, strange in itself – precisely because daydreaming is daydreaming. The crucial point in Bachelard's poetic analysis appears when the elsewhere of daydreaming gives meaning to certain expressions in the visible world. Using the example of what it means to lose oneself in a deep, or immense, forest – as often portrayed in poetry – he shows how phantasmagorical categories and perceptions soon produce a certain form of spatial anxiety; the anxiety of being in a space without limits, that is, without no clear beginning or end. In other words, the forest is gradually becoming strange to us: 'One feels that there is *something else* to be expressed besides what is offered for objective expression. What should be expressed is hidden grandeur, depth' (Bachelard 1958/1994, 186, italics in original).

From Bachelard we can derive a phenomenological sensitivity to the estranging fluctuations of seemingly clear and stable spaces; that is, how strangeness emerges from the elsewhere deep within ourselves, engendering either peace or anxiety. This condition particularly applies to vast spaces, such as forests, plains, plateaus and the depths of the sea, where '*being-here* is maintained by a being from elsewhere' (ibid., 208). In addition, the phenomenological correspondence between the immensity of world-space and the depth of inner space, that Bachelard makes us see, points to the confusion of proportions that saturate daydreaming, so that immensity can be found in the most intimate of spaces. Indeed, such experiences, when interlaced with the proceedings of everyday life, can be classified as bizarre. A complementary example can be drawn from Walter Benjamin's description of how one summer-day in Marseilles, after long hesitation, he took hashish:

> Versailles, for one who has taken hashish, is not too large, or eternity too long. Against the background of these immense dimensions of inner experience, of absolute duration and immeasurable space, a wonderful, beatific humor dwells all the more fondly on the contingencies of the world of space and time. I feel this humor infinitely when I am told at the Restaurant Basso that the hot kitchen has just been closed, while I have just sat down to feast into eternity. (Benjamin 1986, 138)

It is no coincidence that Benjamin, allegedly the most prominent interpreter of modernity as a dream state, explores strangeness through the lens of personal experiences and indeed illusion. Throughout his work, and particularly in his writings on the modern metropolis, there is an inclination to allegory – an ambition to reconstruct modern, urban mythology through reproduction, even imitation. As in the hashish experience, urban space appears as a simultaneously enticing and repulsive, beautiful and grotesque, immense and intimate landscape. Through such overlapping dreamlike experiences, the metropolis is often understood as an alienating force, a strange space, produced through mythology. What appears in the city might rightly be taken for a dreamscape, balancing between ruination and redemption. As Graeme Gilloch (1996, 137) points out in a discussion of Benjamin's work, the allegorical gaze involves precisely this movement. On the one hand, it brings the ruination of things, through a destruction of context. On the other hand, reducing the world to ruins implies that fragments can be recollected and reused. Through allegory, or hallucinatory experiences, strangeness might be brought to the fore, as an overcoming of myth. No wonder, then, that ruins, and especially modern ruins, still lingering in the consciousness of collective memories, often evoke bewildering feelings of nostalgia – as if they were speaking of an historical elsewhere, while at the same time, precisely through their vanishing and fragmentary state, reinforcing mythology 'here and now' (see also Trigg 2006).

Preceding Benjamin's poetic urban explorations, the confounding temporal play of dreaming was also one of the foundational themes addressed in August Strindberg's dramaturgic work around the turn of the twentieth century (see Ericson 2004). As Strindberg stated in the foreword to *A Dreamplay* from 1901 – a surrealistic story of how Agnes, the heavenly daughter of Indra, descends to Earth and experiences the torments of being human – his intention was to imitate the disrupted, but seemingly logical form of the dream, in which illusion creates strange textures of memory, experience, invention and improvization. Such disruptions and illusory reflections saturate the entire play, as in this scene where the old Officer suddenly finds himself in a classroom at school (Strindberg 1918, 279–80, our translation from Swedish).

THE TEACHER *to the Officer.*
Well son, can you tell me now how much is two multiplied by two?
THE OFFICER *remains seated; searching with pain in his memory without finding the answer.*
THE TEACHER.
You are supposed to stand up when you are being asked.
THE OFFICER *in pain, stands up.*
Two ... multiplied by two ... Let me see! That is two two!
THE TEACHER.
Really! You haven't done your homework!
THE OFFICER *ashamed.*
Yes, I have, but ... I know what it is, but I can't say it...
THE TEACHER.
You are trying to dodge! You know it, but can't say it. Perhaps I can help you!
He pulls the Officer's hair.
THE OFFICER.
It's awful, it's awful!
THE TEACHER.
Yes it's awful that such a big boy doesn't have any ambition...
THE OFFICER *in pain.*
A big boy, yes, I am big, much bigger than the others here; I'm grown-up, I've finished school ... – *Like awakening* – I am promoted ... So why am I here? Am I not promoted?
THE TEACHER.
Sure you are, but you must sit and mature, you see. You must mature... Isn't that correct, you think?

Bachelard, Benjamin and Strindberg have in common their notion of strange spaces as the harbours of interlaced dreams and memories – heterotopian spaces of discontinuities, disproportions and dislocations. Strangeness is thus not necessarily tied to any particular spatial form or category, but rather to the co-existence of familiar and phantasmagorical elements.

Starting out from a phenomenology of space, however, there are also other methods of exploring the obscurities of modern life besides poetics and allegory. One may get just as close to the day-to-day experiences of ordinary people if the dreamlike facets of social life are revealed not through allegorical abstraction and recollection, but through a destruction of what phenomenologists (e.g. Schutz and Luckman 1973) call the *natural attitude*. This is to say that what the ethnomethodologists propagated from the 1950s and onwards was an entirely *different kind of awakening*, a social awakening based on defamiliarization through interventions in the lifeworld. Following Harold Garfinkel's (1967/1984) statements in *Studies in Ethnomethodology*, we might conceive of this approach as 'daydreaming in reverse' – a process through which subjectivity is not escaping the 'here and now', but drastically brought back to, and overwhelmed by, that very reality, however strange it might seem:

> Procedurally it is my preference to start with familiar scenes and ask what can be done to make trouble. The operations that one would have to perform in order to multiply the senseless features of perceived environments; to produce and sustain bewilderment, consternation, and confusion; to produce the socially structured affects of anxiety, shame, guilt, and indignation; and to produce disorganized interaction should tell us something about how the structures of everyday activities are ordinarily and routinely produced and maintained. [...] I have found that they [my studies] produce reflections through which the strangeness of an obstinately familiar world can be detected. (ibid., 37–8)

As Garfinkel points out, even the most ordinary spaces involve more or less 'senseless features' that, if multiplied, might distort our basic social trust and understanding. The peculiar thing about this method is that spaces are not entirely transformed, but socially, and/or morally altered, exaggerated – *seen anew*. Since ethnomethodology invokes a clash between deep rooted spatial expectations and strange appearances, as social actors we are taken off-guard, as in a candid camera performance, unable to set aside the established typifications we normally live by. We are forced into a daydream-like elsewhere, unveiling itself here and now, in the midst of ordinariness.

The theoretical legacy of ethnomethodology and the broader stream of social interactionism, is valuable for any exploration of how 'strange spaces' are socially embedded, negotiated and sometimes desired. The tension-field between strangeness and ordinariness is one that not only ethnomethodologists but also 'ordinary' people, seek out and explore. Notably, strangeness is brought about through a breakdown of the codes and conventions normally governing morally and/or aesthetically legitimate social appearances. Such transformations or transcendences tend to balance between the unspeakable attraction of the illegitimate and the aversion towards the obscure, perverse, grotesque or bizarre. What Erving Goffman (1959) has termed *back region behaviour* often involves a routinized estrangement of the legitimate codes and procedures that mark the performances of front regions. As he found in a study of a Shetland hotel, while the kitchen staff, once back-stage, regularly referred to their customers by belittling code-names, the customers, when in their own circles, described the staff 'as slothful pigs, as vegetable-like primitive types, as money-hungry animals' (ibid., 169). Similarly, people in general tend to behave in a rather 'uncivilized', de-disciplined manner once out of reach of public exposure, be it in terms of talk, dress or sexual behaviour. In this way, back regions are always more or less strange spaces, or estranging, when viewed from the outside – that is, from the viewpoint of 'legitimate' forms of activity (which is of course always a relative social construct). They incorporate environments normally hidden to 'outsiders' and expose people in unexpected, even unthinkable ways – sometimes in terms of 'immoral extravagancies', sometimes in terms of plain, but just as shocking, 'ordinariness'. Pierre Bourdieu (1979/1984) has pointed to this estranging moral ambiguity, through which the most common of practices are often judged as de-cultured, decadent and thus strange:

> In the face of this twofold challenge to human freedom and to culture (the anti-nature), disgust is the ambivalent experience of the horrible seduction of the disgusting and of enjoyment, which performs a sort of reduction to animality, corporeality, the belly and sex, that is, to what is common and therefore vulgar, removing any difference between those who resist with all their might and those who wallow in pleasure, who enjoy enjoyment. (ibid., 489)

In a similar way, back regions are spaces of social and cultural de-differentiation, and therefore threatening to moral judgment, order and control. They constitute a kind of omnipresent, ordinary-strange elsewheres – so close but yet so far away. What happens when 'the door is closed', when 'the lights are out', or (in a media age) when the 'camera is turned off', is a general source of human curiosity, fascination and daydreaming.

Communication Geography and Strangeness

In media studies space was, for a long time, treated primarily as context – the context of the production, transmission and consumption of texts. However, new media technologies and cultural forms have in themselves contributed to a problematization of these contexts and processes. The increasing convergence, interactivity and mobility of 'new media' create spatial ambiguities in terms of a relativization and blurring of contexts and text-context relationships. In the era of digital media there are thus few clear-cut 'media spaces', but rather an omnipresent, dynamic and multifaceted realm of communications, saturating our entire lifeworld (see Morley 2000; Couldry and McCarthy 2004). This development partly explains why, since the mid-1990s, media studies have seen a *spatial awakening* taking place, involving not only a new concern with the quite obvious spatial ephemerality of (post)modern media societies, but also an increased understanding of *space as a construct* – a category whose qualities have *always* emerged through representation and communication. Studies of the space-communication nexus, and the emergence of *communication geography* as an interdisciplinary research field (see Falkheimer and Jansson 2006), are thus closely related to a rethinking of the cultural history of media spaces. Such studies may involve our strange and phantasmagorical late modern networked cities, whose buried and vague traces cut across the physical landscape and shape it into a landscape of power with an eerie, invisible yet decisive presence (Graham and Marvin 1996, 2000). On the global market for attention cities also become spaces of subliminal *futurity*, strange spaces replete with the mythologies of capitalism. Digital and technological cities produce endless re-enchantments of a disenchanted urban modernity as they are inscribed with emotions and fuse reason with magic (Jansson and Lagerkvist 2009).

Does this also mean that mediatization is making space increasingly strange? On the one hand, as we saw above, one may, with Lefebvre and others, think of space as something that is strange by its very nature and argue that the strangeness of space largely has to do with its representational dimensions. As Lefebvre (1974/1991, 208) contends, spaces of representation (lived and imagined) constitute 'the obscure counterpart to that luminous entity known as culture'. From this view, mediatization – seen as the multiplication of technologically circulated representations of space – can easily be (mis)interpreted as a process of mythologization and mystification. As spatial representations agglomerate and collapse into our everyday lives, transposing our spatial experience into 'hyper-reality' (Baudrillard 1983a, 1983b; Eco 1986) or 'thirdspace' (Soja 1996), there is indeed something 'strange' happening. But that is just one side of the coin.

On the other hand, from a more phenomenological viewpoint, even the most miraculous and thrilling experiences of our media age are successively transformed into mundane conditions, incorporated as parts of the taken for granted textures of space. Feelings of strangeness and uncertainty may emerge just as well when we find ourselves in spatial pockets of demediation, or in decaying spaces of dying media, where dreams and promises of media society have been broken or put into question. Strangeness does not stem from the 'amount', 'range' or 'impressiveness' of mediation, but from the qualitative relationships between perceived, conceived and lived spaces. One of the central objectives of communication geography is to encircle these *more or less estranging* relationships; to frame and name them by attentively representing them with respect to their vast and ever-growing complexity.

While the general vagueness and opacity of (media) space might be brought to the surface by way of theory, in this book theory is primarily used for interpreting how phenomenologically strange spaces – that is, spaces that (some) people perceive as strange – are produced in and through mediation. In the two following subsections we will further unfold these entwined dimensions of mediated strangeness.

The Confounding Spaces of Representation and Magical Media

Allegedly the most provocative perspective on how mediatization estranges people from society and reality, Baudrillard's writings on postmodern simulations, hyper-reality and implosions cannot be overlooked in a discussion of mediated strangeness. Extending Debord's (1967/1994) post-Marxist notion of the spectacle society and surpassing Jameson's (1991) post-semiotic, yet materialist view of a depthless visual consumer culture, Baudrillard's ironic assertions have profoundly altered our ways of perceiving media culture. His thoughts are ironic because his basic understanding of information society (developed mainly in the 1980s) as a society without sociality, produced through communication without content, kills the very target of what seems like a harsh critique (see Baudrillard 1983a, 1983b, 1987). Indeed, this method of unpacking and reconstructing information society as a state of hyper-reality, disabling all claims for truths, values and critique, unveils a strange state of affairs. It is also fair to say that Baudrillard's thoughts have gained broader relevance and legitimacy in the face of the escalating processes of digitalization and globalization. His postmodern views reveal the most bizarre aspects of a mediatized lifeworld. Regardless of their take on 'post', 'late' or 'liquid' modernity, many contemporary theorists have been united with Baudrillard in this critical fascination with the knowable world sliding into the image (e.g. Eco 1986; Sorkin 1992; Kellner 1995; Ritzer 1999; Clarke 2003).

Still, whether we embrace or refuse postmodernism, what can be theoretically singled out as bizarre and paradoxical is different from people's everyday understandings of the same phenomena. Thus, Baudrillard's most exciting accounts of strangeness are not to be found in his theoretical notion of the hyper-real, but in his personal anecdotes and narratives of what it is to live in a mediatized world – such as his 1980s reflections on America emanating from his own journeys across the continent (Baudrillard 1986/1988). One of the most enchanting things about America, Baudrillard finds, is that the entire country is 'filmic'. The metropolises are like 'a screen of signs and formulas', created as reflections of cinematic images. To grasp the secret of American cities, New York and Los Angeles in particular, one must thus start in the representational realm, understanding that this realm saturates the entire cityscape. In an intriguing passage of the book, Baudrillard, one winter morning in New York, reflects upon the sight of a passing truck with the golden text MYSTIC TRANSPORTATION INCORPORATED painted on its sides, contending that this appearance somehow encircles the very meaning of New York and its mystical view of decadence – the bewildering mixture of filth, promiscuity and special effects – which he sees as the city's *fourth dimension*. Such anecdotes show that, while not understating the relevance of postmodern theory, in order to grasp what is strange about media society we must look at specific *time-space constellations and passing experiences* rather than overarching structural patterns or fixed spaces.

Europeans' journeys in America are particularly apt examples of the preoccupation with the lines between the real and the imagined. Baudrillard was not the first to be confounded by America; the Swedish author and film maker Vilgot Sjöman who travelled in the US in the late 1950s, preceded him by decades. Sjöman was both repulsed and fascinated by *Forest Lawn*, a deeply puzzling burial ground in Los Angeles, which was simultaneously a park filled with media attractions and duplications:

> And Forest Lawn is built on the idea of reproduction – at worst on the idea of recreation. Recreation is when you transfer Leonardo's The Last Supper onto a colored glass window. Or when you make the Sistine Madonna of Rafael into a sculpture – without being at all troubled by the original structure of the art piece. And Forest Lawn does not hide that all the Michelangelo sculptures are reproductions – on the contrary, they tell with pride. If only the copy is 'exact' and made out of Carrar-marble 'from the same quarry that Michelangelo used', then the copy is equal to the original! And Forest Lawn may, at the utter bewilderment of the foreigner announce that they've got 'the biggest collection of masterpieces of Michelangelo outside of Europe.' (Sjöman 1956, 90–91)

The disorienting simulations that European travellers are faced with in America – overlaying the physical geography itself – are also what fascinated and intrigued British author J.G. Ballard and it became the impetus for the writing of his novel *Hello America* (1994). In the introduction to this phantasmagoric literary journey through the US, he underlined that:

> Whenever I visit the United States I often feel that the real 'America' lies not in the streets of Manhattan and Chicago, or the farm towns of the Midwest, but in the imaginary America created by Hollywood and the media landscape. Far from being real, the sidewalks, and filling stations and office blocks seem to imitate the images of themselves in countless movies and TV commercials. Even the Amercan people one meets in hotel lobbies

and department stores seem like actors in a huge televised sit-com. 'USA' might well be the title of a 24 hours a day virtual reality channel, broadcast into the streets and shopping malls and, perhaps, the White House itself ... (Ballard 1994, 4–5).

Ballard underscores Baudrillard's observations about the American simulacrum. This potential for *deceit* in a world saturated by media is another major theme in response to the confounding spaces of representation, shared by Sjöman and Ballard. In the gaudy Hollywood film *The Truman Show* (directed by Peter Weir, 1998) this is taken to extremes. The protagonist of both the film and the show, Truman, lives his whole life in a small town reminiscent of an eternal 1950s America, which is in fact a gigantic TV studio, controlled by the maker of a 24/7 live broadcast, who is a demonic director playing God. Truman is, as the tagline for the TV show repeats, 'On The Air. Unaware.' His spouse, his mother and best friend are actors and the whole world turns out to be a stage with scripted movements and product placements. Here the film represents mediatized spaces as obscure and belittlingly portrays them as profoundly alienating. The film gives full weight to pessimistic views on media society and reality TV. It also thematizes broader cultural tendencies as mediatizations and mediations of space become intimately enmeshed with our inner dream worlds and consequently with experiences of 'real places', amounting to specific knowledge of phantasmagoric psycho-geographies.

Imaginary spaces resting somewhere between the media representations that people consume (and in a context of ICTs, increasingly also produce) and the materiality of the world, are termed *thirdspace* in the geographer Edward Soja's (1996) spatial theorizing. Thinking about space in these terms takes account of the trialectic interplay between three enfolded dimensions: it includes the *firstspace* objectivity that Lefebvre termed 'spatial practices' and *secondspace* mediated qualities, that Lefebvre (1974/1991, 74ff) discussed in terms of 'representations of space'. Strange spaces contain both these aspects – material settings and their 'mappings' or representations of them. But it is only in terms of *thirdspace experience* that they appear to us to retain their full mystery and thrust of sensory-emotive convulsion. Thirdspace equals Lefebvre's lived space, representational space, and is distinguishable from spatial practices and representations of space; still it encompasses both of them. Such poignant and affective spaces bring about expanded and seemingly inscrutable spatial knowledge. According to Lefebvre, who privileged the spaces of representation in his philosophy of spatialization – including the realm of experience and the body – these are constituted by 'a strange interplay between the magical and the rational' (ibid., 203). Hence, they comprise the enigma of space itself.

Edward Soja sees thirdspace as carrying a broad potential. Pervaded by myths and symbols it is also a 'strategic location from which to encompass, understand and potentially transform all spaces simultaneously' (1996, 68).

Thirdspace is an interstitial realm that allows for an expansion of what is possible, thinkable and projectable. Thirdspaces are also more than symbolic spaces. They are

> ... vitally filled with politics and ideology, with the real and the imagined intertwined, and with capitalism, racism, patriarchy, and other material spatial practices that concretize the social relations of production, reproduction, exploitation, domination and subjection. (ibid.)

According to Soja, since they are the dominated spaces, the spaces of the marginalized are also *the chosen spaces for struggle, liberation and emancipation*. We will propose that strange spaces allow our attention to linger on those incompatible facets of space that may potentially foster new projections and directions. It may thus be inferred, not only that audiovisual fantasies generate *spaces in-between*, but that they may entail a capacity of informing and giving rise to *series of ambivalences*: that is, spatial uncertainties. This allows for moving beyond strange spaces understood in terms of fixed contradictory points of reference into embracing the conflicting and incompatible notions and affections implied in a mediatized world. It further calls for opening up the geographical as well as 'communicational' imagination. In highly mediated spaces, the real-and-imagined, the known (knowable) and the unknown (unknowable), are held *in thirdspace suspense* (Lagerkvist 2008). The hypermediated city struck by terrorism is also an uncanny and strange place due to a similar doubling of its imaginary and material actualities and to the strange repetitions and loops of imagery. Hence, as Brian Jarvis argues, New York on 9/11 produced an uncanny effect which occurs 'when something that we have hitherto regarded as imaginary appears before us in reality' (Freud in Jarvis 2006, 49), that is, through the effacement of the real and the imagined that produces, in addition, a crisis of temporality:

> 9/11 is not over. In the case of millennial disaster movies, some of the most important images after 9/11 were replayed before the attacks even took place. 9/11 is a time of uncanny doubling and déjà vu. The attack on the city was foreseen in film, it was seen in the first attack, it was repeated in the second attack. The images of the Twin Towers subjected to a twin attack then endlessly repeated. (Jarvis 2006, 56)

Image dependent terrorism may thus be 'the uncanny of capital itself' (ibid., 62), that is of its logic of waste and destruction, but terrorism is but one drastic example of rendering the urban everyday strange.

No other realm of our contemporary media culture measures up to the complexities of representational space as 'cyberspace'. Digital cultures and 'new' media forms of the information age – the web, virtual realities and game worlds – seemed, particularly in their infancy, quite *strange*. They consequently evoked both astonishment and awe, partly because they brought about complex spatial dislocations. Digital geographies can either be viewed as existing parallel to our material world, or as challenging our perceived, lived and conceived spaces altogether (Crang et al. 1999). One may, for instance, think of computer games that alter spatial scales, causing extended notions of spatiality. The computer game and alternative reality *The Longest Journey* (with the sequel *Dreamfall*) (1999, 2006) is an adventure game that takes place in the year 2209. Players take on the role of April Ryan, who is the heroine with a mission to save the world. The core site within the game – through which the player moves as April explores new worlds, meets strangers and interacts with them – is a place, which is a recreation of Venice. This world, in turn, consists of *two parallel universes*: the so called world of science, or the 'real world' named *Stark*, and the virtual world named *Arcadia*. Ryan is called upon to restore the now unstable boundary between the two worlds of the game.

Here game space is actually a divided place inside another space, to where you log on from yet another place: our physical world (Sundén 2006, 284). There are strange parameters in such game worlds as well as in the user context, where the physical, real, virtual and imagined dimensions of space become complexly jumbled and sometimes inverted. Such obscurity inherent in digital geographies may also account for their appeal and attraction and perhaps for the utopian qualities endowed in them when they are 'new'.

There is also a certain fetishistic strangeness attached to media technologies themselves, at least when they are brand new, as they are often dreamed of as bearers of supernatural powers (e.g. Gunning 2003; Gitelman and Pingree 2003). In our media culture the everyday is not only cognitively or semiotically overloaded. A stream of newly invented technologies constantly appears in our world and in no time they become part of our bodies, inscribed into our lives. Since they are incrementally used but unevenly mastered, media uses and practices are also rendered magical, spooky and strange. The *digital sublime* (Mosco 2004) – 'the ghost in the machine' – comes about through the complexity of media forms. Since we have no way of knowing or fully comprehending how they work, they become 'technologies of captivation' (Gell 1992), potentially sated with agency. Hence these incomprehensible yet lived aspects of our everyday digital cultures contain mundane obscurities that seem enchanting to some of us and outright confusing to others. But as media technologies quickly grow older, or as abstruse navigation in complex worlds of game design is turned into a habit, there is also a movement into a sense of second nature (Gunning 2003), and one day such digital spaces are emptied and left-behind. They are abandoned by both machines (with new, more advanced hardware) and people who use new toys and tools. Media technologies thus bring about *affective movements*. Before they are fully integrated in our daily lives, they even seem *alive* to us, but in the course of time and wear they lose their magical grip, fade away and are potentially met with rejection or neglect – a phenomenon that will be further explored in the next section.

Fading In/Fading Out: Media Spaces of Uncertainty

In their book *Spaces of Uncertainty*, Kenny Cupers and Markus Miessen (2002) discuss the role of public space in cities more and more saturated by information networks and the commercial imperatives of visual culture. They pay particular attention to the seemingly empty, left over, or left behind places that can be found in even the most thoroughly planned cities – empty green areas around communication nodes, decaying zones around industries, cleaned spaces now used as parking lots, etc. While dull and gestureless, these black holes often seem to convey a secret story and therefore easily become the objects of fantasy, creativity and alternative appropriations. In their hiding of something, they also say something. While not loaded with architectural or commercial meaning, they are significant to the socio-cultural circuits of urban life, as well as for marking out the margins and boundaries of architecture and planning. Thus, as Cupers and Miessen argue, they are at the same time outside the city and integral to it.

This kind of urban in-between is a concrete reminder of Lefebvre's general view of spatial opacity:

> Space contains opacities, bodies and objects, centres of efferent actions and effervescent energies, hidden – even impenetrable – places, areas of viscosity, and black holes. On the other, it offers sequences, sets of objects, concatenations of bodies – so much so, in fact, that anyone can at any time discover new ones, forever slipping from the non-visible realm into the visible, from opacity into transparence. (Lefebvre 1974/1991, 183)

The above-mentioned interplay between the visible and the non-visible, between fading in and fading out, cannot be fully grasped without assessing the role of mediation and representation. Urban black holes, for instance, are defined in relation to planning discourses, popular media, social scripts and so on. While grounded in tangible spatial practices, their in-betweenness must be understood discursively and phenomenologically. That is why mediations, and the media, are so important to the study of strange spaces – also when strangeness follows from the absence, attrition and death of such processes.

This way of reasoning also seems relevant if we approach mediation in more institutional and technological terms. What Nick Couldry and Anna McCarthy (2004) have termed *media space*, understood as the conceptual realm through which we can study the intertwining of media circulation and social space, clearly incorporates its own spaces of uncertainty – denoting not hyper-mediated thirdspaces, but spaces beyond, behind or between the dominant structures of control and exploitation. In media society, social norms, expectations and power cannot be uncoupled from the media system. What is 'normal' and taken for granted follows from the mainstreaming of the 'important', the 'picturesque', the 'spectacular', the 'interesting', the 'new' and so on, as well as from the media's self-representation – their continuous efforts to reproduce themselves as the 'centres of society' (cf. Couldry 2003), and the 'magic systems' (cf. Williams 1980) through which social values and norms are spectacularized and sold back to the audiences as something desirable. In this way, the media work as a gigantic machinery of symbolic power through which norms and obscurities are invoked and exploited. What we understand as strange, however, may not be what the media regularly depict as obscure, perverse or deviant, but spatial phenomena and ensembles outside, or in the vague margins of these schemes of classification. As the media must regain or reinvent their centrality and magic, due to the fact that such things tend to become commonplace, the very expressions of that logic – whether the redesign of a television news room, the launch of a new 'provocative' reality show, or the advertising of a 'groundbreaking' digital camera – become the most commonplace of all features of media society.

As the chapters of this volume will show, strangeness is rather to be found in the elsewheres of opacity, figuring as disproportionality, discontinuity, decadence, decay, and so on. Some of these spaces are akin to what Rob Shields (1991) has termed *places on the margin*; that is, spaces occupying peripheral, marginal, or 'low', positions in relation to dominant geographical, social, and/ or moral hierarchies. In an illuminating case study, Shields discusses how the English seaside resort Brighton developed during the 1920s and 1930s into a 'dirty weekend' destination; a place coded as liminal and partially uncoupled from the morality of everyday life. One of Shields' conclusions regards the social significance of 'place images', through which, for example, Brighton was turned into 'a sign of leisure, indolence, and ultimately, illicit sex' (ibid., 25); an issue which since the book was published has become more and more important to study due to the prevailing ideology of branding and place marketing. But again, strangeness always resides elsewhere. Just like spatial planning and architecture, marketing and media exploitation would successively exhaust the hidden secrets of social uncertainty. The beaches of Brighton might have been a strange space for a while, but reputation builds expectations, which in turn make even immoral or illegitimate social appearances seem less strange (in that space).

In this context we must also consider the relationship between social subjects and the more or less mediatized scripts that guide their appropriation of space. Even heavily promoted and simulated environments may seem strange when the relationship between representations, social practice and imaginary geographies are altered or taken apart. Here we might recall the above discussion of Goffman's back regions and the strangeness one might experience when encountering mundane or illegitimate signs or practices in spaces otherwise appearing as very formal or representative. For most of us, the back regions and underworlds – typically the spaces of infrastructure and maintenance, of hotels, shopping malls, cruise ships, broadcasting centres and so on – seem strange, inseparable as they are from their dominant representations. An interesting parallel can be drawn to Couldry's (2000) analyses of how 'ordinary people' experience their encounters with the stages of media production. Leaving one's ordinary position as audience, finding oneself in the setting of actual recording or performance work is a transition marked by contradictory, even bewildering experiences. While entering into the sanctum, the glowing realms of holiness, implies an experience of wonder and celebrification, the sense of 'being there' is paralleled by a sense of 'being the wrong subject in the wrong space and time'. Accordingly, since there are no common scripts as to the appropriation of these normally concealed spaces of media power, one might paradoxically think of them as black holes – sometimes even as back stages, hidden behind mediation itself – spaces that escape the classificatory forces of dominant meaning circulation.

Finally, many uncertainties related to media space can be understood as a function of time. As in the opening example with abandoned industrial wasteland or cleaned urban areas waiting for exploitation, media spaces may be affected by stagnation, cultural lags and processes of obsolescence and decay. Such temporal incongruence is typically defined at the intersection of technology and cultural form (Williams 1974). Flea markets and garage sales, for instance, where one can find once sophisticated media equipment and outmoded records together with old ceramics, toys, sports equipment and second hand clothes, are not only phantasmagorical, multilayered time-spaces in themselves, but also environments that remind us of the somehow bizarre circuits of goods characterizing media society, and how these circuits are functionally and emotionally interwoven with our own life biographies. Objects that were recently a source of personal pride and confidence can be re-encountered at the dusk of their lifespan, evoking feelings of contemplation and nostalgia (cf. Straw 2005).

However, it can be just as strange to encounter spaces where old media are in functional use. Who would today expect to hear the sound of a typewriter, a turning dial, or even a connecting telephone modem, in a high-tech office building? Still, certain institutions, in particular bureaucratic organizations marked by rigid structures, just slowly adopting new media technologies and forms, may incorporate a multilayered order of administrative techniques – combining rationality with slowness. Such contradictory patterns may seem odd to clients who are getting more and more accustomed to the instantaneity and impersonality of digital media infrastructures, such as on-line shopping and banking. Waiting in line outside different offices, answering personal questions and filling out complicated paper forms may be a more dreamlike experience than wandering through the media monitored consumer phantasmagoria of shopping arcades and international airports. Once again, this is to say that the strangeness of strange spaces is a relational, phenomenological and discursive matter.

This Volume – Three Fields of Mediated Obscurity

In this introductory chapter we have outlined some key theoretical aspects from which the strangeness of strange spaces can be analysed and understood. We have also introduced several distinctions that encircle the more empirical terrain of *Strange Spaces* that we have chosen to present under three thematic headings – parts that will be more meticulously presented in this book in relation to the respective chapters.

In the first part of the book, 'Scales of Opacity' we approach the strangeness of mediaspaces with a focus on some of those miscellaneous scales involved in media geographical experience. The contributions move from a kind of phenomenological and transformative *condition of vagueness* in certain oppositional or alternative spaces challenging social norms of modern and mediatized spaces (Carney and Miller), to the milieu in which we all live and sleep and eat: the *home* in which our cherished, forgotten and obsolete media forms die and are resurrected (Löfgren). From there occurs a move though visual representation *into the body* – gauging the interior scales and strangenesses of our selves and bodies in modernity (Åhren and Sappol). Then the final chapter *lifts from the planet* into satellite spaces investigating media forms that escape our ordinary mundane knowledge and experience, while thoroughly interpenetrating our lifeworlds (Parks). Lingering on scales makes this variation and complexity visible and researchable.

The second part 'Dislocation, Disruption, Disobedience' focuses instead on moments of inaugurated strangeness in trying to answer the question of when, how and through what means spaces become strange, and for whom? Our approach here is that a sense of dislocation occurs when a spatial code is lost or challenged through some kind of interruption or disruption, either through violence and noncompliance or a breaking off from general expectations of what normally occurs/should be in that space. The contributions interrogate mediated obscurity at an exhibition visited by the 'wrong person' (Habel); through the spatio-temporalities of a building where materialities of old media apparatuses and gender representations estrange visitors (Lagerkvist); within representational styles and aesthetics of sinful print cultures (Straw); in the overlayering of urban space with mediated murder stories (Wilbert and Hansen), and by emphasizing the communicative aspects of architecture (Hammond) – examples that form the media-spatial backdrop against which transformative or dislocational moments of strangeness occur.

In the final part, 'Secrets and Wonders of Media Space', the chapters focus on how mediaspaces themselves become strange, magical and alluring due to aspects of their secretive yet powerful omnipresence as mediators of flows of images and texts, as technologies that define the everyday, and as institutions with the power to pervade our modern lives, practices and values. The chapters in this section show that there is something sacred to the media. New phenomena may highlight the many wondrous aspects of for example the media houses and machineries themselves (Ericson; Jansson), that the cultural history of representational techniques, such as the postcard, can teach us about the complex imaginaries of new machines and aerial space (Sloan), and that certain environments of mediated obscurity seem especially conducive to the film medium, such as museums (Jacobs) and the underground (Pike).

While opening up a highly dynamic and multifaceted research field, implicitly asking for interdisciplinarity and theoretical anti-orthodoxy, we aim to show that strangeness is reliant upon a cultural and social *situation* and that it is thoroughly *relational*. The empirical chapters comprise piecemeal and probing explorations into mediated obscurities – obscurities that may reside *beyond* mediation; *inside* its very core, or even *between* the mediated and the 'real'. They open up a terrain where there is no neat, linear or consistent narrative to follow: it is not mediation as such that is strange, but also the spatial relationships and situations it produces. Strangeness may belong to one discursive order or another, to the body or to its habitat, to light or shadow, to the city or the countryside, to the streets or to the sky, to violent or extraordinary action, or to tedious, humdrum stillness.

References

Bachelard, G. 1958/1994. *The Poetics of Space*. Boston, MA: Beacon Press.
Ballard, J.G. 1994. *Hello America*. London: Vintage.
Balshaw, M. and Kennedy, L. 2000. *Urban Space and Representation*. London: Pluto Press.
Baudrillard, J. 1983a. *Simulations*. New York: Semiotext(e).
Baudrillard, J. 1983b. *In the Shadow of the Silent Majorities, or, The End of the Social, and other Essays*. New York: Semiotext(e).
Baudrillard, J. 1986/1988. *America*. London: Verso.
Baudrillard, J. 1987. *The Ecstasy of Communication*. New York: Semiotext(e).
Bell, D. and Binnie, J. 1994. 'All Hyped Up and No Place to Go', *Gender, Place and Culture: A Journal of Feminist Geography*, 1:1, 31–47.
Benjamin, W. 1986. *Reflections: Essays, Aphorisms, Autobiographical Writings*. New York: Schocken Books.
Bourdieu, P. 1979/1984. *Distinction: A Social Critique of the Judgment of Taste*. London: Routledge.
Butler, J. 1990. *Gender Trouble: Feminism and the Subversion of Identity*. London: Routledge.

Certeau M.de1984. *The Practice of Everyday Life*. Berkeley: University of California Press.

Couldry, N. 2000. *The Place of Media Power: Pilgrims and Witnesses of the Media Age*. London: Routledge.

Couldry, N. 2003. *Media Rituals: A Critical Approach*. London: Routledge.

Couldry, N. and McCarthy, A. (eds) 2004. *Mediaspace: Place, Scale and Culture in a Media Age*. London: Routledge.

Clarke, D.B. 2003. *The Consumer Society and the Postmodern City*. London: Routledge.

Crang, M., Crang, P. and May, J. (eds) 1999. *Virtual Geographies: Bodies, Space and Relations*. London: Routledge.

Cupers, K. and Miessen, M. 2002. *Spaces of Uncertainty*. Wuppertal: Müller and Busmann.

Debord, G. 1967/1994. *The Society of the Spectacle*. Cambridge, MA: Zone Books/MIT Press.

Eco, U. 1986. *Travels in Hyper-Reality*. London: Picador.

Ericson, S. 2004. *Två drömspel: Från Strindbergs modernism till Potters television*. Stockholm: Symposion.

Falkheimer, J. and Jansson, A. (eds) 2006. *Geographies of Communication: The Spatial Turn in Media Studies*. Göteborg: Nordicom.

Foucault, M. 1967/1998. 'Of Other Spaces', in Mirzoeff, N. (ed.) *The Visual Culture Reader*. London: Routledge.

Garfinkel, H. 1967/1984. *Studies in Ethnomethodology*. Cambridge: Polity Press.

Gell, A. 1992. 'The Technology of Enchantment and the Enchantment of Technology', in Coote, J. and Sheldon, A. (eds) *Anthropology, Art and Aesthetics*. Oxford: Clarendon Press.

Gilloch, G. 1996. *Myth and Metropolis: Walter Benjamin and the City*. Cambridge: Polity Press.

Gitelman, L. and Pingree, G. (eds) 2003. *New Media 1740–1915*. Cambridge, MA: MIT Press.

Goffman, E. 1959. *The Presentation of Self in Everyday Life*. Harmondsworth: Penguin.

Graham, S. and Marvin, S. 2001. *Splintering Urbanism: Networked Infrastructures, Technological Mobilities and the Urban Condition*. London: Routledge.

Gunning, T. 2003. 'Re-Newing Old Technologies: Astonishment, Second Nature and the Uncanny in Technology From the Previous Turn-of the-Century', in Thorburn, D. and Jenkins, H. (eds) *Rethinking Media Change: The Aesthetics of Transition*. Cambridge, MA: MIT Press.

Jameson, F. 1991. *Postmodernism, or, the Cultural Logic of Late Capitalism*. Durham, NC: Duke University Press.

Jansson, A. and Lagerkvist, A. 2009. 'The Future Gaze: City Panoramas as Politico-Emotive Geographies', *Journal of Visual Culture* 8:1, April, 25–53.

Jarvis, B. 2006. 'New York, 9/11', in Lindner, C. (ed.) *Urban Space and Cityscapes*. London: Routledge.

Kellner, D. 1995. *Media Culture: Cultural Studies, Identity and Politics between the Modern and the Postmodern*. London: Routledge.

Kristeva, J. 1991. *Strangers to Ourselves*. New York: Columbia University Press.

Lagerkvist, A. 2008. 'Travels in Thirdspace: Experiential Suspense in Mediaspace – the Case of America (un)known', *European Journal of Communication*, 23:3, 343–63.

Lauretis, T. de 1987. *Technologies of Gender: Essays on Theory, Film and Fiction*. London: Macmillan.

Lefebvre, H. 1974/1991. *The Production of Space*. Oxford: Blackwell.

Lefebvre, H. 1992/2004. *Rhythmanalysis: Space, Time and Everyday Life*. London: Continuum.

Massey, D. 1994. *Space, Place and Gender*. Cambridge: Polity Press.

Massey, D. 2005. *For Space*. London: Sage.

Morley, D. 2000. *Home Territories: Media, Mobility and Identity*. London: Routledge.

Mosco, V. 2004. *The Digital Sublime: Myth, Power, and Cyberspace*. Cambridge, MA: MIT Press.

Ritzer, G. 1999. *Enchanting a Disenchanted World: Revolutionizing the Means of Consumption*. Thousand Oaks, NJ: Pine Forge Press.

Rose, G. 1993. *Feminism and Geography: The Limits of Geographical Knowledge*. Cambridge: Polity Press.

Schutz, A. and Luckmann, T. 1973. *The Structures of the Life-World*. Evanston, IL: Northwestern University Press.

Sheringham, M. 2006. *Everyday Life: Theories and Practices from Surrealism to the Present*. Oxford: Oxford University Press.

Shields, R. 1991. *Places on the Margin: Alternative Geographies of Modernity*. London: Routledge.

Sjöman, V. 1956. *Flygblad. Resereportage*. Stockholm: P.A. Norstedt och söners förlag.

Skeggs, B., Moran, L., Tyrer, P. and Binnie, J. 2004. 'Queer as Folk: Producing the Real of Urban Space', *Urban Studies*, 41:9, 1839–56.

Smith, A. 1996. *Julia Kristeva: Readings of Exile and Estrangement*. London: Macmillan.

Sorkin, M. (ed.) 1992. *Variations on A Theme Park: The New American City and the End of Public Space*. New York: Hill and Wang.

Soja, E. 1996. *Thirdspace: Journeys to Los Angeles and other Real-and-Imagined Places*. Oxford: Blackwell.

Straw, W. 2005. 'Pathways of Cultural Movement', in Andrew, C., Gattinger, M., Jeannotte, S. and Straw, W. (eds) *Accounting for Culture: Thinking Through Cultural Citizenship*. Ottawa: University of Ottawa Press.

Strindberg, A. 1918. 'Ett Drömspel', in *Samlade skrifter 36*, 213–330. Stockholm: Albert Bonniers förlag.

Sundén, J. 2006. 'Digital Geographies: From Storyspace to Storied Places', in Falkheimer, J. and Jansson, A. (eds) *Geographies of Communication: The Spatial Turn in Media Studies*. Göteborg: Nordicom.

Trigg, D. 2006. *The Aesthetics of Decay: Nothingness, Nostalgia, and the Absence of Reason*. New York: Peter Lang.

Williams, R. 1974. *Television: Technology and Cultural Form*. London: Fontana.

Williams, R. 1980. *Problems in Materialism and Culture*: Selected Essays. London: Verso.

Other Sources

The Longest Journey (1999), Funcom, IQ Media Nordic.
Dreamfall: The Longest Journey (2006) Funcom, IQ Media Nordic.
The Truman Show (dir. Peter Weir, 1998).

PART 1
Scales of Opacity

Introduction to Part 1

André Jansson and Amanda Lagerkvist

In one of his 1920s Weimar essays, *The Hotel Lobby*, Siegfried Kracauer (1963/1995) provides an early account of a non-place (cf. Augé 1995). In the essay, which was part of his book-length study of the detective novel, first published as a chapter in *The Mass Ornament*, Kracauer sketches the socially alienating character of a place whose (lack of) identity is based on 'a relation to the nothing' (1963/1995, 179). The hotel lobby, Kracauer contends, is an inverted church. It is a space where people in transit, engaged in nothing but 'lounging', become equal to one another because they have 'dispossessed themselves of themselves', not in front of God, but through the anonymity of the modern crowd. In the hotel lobby there is silence, namelessness and pure exterior: 'Remnants of individuals slip into the nirvana of relaxation, faces disappear behind newspapers, and the artificial continuous light illuminates nothing but mannequins' (ibid., 183). Like most other non-places the hotel lobby is an opaque space. Coming as visitors we know what it is and what to expect. Still the lobby does not keep us there for anything other than *waiting* – for encounters, a taxi or the time to fly. It holds a merely transitory, disembedding function in-between the more concrete events of what Kracauer calls the 'actual life'.

Kracauer's analysis reveals that modernity in spite of its obsession with rationality and reason has not abolished the social presence of opacity and vagueness – quite the contrary. Almost a century after their creation there is still a peculiar actuality to Kracauer's writings, notably as to their identification of (post)modern spaces beyond or in-between the spatial and temporal stages of actual life, or spaces that fleet into one another or dissolve into the mists of uncertainty and forgetfulness. Similar themes were present, as we saw in Chapter 1, in the writings of Kracauer's contemporary Walter Benjamin, sometimes regarded as an early advocate of postmodern thinking (see Gilloch 1996). Both authors hold that it is precisely through the dynamism and progressivism of modernity, its continuous refinement of productive and consumptive functions, that spaces become opaque – either through alienation or through the dreamlike transmutation of spatial forms (such as the problematic confusion of interiors and exteriors in a hotel lobby or a shopping arcade).

Today it goes almost without saying that the mediatization processes of our times, through their very fluidity, contributes to these instances of spatial uncertainty. And the scholarly debate has contributed to both insights and mystifications related to this matter. What is hardly covered in the literature on mediated spatial opacity, however, is the full breadth of the question – the various *scales* in which opacity operates. In this part of the book our aim is to provide a multifaceted illumination of how mediation (the representation of space) and mediatization (the saturation of media in space) not only affect spaces of different size or scale, but also, and more significantly, cause ambiguities in terms of spatial proportions, positions and meanings (cf. Couldry and McCarthy 2004; McCarthy 2006).

In the first chapter Phil Carney and Vincent Miller introduce the concept of *vagueness*, aiming to highlight its positive implications for the study of modern space and culture. To understand vagueness properly, the authors argue, one must engage in spatial thinking from the very beginning, since the vague is always a matter of wandering, an activity that occurs in space and which produces uncertainties through its undermining of fixity. Carney and Miller apply two illuminating cases; firstly, how photographic practices in inter-war Paris opened up the *terrains vagues* of urban wastelands, and, secondly, how the representational vagueness of Jewish *eruv territories* in London articulates the traits of nomadic identity.

The following three chapters explore various scalar transformations in the geographies of media and communication. Orvar Löfgren goes on an archeological exploration in the back regions of his own home – the storages of obsolete, dying or marginalized media forms. His analysis exposes an often neglected scale of media circulation, the system of domestic spaces that we create for objects that wait for a new stage in their life-cycles. This is a scale of hidden cultural transformation, where time mercilessly kills or revives material objects. In the following chapter, in the form of two dialogues, Eva Åhrén and Michael Sappol discuss medical representations and ways of seeing the body. The first dialogue deals with modern see-through illustrations of the bodily 'interior' and problematizes the discursive regimes of revealing and magnifying the otherwise vague organic architectures and infrastructures behind the human shell. The second dialogue actualizes a different scalar transformation; the conservation and exposition of human body parts in the Mütter Museum of the College of Physicians of Philadelphia. Finally, in Chapter 5, Lisa Parks provides a critical perspective of the technological, economic, cultural and political geographies of communication satellites. Through a contextualization of their functions, charting for instance their media transmissions and abstract footprints on earth, Parks makes visible a scale of media space that normally escapes the realm of everyday attention, but nonetheless constitutes people's lived environments. The geography of satellites is a case in point of how modern communication ideologies operate through opaque infrastructures of cultural dissemination and control, which also contest the pre-established modern order of nation states.

Altogether, the four chapters of this section teach us that what we see and sense is not always, and perhaps not even very often, the 'true' face of a place, but rather a distorted surface or phantom image, while the logic of its production may stem from (communication) processes occurring within other scales. It is emblematic that Siegfried Kracauer paid particular attention to the hotel lobby in his analysis of the detective novel, perhaps the genre *par excellence* of modern literature. Like in a detective story, the aestheticized 'pseudo-life' of the hotel lobby refers to nothing else than its own 'strange mysteries', the slippings between the masks, concealing what might be going on underneath or beyond its own exteriorized immanence (Kracauer 1963/1995, 184–5). Something similar could be said about a conserved head in a glass jar in a museum. What goes on goes on elsewhere.

References

Augé, M. 1995. *Non-Places: Introduction to an Anthropology of Supermodernity*. London: Verso.

Couldry, N. and McCarthy, A. 2004. 'Orientation: Mapping Media Space', in Couldry, N. and McCarthy, A. (eds) *MediaSpace: Place, Scale and Culture in a Media Age*. London: Routledge.

Gilloch, G. 1996. *Myth and Metropolis: Walter Benjamin and the City*. Cambridge: Polity Press.

Kracauer, S. 1963/1995. *The Mass Ornament: Weimar Essays*. Cambridge, MA: Harvard University Press.

McCarthy, A. 2006. 'From the Ordinary to the Concrete: Cultural Studies and the Politics of Scale', in White, M. and Schwoch, J. (eds) *Questions of Method in Cultural Studies*. London: Blackwell.

Chapter 2

Vague Spaces

Phil Carney and Vincent Miller

One might say that the concept 'game' is a concept with blurred edges. – 'But is blurred a concept at all?' – Is an indistinct photograph of a picture of a person at all? Is it even always an advantage to replace an indistinct picture by a sharp one? Isn't the indistinct often exactly what we need?

Wittgenstein 1972, 34

Introduction

In this chapter we engage with the complex relationships between space, knowledge and power through a consideration of vagueness, vague practices and vague spaces. We argue that interlinked modern processes of the state and capital constitute hegemonic power through processes of fixing and enclosure of space, meaning and practice. In such operations, the strange and the vague are represented in a pejorative or marginalized manner, and become the target of order, control and rationalization. Thus we see the possibilities in strangeness and vagueness, and the practices associated with them (such as wandering, rambling, borderless existence), as political activities that run counter to the hegemonic powers of modernity, opening up possibilities for other forms of space and practice.

In the five substantive sections which follow, we will first examine the notion of vagueness and its relationship to spatial practices through its etymological origins. We will then look at the relationship between vagueness, representation and modern capitalism primarily through an examination of the work of Henri Lefebvre. The third section will consider vagueness and the practice of everyday life, seeing the vague as an inherently pragmatic understanding of the world with radical possibilities. We will then illustrate the concept of vague spaces through two examples: representations of the strange possibilities of *terrain vague* in urban photographic practice, and a discussion of Jewish 'eruvim' as a form of re-enchantment in contemporary urban space.

These two examples indicate the key characteristics of vague spatial practices. First, they both demonstrate the temporal spatiality of the vague by working within the interstices of modern urban landscapes. Secondly, they actively work against modern practices of enclosure by emphasizing multivalency. That is, they are spaces of additional meaning or practice, and in that sense are spaces of openness and possibility in meaning and in use. Thirdly, such spaces work against the rationalized urban landscape and its hegemonic order by inserting otherness and irrationality into the urban fabric. They of course differ in the sense that one is an artistic re-articulation critical of modernity, yet born within it, and the second is a landscape inspired by pre-modern religious tradition and cultural heritage. Yet it is also reconfigured according to the demands of contemporary urbanism, and is not, therefore, a backward-looking exercise in pre-modern nostalgia but the reinvention of a religious practice for modern times. Lastly, both practices are also capable of *estrangement*, not in the sense of alienation but in their capacity to *create* what might be called surreal effects, which work in excess of an 'otherness' to rationality. This is a process that combines interruption with production, the interruption of the modern matrix with the tactical elaboration of imaginative mobility.

Playing with Words, Playing with Power: An Etymological Introduction to Vague Spaces

In what follows we will wander across the domains of etymology, genealogy, politics, ontology and epistemology in an attempt to follow the vagaries of the vague. Etymology must always be a vague pursuit because traces blurred by the mists of time will leave us with an inevitable sense of the opaque and the indistinct. So it is with the word vague itself where there is a network of significations across which meaning roams, observing no borders and with no fixed points of origin or destination.[1]

We seem to have a fairly good idea that the word vague comes from the Latin verb *vagari*, to wander or roam, and the related noun *vagus* that means, variously, wandering, roaming; fickle; diffuse, aimless. There is a noun in English meaning an indefinite expanse so that 'a vague', perhaps close to its 'wandering' origin, relates to terrain and is thus immediately connected to considerations of space and geography. We feel it is better to examine the *dynamics* of the word, that is to say what it indicates as an *action*. Thus 'a vague', as an indefinite terrain, is the 'place' where wandering or roving occurs, a place of the nomad. The verb of wandering comes before the noun of space, the activity before the thing. More opaque, vague indeed, is the etymology of 'vagrant', which is not *necessarily* a cognate word ... but it *might* be. There is an Anglo-French word *wakerant* of Germanic origin (the origin of 'walk') that has been assimilated to the Latin *vagari*. *Wakerant*, not that far from 'walker', may have very readily shifted to 'vagrant' because of the similar sound *and* meaning attached to the first three letters *vag-*. Of course this is interesting

---

1   We will use the *OED*, Chambers, Onians (1966) and Partridge (1966).

from a linguistic view in that we might thus connect 'vagrant' to *vagari*, not because of its *origins* but because of its *destination*.

Vagabond is closely related though it appears later than its synonym 'vacabond': both seem to be traceable to *vagari*. It may be the case that the 'c' in vacabond appeared as a result of some kind of homophonous association between *vagari* and the Latin *vacare*, to be empty or free. In both vacabond and the subsequent vagabond such 'freedom' was to develop strong connotations of an immoral idleness.

Vague and its cognates have many 'vagaries' freighted with negative connotations. Thus we find 'lacking in character and purpose', or 'haziness in thought' in the word vague, or something that could be 'devious', with similar negativity in the word *vagary*. A not dissimilar word denoting mobility, to *deviate* or to be a *deviant* or *devious*, is, properly again, just a kind of wandering, though is largely negative *except* in the word 'deviate' which may retain a balance of positivity. Likewise to 'ramble', while retaining something positive in walking, is definitely negative in a speech, even though we know that the dialogue we call 'good' conversation will, indeed should, *ramble*.

There are two French words, feminine and masculine, *la vague* and *le vague* . Related to the German *Woge*, the French *la vague* is a wave and thus a 'movement' or a trend, something that arrives, or something that comes and goes, something that, being *in* movement, is thus, in the instability and fluctuation of the wave, related to the Latin *vagari*. *Le vague* takes in all the English meanings of 'vague' in addition to certain French metaphorical uses not found in English

We may thus argue that it is not a phenomenological notion of the indeterminate that gives us the activity of free mobility, but the active process of wandering that is crystallized in the indeterminacy of the phenomenological vague. The German *Woge* and the French *vague* may denote a wave or a sea-swell, partaking in an activity that cannot be bordered, territorialized or occupied in the same way as land. The wave and the swell are associated with the mobility of the boat, of the movement of rivers and the dangerous indeterminacy of marine weather in its squalls and storms. Are not sailors wanderers, and have they not at times been stigmatized with the rogue qualities of vagabonds? Rivers and the sea are not places of *being* or 'identity' despite the static names we give them. As flows they are always in movement, always *becoming*. No wonder that Herakleitos – the pre-Socratic philosopher of becoming, whom we should oppose to the mainstream Platonic philosophy of being – is best known for his questioning of identity in the statement 'One cannot step twice in the same river, for the water into which you first stepped has flowed on' (Davenport 1979). Vague as motion resists categories, boundaries, calculations and identities. Vague is what moves to escape these determinations and gives its name to the spaces through which such resistance occurs.

Enclosure: Vagueness, Wandering, Modern Capitalism and the State

Meanwhile, in our laws today, 'vagrancy' remains illegal, not for any other reason than wandering *itself* is seen as an inherently suspicious activity so that the act of spatial errancy is sufficient to be an object of criminal prosecution. According to established power, the vagrant, the tramp and the traveller must be regarded, aside from anything else they might do, as somehow criminal purely as a result of their wandering practice. (see, for example, Chambliss 1964; McNeel 1972). Cresswell (1997, 2006) has examined mobility as a problem not only in modernity but also for the modern and the so-called 'post-modern' imagination. Through his study of the figure of the tramp, the hobo and the drifter, he has shown how modernity produces 'others' in its discourses and practices of space (Cresswell 2001). Such exercises of power also aid in the construction of the important contrast between ordered domesticity and potentially disordered public space, between fixity inside the ideal bourgeois home and movement in the dangerous outside. In this way power also orders on the level of gender and age. This is evident not only in the problematization of the female tramp (Cresswell 1999), a kind of double other, but also in panics about unruly young people in public space (Pearson 1983). In all these developments that target aimless wandering, we see two major interpenetrating imperatives: a state that fixes people to their proper territory and a will of capital to put them to good use in an efficient division of labour.

Thus modernity will turn a wandering with a certain potency in its strangeness, into that which is moralized as unnecessarily profligate and uneconomically exorbitant or *extravagant*. Here we find the operation of a carefully calibrated, calculative rationality of a political economy. It follows a logic of quantitative equivalence in the market – spoken about variously by the political economists (as exchange value) and criticized in various ways by Marx (1976), Adorno and Horkheimer (1997), Bataille (1988), Deleuze and Guattari (1984, 1988), among others – that is also a rationality of precision in science, definition in knowledge and measurement of space.

The spaces of our late-modern cities are constructed through such an axiomatic. Capital flows through the micropolitics of our lives, interacting with the other modalities of power and desire that produce the everyday. Employees and customers, citizens and flaneurs alike are encouraged to inhabit a docile world designed as a rational eutopia (sic), a perfectly functioning, predictable, measurable, calculated and secure space, fit for the Last Man (Nietzsche 1969).

Such relationships between a functioning, reasonable and calculated space, and the vagaries of everyday life are the major starting point for Henri Lefebvre's analysis of space and everyday life under capitalism. Lefebvre sees the state, bureaucracies and, fundamentally, the practices of the *spatial sciences* as the primary players in the reproduction of the capitalist state. Planners and engineers 'rationalize' the landscape, divide it into parcels and routes for the efficient production and flow of goods and people. Geographers and other social scientists also legitimate these ambitions in their will to measure and arrive at a calculated truth of space. Lefebvre argues that the rationalized 'system', in the form of spatial science, produces everyday life through the use of space. In other words, the spatial operation of power occurs in the spatial practices of everyday life, not just in the workplace, and it is in the realm of everyday life where struggles against such rationalized spatialization must occur.

Lefevbre conceptualized the relationship between capitalist rationality, everyday life and the production of space as a trilectics of three 'moments'.

Spatial Practice. In this arena of power is action, in the sense of both the medium and outcome of human activity, behaviour and experience (Shields 1999). As such, it is the quantitative aspect of social life, rendered open to accurate measurement and description (Soja 1996). This 'moment' is often referred to as the *perceived*, meaning that spatial practices are perceived in the commonsensical mode and therefore involve daily life and routines under the logically rationalised urban (Shields 1999). The cognitive consequences of spatial practice could be compared to the 'natural attitude' or the 'paramount reality' of phenomenological sociology (see also Miller 2007).

Representations of Space are linked to the expert knowledges and discourses on space, which *conceive* of space in relation to production, reproduction and order within the capitalist state (Shields 1999; Soja 1996). They 'theorize' space and present an abstract form of lived experience in space. These are 'plans' conceived to impose order and make things run smoothly. Such forces are central to the forms of knowledge and claims of truth made in the social sciences. They ground the rational/professional power structure of the capitalist state (Shields 1999).

Spaces of Representation is perhaps the most complex term in Lefebvre's trilectics. Loosely, Spaces of Representation can be considered 'discourses' *of* space as opposed to discourses *on* space (Shields 1999). However, there are several possible themes to this 'moment' in the production of space:

1. A utopian element where Spaces of Representation are characterized as being at the heart of 'fully lived' spaces (ibid.) or 'space as it might be'. Spaces of representation involve what is thought to be potential, possible or achievable. Thus Lefebvre also refers to them as *the imagined*.

2. Through a space of the other, Spaces of Representation may be sites of resistance to the dominant social order (Soja 1996), seen in many different ways, and in radically different situations. For example the spaces of the slums, barrios and favellas embody and enact a certain resistance to the rationalized order of the city, even as they are at the

same time products of exploitation, exclusion and othering. The work of the Situationists which upset the rational everyday use of space at strategic sites (Shields 1999). It is also linked to the 'clandestine or underground side of social life' (Lefebvre 1991, 33).

3. Spaces of Representation are linked to the qualitative aspects (or the 'art' as opposed to the 'science') of social life and as such may be partially unknowable (Soja 1996).

4. Spaces of Representation are linked to a social imaginary (Shields 1999) full of myths and legends laid down through history and embedded in symbolism.

5. Elden (2004), has recently emphasized the influence of Heidegger on Lefebvre's work, and sees Spaces of Representation as *connaissance*: less formal or more local forms of knowledge. This is similar to Heidegger's notion of poetic dwelling, a lived experience of everyday life, knowledge and imagination through practical engagement. Thus Elden, focuses on the concept of spaces of representation as an 'authentic' space of 'doing' or 'dwelling'.

Thus Spaces of Representation have many dimensions: the possible or potential, the non-paramount or 'other', the qualitative and unknowable, the 'authentic, practical, and 'fully lived'. In Lefebvre's trilectics of spatiality, this *potential* space – the practice of a Space of Representation – battles with the production of dominating representation and abstraction. The latter are attempts at enclosure through conversion into property, the enforcement of order (in, for example, planning law), the imposition of homogeneity and rationality (through separation of land uses), grounding the spatialized power structure of capital.

In the modern era, Lefebvre argues, Spaces of Representation have been overthrown by Representations of Space as space has been 'pulverized' by modern demands for exchangeability and homogeneity through commodification and control. In short, 'colonization' occurs when conceptions or representations precede space that he calls 'fully lived' and space itself becomes a representation. All this is summed up in the term *abstract space*:

> This space relies on the repetitive – on exchange and interchangeability, on reproducibility, on homogeneity. It reduces differences to induced differences: that is, to differences internally acceptable to a 'set' of 'systems' which are planned as such, prefabricated as such – and which as such are completely redundant. (Lefebvre 1991, 396)

It seems that Lefebvre is a humanist in his vocabulary of alienation, and a Heideggerian in his appeal to authenticity, relying on a past of authentic, unalienated existence prior to the arrival of domination. A different perspective would see 'living' and 'conception' as inextricably bound together:

hand in hand the state and capitalism thus produce a new form of living and conception, practice and representation, characterized not so much by power as such, but by a different kind of power, with new characteristics, stabilities and instabilities. The order of the state-capital nexus need not overwhelm us and its weaknesses, interstices, and volatility create the potential spaces of resistance.

De Certeau (1984) would argue that the relationships between time and space are integral to our understanding of everyday practices, and in particular the practices of the marginal versus the more powerful. He characterized the pragmatics of power through the distinction between the strategic and the tactical. Strategic operations are designed to inform or control the social order. They are actions of the powerful that elaborate theoretical places as a centre of power or a vantage point from which to realize projects. Strategies are the practices of actors who have the capacity of space, and the ability to represent a situation through abstraction, to produce hegemonic discourses, to stabilize and to 'plan'. In this sense, the link with Lefebvre's 'representations of space' becomes clear. However, this is only half the story. Strategic actions are liable to become overrun by a multiplicity of other kinds of operations called tactics. Where strategies are the result of a control over space, tactics rely on time:

> I call a 'tactic', on the other hand, a calculus which cannot count on a 'proper'
> (a spatial or institutional localisation) ... The place of the tactic belongs
> to the other. A tactic insinuates itself into another's place, fragmentarily,
> without ever taking over its entirety, without being able to keep it at a
> distance... The 'proper' is a victory of space over time. On the contrary,
> because it does not have a place, a tactic depends on time – it is always on
> the watch for opportunities that must be 'seized on the wing'. (de Certeau
> 1984, xix)

In this relationship, time is associated with the marginal, the subversive, the subcultural. Marginal groups cannot take up a 'proper' position so they are obliged to take advantage of situations as and when they present themselves: they must respond to their environment. For example, colonizing disused urban buildings and spaces that are mostly outside hegemonic control (*terrain vague*), or flourishing during night-time, or other times when commercial use is at a low ebb. Thus the space of the marginal is as much about time as it is about space, it is fleeting or interstitial, giving it a vague nature. Not so much spaces, they are 'interruptions' or 'events' within dominant land uses and hegemonic spatial discourses.

The Chicago School associated marginalized subcultures with what they called the 'zone in transition', a space that took in immigrant communities, transient populations, 'bohemians' and sexual subcultures. It was also a space of the strange. The term itself implies a link with time and movement: influx and abandonment. It was seen as an unstable, itinerant space susceptible to the temporal ebb and flow of mobile populations. In short, it was the zone of the city which, in many ways, resisted hegemonic order because of spatial and social wandering. And it was for this reason that it was targeted by the police and the courts, who were, of course, involved in the greater crimes of the central Loop zone: crimes ignored because they were committed in the suites, chambers and offices of power.

Vagueness and Everyday Life: Ontology, Phenomenology and Practice

We have tried to show that in its provenance and its genealogy the *philosophical* concept of the vague must acknowledge in some way the historical dynamics of power. It would be idealist to attribute to the abstract notion of 'vague' a greater priority than the 'vague' as concrete practice or the product of concrete practice in the matrix of power. Therefore to speak of an *ontology* of the vague is to miss its meaning as a process or activity or practice of wandering, as opposed to staying still or settling. It may be here that we encounter Lefebvre in a Heideggerian desire to evacuate from ontology its metaphysical stasis and give it some root in active practice. Marx (1975) had already indicated the priority of social practice over philosophical contemplation, and it may be the case that Lefebvre offers a gloss not just on the passive nature of contemplative representation but on its operation as a practice of domination, to which what we would call vague and strange social practices offer the possibility of resistance.

From a radical phenomenological standpoint abstract space is the attempt to produce a 'paramount reality' (of a nexus of rationalization, of capitalism, the world of work and the state). It is the imposition of the 'natural attitude', with all of its mechanical repetition, rationality and blasé-ness, on our experience of space. Space becomes knowable, predictable, and manageable (Miller 2007). The dynamics of the vague – openness, possibility, ephemerality, unpredictability – are squeezed from its potential. If we can see Spaces of Representation as part of the practical experience of space as 'doing' or 'dwelling', then this practice may also emerge from that phenomenology of experience. Schutz's (1962, 1964, 1970) more general phenomenological view of space takes a *practical* perspective so that the typifications, categories and recipes we use in daily life 'work' in the social practice of the world but, translated into the cognitive, they are also essentially uncertain in the sense that they cannot function as exact understandings. At the level of representation they are vague generalizations that through *use* are pragmatically 'good enough'. Thus, Schutz's phenomenological approach to the social world rests on vagueness (Natanson 1986; Stoltzfus 2003; Miller 2007). In this way, Schutz allows us to see how some of our pragmatic understandings of the world may act in Lefebvre's Spaces of Representation.

In his *Philosophical Investigations* Wittgenstein (1972) found similarly vague epistemological concepts in certain practices of language in everyday life:

> Compare *knowing* and *saying*:
> How many feet high Mont Blanc is –
> How the word 'game' is used –
> How a clarinet sounds.
> If you are surprised that one can know something and not be able to say it,
> you are perhaps thinking of a case like the first. Certainly not of one like the
> third. (Wittgenstein 1972, 36)

Yet while we might perhaps speak about an epistemology of the vague, it should really only occur in the form of a nomadic knowledge. Here we have not so much an epistemology of the vague, as a vague epistemology, and even then we are left with a rather dissatisfying abstraction. A vague knowledge is, first of all, a practical knowledge. Perhaps we can go further than Schutz and argue that there is no 'epistemology' or meta-discourse of knowledge outside the specific practice in question.

Using the term more actively vagueness is a practice of wandering in a vague terrain: it is an activity in space. It is only by a metaphorical shift that we apply it to an epistemology or an ontology, a shift that takes us away from the concrete notions of space that we should be invoking. If we speak about a vagueness in theoretical writing, that is to say a positive vagueness, then it is only by thinking of such writing as a kind of spatial practice – or at least a material practice that occurs in the 'spaces' of institutions and of discourses – that we understand what 'vague' theory should be.

This creates room for the multivalency of urban spaces through an emphasis in the vagueness embedded in the moment or the 'doing' of life. Ultimately, the construction of intersubjective 'reality' is seen as an open-ended pragmatic process of living. The emphasis on space as a construction (or 'becoming') in practice links the concept of intersubjective vagueness with space as an ephemeral, multivalent, unknowable 'event'.

Practices of Photography in the *Terrain Vague*

We know how photography may fix, capture and control through various identification and surveillance practices (Tagg 1988). It is also a servant of commodification in advertising and packaging, product placement and marketing (Goffman 1969). In those asymmetries of power constituted by a highly centralized mass media (Couldry 2003), photography is also a vehicle and frame of news, political performance and military display all of which have their spatial dynamics. Restated in Lefebvrian terms, photography may be complicit with practices of state and capital enacting the 'Representation of Space'. However, photography can also be used to allow space to present itself in a way that attaches itself to, interrupting and fracturing power: it may engage in a certain creative practice of the real, and in the dynamics of a space that resists commodification, categorization, calculation and control. Therefore this mode of 'representation' opens itself up to vague and strange spaces that defy more *formalized* representation. In Lefebvrian terms the resistant, subversive and creative use of photography involves practices in 'Spaces of Representation'.

A consideration of photography highlights how Lefebvre interchanges the terms representation and space. To oversimplify: if a photography of control is the Representation of Space, in which an attempt is made to submit space to the force of representation, then a photography of resistance is the Space of Representation, where representation is interrupted by the force of space. In other words Lefebvre might be able to tell us that the photographic relationship to 'the real' (in the sense of Barthes, 1981) is not so much the representation of space as opening up representation to space. This is where photography encounters Henri Lefebvre on the terrain of the vague and the strange.

The oscillation of city spaces between the rationalized wealth-producing landscape and indeterminate terrain is worth noting, and has been considered within architectural discourse by Ignasi de Solà-Morales Rubió (1995) who has used the French phrase *terrain vague* to theorize spaces in the city that are empty, abandoned, derelict, in which often a series of land uses has taken place. Encompassing some of the etymology of the French *vague* (absence of use, the indeterminate, blurred and uncertain, and 'wave' of movement), terrain vague is seen as space that is free, available, unengaged, limitless, uncertain, roving and temporary.

The wider context of his discussion should begin with mid-late nineteenth century Paris where the paradigm modern urban renewal by Haussmann and his successors gives rise to *terrains vagues*, vacant wastelands around the railway stations, at the edges of the new developments and on the periphery of the city between the urban and the suburban. They were vague in their boundaries, in their future, in their ownership and the rules governing those who occupied them, whether ragpickers, beggars, tramps, gypsies, or other shanty-dwellers. Eugène Atget, later beloved of the surrealists, patiently documented these and many other spaces of disappearing 'old Paris' (Walker 2002). Today we may find not dissimilar spaces – bohemian, illicit, criminal, squatting, vagrant – that escape controlled consumerism, that are 'non-productive' and strangers to the calculative rationality of security and the market. What is interesting for the purposes of this chapter, is how Solà-Morales Rubió makes explicit the connection between 'Terrain Vague' and 'Strange Spaces'. While we might doubt both the specificity of his description of urban estrangement to late modernity, the precise nature of this estrangement, as well as his psychoanalytic attempt to see the response to these spaces as projections of our estrangement, fear and anxiety, it is nonetheless true that our urban spaces embody a tension between order and disorder, determinacy and the vague, the familiar and the strange.

In that sense, *terrains vagues*, and their representation in photographic art, bring us to the fundamental problem of contemporary social life: a fear of disorder, indeterminacy and the strange, and of a call for the 'management' of forms of resistance to hegemonic structures of control designed to impose order. Thus, Solà-Morales Rubió argues that the eye of the photographer is drawn to terrain vague because of its potential exception from forms of power. *It connotes expectation, of the other or the alternative.* It possesses possibility in a landscape otherwise colonized by the structural violence, rationality and order of architecture and planning. The subject of architecture is the citizen: the subject of terrain vague is the nomadic stranger, subject, however, to forces that 'other' and estrange.

Examining the surrealist possibilities of documentary photography in interwar Paris, Ian Walker (2002) turns his attention to the *terrain vague*. Going beyond Solà-Morales Rubió's narrow periodization he discovers how photography is attracted to the surreal – read 'strange' – possibilities of these spaces from the time of Eugène Atget in the late nineteenth century, to the surrealists of the inter-war period. Indeed, unnoticed by Rubió, Man Ray had given the title 'terrain vague' to a photograph of a strange, sloping, wasted space occupied by an odd remnant of ironwork next to a tree, a space traversed by the traces of human movement in the imprinted steps (Figure 2.1). In one of his most famous photographs, Cartier-Bresson the surrealist shows another strange scene of wasteland (Figure 2.2). This time we recognize it as behind the Gare St Lazare in which a man balletically leaps from a half-immersed ladder to a point on a shallow pool of water, the leap both reflected in the water itself and in a decaying poster in the background, all enveloped in a faint, vague mist of the type to which cities can succumb. This is a little miracle of a leap in which the blurred mobility of the silhouetted stranger, the vibration of a transient fall of water, a ripple across its surface, the happenstance resonance of figures and shapes, the misty scene in which light begins to dissolve the St Lazare sheds, the chance symbolization of man and poster with the possibility of something else happening, all combine to constitute a strange, vague dream that erupts from a disordered vacuole in ordered city space. All the time the surrealists sought to interrupt the rationalized world of modernity with the dream. They also found the dream in photographed reality, including the dream of certain spaces of representation existing in the midst of the hegemonic representation of space.

Sheridan (2007) fruitfully develops and revises Solà-Morales Rubió's use of *terrain vague*, broadening the term from the indeterminacy associated with dereliction and abandonment to any area, space or building where the city's normal forces of control have not shaped how they are currently perceived or occupied. This brings us back to Lefebvre's description of colonization,

Figure 2.1   Man Ray (1932) *Terrain Vague* © Scala (for MoMA in New York)

where abstracted forces and representation assert order and control on the imagined and the perceived. Sheridan rightly points out that indeterminate territories are, just as the term 'Spaces of Representation' implies, also the space of subculture within the city. These are spaces drifting outside, and interrupting the inside of, the hegemonic control of capital and state, the market, planning and authority. Such an observation captures the 'possibility' of vagueness and indeterminacy associated with Spaces of Representation, whether the uncolonized existence associated with phenomenological poetic dwelling, or the practice of forging new dreams, possibilities and forms of life out of fractures in rationalized space.

Rubió, in his conception of the photography of *terrains vagues*, Walker, in his description of surrealist documentary photography, and Sheridan, in his discovery of subcultural possibilities (with a text suffused by photographs), allow us to see how both photography and space might work in more creative ways. Indeed photography itself, as Rubió hints and Walker elaborates, is an inherently vague and estranging practice. It may have a dynamic of coding in its conventions but it also escapes, as Roland Barthes' last work (1981) argues, from the code into the affective real. The photographic real pierces the code,

Figure 2.2   Henri Cartier-Bresson (1932) *Behind the St Lazare Station, Place de l'Europe* **© Magnum**

but we can also say it strays outside it. So in this sense we might allow Lefebvre to tell us that 'documenting' the real of the terrain vague is not so much the Representation of Space as opening up representation to space. It is in this light that we can situate, so to speak, the recent 'terrain vague' expeditions not only into abandoned spaces in the interstices of the city, but also into the hidden underbelly of the 'functioning' city, where photography turns the urban inside-out and upside-down, rendering strange its most mundane functions (for example, the practices seen in http://www.sub-urban.com/ and http://www.urbansickness.co.uk/). Here the photographic image is not just the *record* of a trangressive expedition, it is also its *practice*. These images bring back traces, souvenirs, of the previously unseen, but they also indicate – not infrequently through self-conscious lighting effects – the possibility of another resistant practice of photography.

Vague Spaces and Subculture: Eruvim as Spaces of Nomadic Possibility

> I just wanted to make sure that the whole thing is about presentation and
> explanation, and about this silly idea that you have to find a word, an English
> word, for eruv. You don't. You don't have to find an English word for kosher,
> you don't have to find an English word for halal, you don't need an English
> word for chuztpah and you don't need to find an English word for eruv. And
> they're rushing to find an English word for eruv and came up with boundary,
> and as soon as you talk about boundary people start panicking, and frankly
> speaking Jews don't want to be in a boundary either. (North West London
> eruv user, interview by author, 2006)

An interesting, perhaps strange, example of what might be regarded as a
'subcultural' vague space is provided by eruvim. Unlike the practice of *terrains
vagues* photography, which actively seeks out spaces which have become
exceptional to rationalized hegemonic order, because of their tenuous,
interstitial nature, eruvim are tied more directly to a specific community.
Thus, as opposed to poetic happenstance or an opportunity 'seized on the
wing', eruvim are more intentional spaces, created with purpose. Their
'vagueness' relies more on the temporal nature of their existence, and the
nomadic circumstances of their creation. In many ways they provide a more
direct challenge to the hegemonic forces of enclosure and a rationalized urban
order.

Recently, the eruv has captured the imagination of geographers, architects
and others because they are non-material, ephemeral spaces that demonstrate
the plural forces and multiple meanings possible in the terrain of the city
(Cooper 1996; Herz and Weizman 1997; Vincent and Warf 2002; Watson
2005; Stoker 2003; Siemiatychi 2005; Smith 2007). An eruv (or, more properly,
an *eruv chatzerot*) is a space designated under Talmudic law. Its use comes
from the set of restrictions placed on observant Jews during Sabbath days
(Friday sunset to Saturday sunset), most importantly the 'carrying law'.
Basically this states that one is not allowed to carry, pull or push *any* object
in the public domain during the Sabbath. The word 'eruv' essentially means
'mixing' or 'blending' and what the eruv does is effectively negate or blend
separate domains of space by creating a new legal context. Public and private
space is mixed by officially creating one large private domain outside specific
dwellings, encompassing streets, parks and other areas to the limits of the eruv
boundary.

By creating a totally enclosed space, an eruv is able symbolically (and in
Talmudic law, legally) to turn a collection of houses (or private spaces) into
one large house or private space during Sabbath days. Therefore, observant
Jews residing in the eruv only have to face the lower level of restrictions that
apply to private domains. Practically, this means that they are allowed to carry
house keys, push prams and wheelchairs, use a cane or carry a book from place

to place within the eruv, and not violate the spirit of the Sabbath. Families with young children are able to go to synagogue together, and older and infirm people are not isolated in their homes. The celebration of the Sabbath is made more inclusive and enjoyable, and yet still kept holy.

In a challenging example of vagueness, the eruv boundary is a symbolic wall, which may be made up of existing structures, fences, streets, some forms of topography (natural barriers such as rivers for example), as long as these are continuous. Where they are not continuous – say, where a boundary crosses a street – a (symbolic) doorway has to be created to allow for the continuity of the wall. There is no limit on the number of doorways or entrances but they have to observe certain rules. First, like any doorway, they need to have vertical supports on either side which act as the frame. According to Talmudic law, these need to be a minimum of ten cubits in height. These supports are usually accomplished through the use of existing utility poles, but sometimes new poles are erected. For example, the North West London eruv established in 2003 uses boundaries such as the M1 motorway, a footpath through Golders Hill Park, the Finchley Central train line, existing utility poles, thin wires across streets, and a small number of new poles, to maintain its borders (see Figure 2.3).

Secondly, on top of the vertical supports any door must have a lintel: often fishing line, or cable, or some sort of wire is stretched from pole to pole across a street or open space. The lintel effect is further sustained by the addition of a *lechis*, a small post or rod which is added to the top of the pole so that the cable sits 'on top' of the vertical support.

Eruvim are a relatively unknown yet increasingly common feature of modern Western cities. There have been over 200 established in cities all over the world and indeed many cities, such as New York, Toronto, Chicago and Los Angeles have several eruvim (Vincent and Warf 2002). Many have been established within the last 15 years, the result of a recent surge in those adhering to more orthodox Judaism. Thus, on Friday afternoons in cities all over the world, a new space is created overlaying the existing urban fabric, unbeknownst to the vast majority of their inhabitants. For those who do observe, websites are checked (using a traffic light system), or text messages are received to let them know that the eruv border is intact and working.

Of all the recent work on eruvim, the architects Herz and Weizman (1997) have arguably the most interesting account of what an eruv is, and the historical condition from which eruvim have emerged. They argue that to understand eruvim, one has to understand the two poles of existence that define Jewish history and the relation to space: 'kingdom' versus 'placelessness', or 'city' versus 'desert'. Following the exodus from Egypt, the Jewish people were essentially nomadic tribes wandering through the desert before it was revealed to Moses that a 'promised land' awaited them. They were in a continual situation of pursuit and preparation, looking for place and stability. Thus,

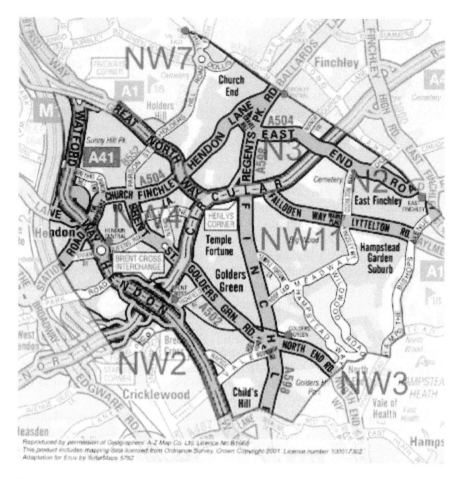

Figure 2.3 **The North West London eruv (http://www.nwlondoneruv.org/images/masterwebmap.jpg)**

laws with regard to place had implicitly to refer to places that were, as yet, unattained: imagined places in a future.

A second major displacement took place following the destruction of the First Temple and the Babylonian exile. After the return to Judaea and the building of the Second Temple, a substantial diaspora remained behind and spread over the Eastern Mediterranean. In this way a kind of alternative, stateless, nomadic state was chosen by many, with 'home' more as an imaginary than a loss. A nomad is not tied to a place, and does not claim rights to a fixed location, so the 'places' of the Jewish people in a symbolic desert had to become portable and tied only to the temporary state of *'encampment'*. Like all nomadic forms of existence, this experience is in marked contrast to the 'state' form of fixed settlement, boundaries and 'enclosure'. Eruvim thus carve a temporary, portable space out of the placelessness, with a vision, promise or utopia of a settled future (Herz and Weizman 1997).

Displacement was renewed after the Roman destruction of Judea in the first century AD. The Second Temple was destroyed and Jewish people themselves were banned from entering Jerusalem under pain of death. A second Jewish Diaspora occurred, with new migrants joining old in settlements that would be scattered throughout Europe, North Africa and the Middle East.

As Herz and Weizman point out, Jewish thinkers of the new diaspora were once again put in the position of revising laws that place an emphasis on a future constituted by an ideal reappropriation of the 'place'. Within this context, they argue, it was possible to interpret the eruv, at least in part, as a symbolic reconstruction of the temple: 'Walls' and 'doorways' imply a roof, the implication of a 'roof' encompasses all space within it into the realm of the private.

In that way, the eruv is built on a chain of signifiers: borders imply walls; wires signify doorways; walls and doorways signify a roof; a roof signifies the Temple. This ultimately leads the eruv back to Jerusalem: a new emblematic centre of diaspora and the possible focus of an ideal settled 'place', perhaps interpretable as the land promised to a nomadic people of the desert, or perhaps a symbolic point of departure and/or destination for the diaspora. A complex set of meanings is at play, since the nomadic state refers both to a Biblical past, a possible symbolic present, as well as a relationship to an ideal future in a world that hovers between this and the next, the present world and the one to come. In this respect, the rationalized urban landscape becomes overlain, indeed re-enchanted with a complex web of historical, Biblical and Talmudic meanings, combining significances and allusions to past, present and future, all in the interacting registers of the actual, the potential, the imaginary and the ideal.

Not without controversy, the recent establishment of eruvim in north London, Tenefly, New Jersey and Outremont, Quebec have seen high profile conflicts in which they have been bitterly resisted by some local groups (Cooper 1996; Watson 2003; Stoker 2003; Siemiatychi 2005; Smith 2007). Cooper (1996) agues that the North West London eruv controversy was built around objections derived from modernist perspectives of space and property. Here the eruv was seen to challenge the principles of universalism, public secularism, and the division between public and private space. Eruvim challenge modern state and capitalist notions of space as 'territory' or 'enclosure', or what Cooper refers to as 'zero-sum' ownership of space, in which rights to or authority over a space for one group or individual withdraws rights to or authority over space for another.

This is also recognisable as the logic of mutual exclusivity that also determines the operations of identity and difference in forms of knowledge, disciplinary division as well as most concepts of nationality. In modernity, borders are identical with regions, territories, or enclosures, and imply, to a certain extent, a singular monopoly or homogeneity on material ownership, jurisdiction, law, or meaning. This is dependent on the presumption that space is a singular in terms of its use, jurisdiction, authority or ownership. This jurisdiction, we assume, is dividable into subregions, like continents being divided into countries, countries into provinces or states, states into counties, (and so on), with each domain nested within a hierarchy of authority (Smith 2007, 43).

Much of the controversy surrounding the establishment of eruvim has revolved around this slippage between modern capitalist conceptions of use, rights and ownership and the idea that an eruv boundary would only affect those who 'observe' or 'believe' in it, that it would only be relevant to certain people and have no impact on others. Opponents to the eruv could not see the eruv as anything but an attempt to claim territory and impose a dominant meaning on the space coincident with it (Cooper 1996).

> I run a public relations business so I blame the people from the very beginning who got the messaging wrong. They translated eruv as 'boundary', they shouldn't have. They shouldn't have but they did. And so it became the Ham and High starting writing about the boundary, the Jews are building a boundary, and they shouldn't. (North West London eruv user, interview by author, 2006)

Eruvim do not work with this modernist spatial ontology. First, they are, a combination of real and imagined space (Siemiatychi 2005), using real physical borders and a fully enclosed space in a physical sense, but with a non-demarcated symbolic meaning. They give existing material structures an *additional* symbolic function. Secondly, they are ephemeral, and not universally identical with any region of space (i.e. they have a time-based relevance). And lastly, their meaning and legal status are only relevant to one specific group of people (Orthodox Jews). Those outside Judaism would have no idea if they were in an eruv or not.

In that sense, eruvim challenge the notion of a secular, rational modern public space with a demonstration of polyvocal, multivalent, and religious meaning (brought into a modern context and a modern set of problems). As Cooper (1996) and Siemiatychi (2005) have pointed out, it is a good example of how space can, in some sense, be 'used by' or 'belong to' more than one group or set of persons at the same time, thus allowing for the display and acceptance of difference in public contexts in demographically diverse, multicultural cities.

Conclusion

'I'm going to the countryside ...', 'where are you going?' Where does the countryside begin and the city [end]? You don't really have those sort of boundaries. It is not an unusual thing. We're not talking about Canary Wharf here where you drive up, there's a barrier and you're in. We're not talking about a border where your passport is stamped and now you're in ... So it's not about going in and going out, we live within an eruv, feel comfortable within ... before the eruv was erected we still felt really good within our community. The border of our community doesn't exist, there isn't one, it's notional. (North West London eruv user, interview by author, 2006)

When we speak about the vague in the everyday it is often morally negative in its association with blurring, haziness, the erratic, the restless, a lack of character, an absence of purpose. Similarly, when we philosophize about the vague, we only seem to be able to speak about it in terminologically negative words, of the in-definite, the im-precise, all of which bear traces of the morally deviant and the contemptibly transgressive. In this chapter we have emphasized the vague and the strange as forms of social practice. Along the way we have encountered the related, even at times concurrent, actions of 'wandering' and 'estrangement'. We maintain that the interlinked statuses of the strange, the stranger and the estranged are linked to the *vague* as a wandering practice.

Two main forms of power have confronted wandering practice: first, the state and, secondly, capital. Modernity is characterized by a state-capital nexus, which is particularly ruthless in its stigmatization, marginalization and exclusion of vague practice. The development of the state has been associated with ever more secure borders and barriers, as well as the fixing of citizens in physical space and in the bureaucratic archive. As a result, space becomes measured, precise, knowable, exchangeable and repetitive, encouraging the unreflexive blasé nature of the natural attitude. Land is enclosed and territorialized by the coded order of the market and the state, with its laws, violence and practices of bordering, categorization and representation. These forces contribute to a calculated spatialization of cities, rendering them fit for secure and controlled consumption.

Modernity seeks to eliminate the unknown, the obscure, the opaque, the strange and the vague. It is an imperative that seeks not only to extinguish the vaguely known and unknown but also the unmappable and the unrepresentable in space. Henri Lefebvre helps us both understand these basic dynamics and appreciate how they produce space in modernity, as well as making space for counter-hegemonic possibilities amid control.

In Lefebvre's Spaces of Representation there is another kind of theoretical practice that resists the Representation of Space: it discovers not only what is left behind but also the possibilities emanating from within the unstable practice of domination. Practices of the strange and the vague, not only detect their resisting movement but help to create it. Vague spatial practices have the following characteristics. First, they work in the spatial or temporal *interstices* of modern urban landscapes. Second, they are *multivalent*, permitting *openness* and *possibility*. Third, among their possibilities is the chance that they may become *counter-hegemonic*. Arts of resistance may take the form of spatial practices and we have encountered possibilities in the very different cases of photography and the eruv.

We have seen how the eruv moves in time what would otherwise be a fixed space, and how it shifts uncertainly and indeed vaguely through the registers of the actual, the virtual, the imaginary, the symbolic, the ideal and the real. If it refers back to the practices of desert nomadism and diaspora, then its wandering in time is also a kind of roaming in space. At the cognitive level it defies the identity of easy definition and categorization; at the spatial level it interrupts the normal way in which places are parcelled out in the modern city. Calling up the idea of hybridity in order to subject a religious rule to a different perspective, the eruv is traversed with multiple possibilities. In one sense it is a practice of nomadic evanescence, in another it is an attempt to fix space, albeit temporarily. But as a kind of encampment, the eruv passes through time and space without possessing either. It might be seen as a 'rationalization' of a religious practice adapted to the demands of the city, or an eruption of strange, indeed surreal enchantment in the disenchanted, rationalized spaces of the city.

Surrealist and *terrain vague* photography demonstrate how the vague and the strange may constitute an art in the interstitial spaces of the city. We have seen how Sheridan looks at vague living in the subcultural spaces of Berlin, and how in the vague image Walker discovers an artistic practice of estrangement of and in the 'real' through proto-surrealist, surrealist and post-surrealist photography. Solà-Morales Rubió gives us an account of the practice of photography in the space of the terrain vague that continues to this day. In this way, photography may act as both a 'representational' practice in the Spaces of Representation, and as a creative production – perhaps linked to forms of living, at least potentially – in the interstices left behind, or erupting from, the modern, urban, calculative representation of space by the nexus of state and capitalism. Here Cartier-Bresson's estranging 'decisive moment' is a Space of Representation that is, in de Certeau's words, 'seized on the wing'.

Used outside the confines of commerce and control, the photograph allows the eruption of an affective real amidst the coded conventions of everyday life. Strangeness is returned to spaces that might otherwise be unseen. As a tool of urban exploration, it crosses boundaries, disrupts the ordered signification of space, and makes visible the unconscious of the city. Disappearing as soon as it has appeared, the nomadic photographic image may set up camp for a brief time in urban space, producing for us the souvenirs of an evanescent contact with the vague

Thus in the very different possibilities of photography and the eruv, resistance may occur not only by discovering and adhering to what has been left behind by the relentless processes of modernization, but also through a practice of estrangement, of rendering vague, of creating interstices out of unstable spaces of the city. In this way, the connection between the alien and the strange has less to do with a representation of what we are losing, but more as a set of possibilities for the creation of a counter-hegemonic practice of the vague.

References

Adorno, T.W. and Horkheimer, M. 1997. 'The Concept of Enlightenment', in *Dialectic of Enlightenment.* London: Verso.

Barthes, R. 1981. *Camera Lucida: Reflections on Photography.* London: Jonathan Cape.

Bataille, G. 1988. *The Accursed Share: An Essay on General Economy (Vol 1: Consumption).* New York: Zone.

Certeau, M. de 1984. *The Practice of Everyday Life.* Berkeley: University of California Press.

Chambers Twentieth Century Dictionary 1972. Edinburgh: Chambers.

Chambliss, W. 1964. 'A Sociological Analysis of the Law of Vagrancy', *Social Problems*, 12, 67–77.

Cooper, D. 1996. 'Talmudic Territory? Space, Law, and Modernist Discourse', *Journal of Law and Society*, 23:4, 529–48.

Couldry, N. 2003. *Media Rituals: A Critical Approach.* London: Routledge

Cresswell, T. 1997. 'Imagining the Nomad: Mobility and the Postmodern Primitive', in Benko, G. and Strohmayer, U. (eds) *Space and Social Theory: Interpreting Modernity and Postmodernity*. Oxford: Blackwell.

Cresswell, T. 1999. 'Embodiment, Power and the Politics of Mobility: The Case of Female Tramps and Hobos', *Transactions of the Institute of British Geographers*, 24, 175–92.

Cresswell, T. 2001. *The Tramp in America.* London: Reaktion.

Cresswell, T. 2006. *On the Move: Mobility in the Modern Western World.* New York: Routledge.

Davenport, G. 1979. *Herakleitos and Diogenes.* Bolinas, CA: Grey Fox.

Deleuze, G. and Guattari, F. 1984. *Anti-Oedipus: Capitalism and Schizophrenia.* London: Athlone.

Deleuze, G. and Guattari, F. 1988. *A Thousand Plateaus.* London: Athlone.

Elden, S. 2004. *Henri Lefebvre: A Critical Introduction.* London and New York: Continuum.

Goffman, E 1969. *Advertising Photographs.* London: Macmillan.

Herz, M. and Weizman, E. 1997. 'Constructing the North London Eruv', *AA Files* 34 (also as *Between City and Desert*, http://www.manuelherz.com/t-eruv.html, accessed on 17 June 2008).

Lefebvre, H, 1991. *The Production of Space*. Oxford: Blackwell.

Marx, K. 1975. 'On Feuerbach', in *Early Writings*. Harmondsworth and London: Penguin and New Left Review.

Marx, K. 1976. *Capital: A Critique* of *Political Economy. Vol. 1*. London: Pelican and New Left Review Books.

McNeel, N. 1972. 'Vagrancy Ordinance Is Void for Vagueness', *Mississippi Law Journal*, 43, 403–5.

Miller, V. 2007. 'The Unmappable: Vagueness and Spatial Experience', *Space and Culture*, 9:4, 453–67.

Natanson, M. 1986. *Anonymity: A Study in the Philosophy of Alfred Schutz*. Bloomington, IN: Indiana University Press.

Nietzsche, F. 1969. *Thus Spoke Zarathustra*. Harmondsworth: Penguin.

Onians, C. T. (ed.) 1966. *The Oxford English Dictionary of English Etymology*. Oxford: Clarendon.

The Oxford English Dictionary, 2nd edn, 1989. Oxford: Clarendon.

Partridge, E. 1966. *Origins: A Short Etymological Dictionary of Modern English*, 4th edn. New York: Macmillan.

Pearson, G. 1983. *Hooligan: A History of Respectable Fears*. London: Macmillan.

Schutz, A. 1962. *Collected Papers I: The Problem of Social Reality*. The Hague: Martinus Nijhoff.

Schutz, A. 1964. *Collected Papers II: Studies in Social Theory*. The Hague: Martinus Nijhoff.

Schutz, A. 1970. *On Phenomenology and Social Relations: Selected Writings*. Chicago: University of Chicago Press.

Sheridan, D. 2007. 'The Space of Subculture in the City: Getting Specific about Berlin's Indeterminate Territories', http://www.field-journal.org, 1(1).

Shields, R. 1999. *Lefebvre, Love and Struggle: Spatial Dialectics*. London: Routledge.

Siemiatychi, M. 2005. 'Contesting Sacred Urban Space: The Case of the Eruv', *Journal of International Migration and Integration*, 6:2, 255–70.

Smith, B. 2007. 'On Place and Space: The Ontology of the Eruv', in Kanzian, C. (ed.) *Cultures: Conflict Analysis Dialogue*. Frankfurt: Ontos.

Soja, E. 1996. *Thirdspace*. Oxford: Blackwell.

Solà-Morales Rubió, I. 1995. 'Terrain Vague', in Davidson, C. (ed.) *Anyplace*. Cambridge, MA: MIT Press.

Stoker, V. 2003. 'Drawing the Line: Hasidic Jews, Eruvim, and the Public Space of Outremont, Quebec', *History of Religions*, 43:1, 18–49.

Stoltzfus, M. 2003. 'Alfred Schutz: Transcendence, Symbolic Intersubjectivity, and Moral Value', *Human Studies*, 26, 183–201.

Tagg, J. 1988. *The Burden of Representation: Essays on Photographies and Histories*. Amherst, MA: University of Massachusetts.

Vincent, P. and Warf, B. 2002. 'Eruvim: Talmudic Places in a Postmodern World', *Transactions of the Institute of British Geographers*, 27, 30–51.

Walker, I. 2002. *City Gorged With Dreams: Surrealism and Documentary Photography in Interwar Paris*. Manchester: Manchester University Press.

Watson, S. 2005. 'Symbolic Spaces of Difference: Contesting the Eruv in Barnet, London and Tenafly, New Jersey', *Environment and Planning D: Society and Space*, 23:4, 597–613.

Wittgenstein, L. 1972. *Philosophical Investigations*. Oxford: Basil Blackwell.

Chapter 3

Domesticated Media:
Hiding, Dying or Haunting

Orvar Löfgren

A Domestic Geography of Media Life and Death

Where do old media go to die or hide? Why are some resurrected or just keep haunting the present by their absence? During the last decade there has been a rising interest in the ways media age, often as an antidote to all the spin surrounding so-called 'new media' (see, for example, Gitelman and Pingree 2003; Hui Kyong Chun and Keenan 2006). Following scholars like Williams (1980), Kittler (1986/1999) and Marvin (1988), there have been discussions of the residual (Acland 2007), of remediation and of the ways in which old and new media coexist and intersect (Bolter and Grusin 1999; Jenkins 2006). Inspired by material culture studies, there are also studies of the life cycles of media, linked to the materialities and practicalities of everyday life.[1]

I am interested in this materiality of what I will call 'media stuff', a suitably vague term that includes media as well as all kinds of media technologies, products and props that facilitate or clutter up daily life. Dead or hiding media are found all over the domestic landscape, in the bottom drawer, up in the attic or even right in front of you in the corner of the desk. From here they may continue to influence us in strange and devious ways.

This chapter deals with such transformations of media in their movements inside and out of the home. I focus both on the emotional geography of such movements and on their moral economy, to illustrate how cultural investments and emotional commitments are changed over time and in space.[2] The analytical approach comes from an ongoing study of the cultural organization of the disappearing (see Ehn and Löfgren 2007; Löfgren 2006), a theme that is strangely underdeveloped in contemporary cultural analysis.[3] I will explore

---

1   See for example Appadurai (1986), Attfield (2000) and Miller (2001, 2008).

2   For a discussion of media moral geographies see Silverstone and Hirsch (1992) and for emotional geographies, Bruno (2002) as well as Falkheimer and Jansson (2006).

3   The study on which this chapter is based is part of the project *Home Made: The Cultural Dynamics of the Inconspicuous*, financed by the Swedish Research Council, see www.etn.lu.se/homemade. Special thanks to my project colleague Robert Willim for his constructive comments on an earlier version of this chapter.

how processes of vanishing, waning or fading can be organized around different trajectories, taking my own media history as a starting point. How does media stuff age, become obsolete, go out of fashion, or turn invisible by becoming what Nigel Thrift (2004) has called 'epistemic wallpaper' – drifting into the technological unconscious? What happens when the disappeared reappears, recharged or recycled, sometimes in surprising places and unsettling ways?

Hibernating Media

> Sony portable radio, Olympus XA2 35 mm camera with flash, VHS tape of three boxing fights, Lou Reed Live LP, Facit 9411 typewriter, Nirvana CD Case (empty), plastic toy stereo radio, roll of 35 mm Fuji film, Jim Reeves cassette tape, Sony widescreen colour television, Nintendo Game Boy cartridge 'Men in Black', 18 assorted pictures of suburban homes, Blockbuster video hire card, 2 AA batteries ...

These examples come from a project by the artist Michael Landy. In 2000 he embarked on a quest that started with systematically making an inventory of all his belongings, from his white Saab 900 Turbo MS to forgotten paperclips wedged into the corner of the desk drawer. He then transported all his belongings to a department store in Oxford Street. Passers-by could either peer through the shop window or step right in and inspect Landy's possessions at close quarters. One by one the goods were placed on a conveyor belt that led towards total destruction, split up, sorted, shredded and finally pulverized into a grey powder. Every single thing that he once owned was to be buried, most likely in a car park in front of a shopping centre in this anti-consumption manifestation (Landy 2001).

When I leaf through his catalogue of thousands of items I am struck by the many media gadgets and products that surface. Strikingly enough, many of them are not found under the heading of 'home electronics', but under other labels. Lists like his make us aware of how definitions of media technologies and products can be narrowed down in ways that exclude more mundane or too well-known media, from yellow Post-it notes to answering machines.

Landy had long been obsessed by the ambiguities and paradoxes of consumer society: the joys of shopping and its tedium. He took the final step and chose to display the totality of his chaotic life as a consumer. He realized that this also meant exposing the whole gamut of consumption – all the broken, clumsy and old-fashioned stuff that people collect in their homes. He also became aware of just how difficult it can be to condemn some things to destruction. Is everything to be obliterated? Will he really let go of personal letters, holiday videos and wedding photographs? It is no coincidence that some media stuff is harder to throw away than others.

Realizing that he owned more than 5,000 articles (many Westerners own considerably more) came as something of a shock to Landy – not to mention the fact that he had forgotten so many of his possessions. He found all kinds of bits and pieces hidden at the back of a cupboard or in the basement that he did not even know the purpose of. He discovered well-meaning presents that had never even been used or ingenious gadgets that had never worked or fitted in. He became acutely aware of all the things he turned out to possess, although it was a collection that, to all intents and purposes, he had lost control over. It was as if his former life had come back to haunt him, with reminders of impulse buying, unfinished projects, disappointing acquisitions as well as changes in lifestyle and personal interests.

Landy's project is a reminder of how many of our own possessions are condemned to a shadowy or completely forgotten existence in basements and attics, or drifting between the kitchen and the subconscious. They can suddenly reappear to a shout of joyful recognition – or annoyance. Should we really keep those old loudspeakers in the attic? Has anybody seen my old Tina Turner records? I think I have found the charger to your old mobile phone, and hey, look over here, what is this, did you buy that?

Reading Landy's report, I started thinking of my own abandoned media stuff and climbed down into the basement, where most of it is still resting. Rummaging through the boxes and searching along the shelves is like going through a media history. Here are all the technologies and products gathered, waiting to be salvaged and brought to life again or just carried away to the recycling station in the next spring clean. Old photo albums, boxes of slides, reels of home movies, video cassettes, an old tape recorder, bunches of music cassettes, children's drawings, a large collection of vinyl records, but also teenage diaries and scrapbooks, some computer games and a Nintendo games machine. I return upstairs to continue my hunt for dead or hiding media in cupboards and desk drawers, finding pre-digital cameras, a Sony Discman next to a cheap Walkman copy, and everywhere old cords, transformers, batteries.

All over the Western world there are basements, cupboards and garages waiting for such kinds of media ethnographies. Why is it that homes tend to become cluttered with old media stuff, in all kinds of shapes and forms? Is there some magic or sentimental aura that protects them from the garbage can? Going through my own materials, I try to grasp what these collections stand for. A family archive or shrine, a catalogue of media life histories, some fossilized technology, a burial site or just unspecified junk?

In and Out of the Home

Look at the photo of a typical Swedish rural living room around 1970 (Figure 3.1). In what was then called 'the best room' of the house the TV set had an altar-like position, decorated by framed family photos, arranged like a kinship

Figure 3.1   'The best media', the new TV-set and the silver framed family
photos, in the best room of a Swedish rural home in 1970. Photo:
Folklife Archive, Lund University

shrine with flowers on a small lace cover. A decade earlier the radio had had
that same central position, but by 1970 it had been dethroned and moved over
to the bookcase and into the kitchen.

A life history approach to media stuff may start by looking at the stations
through which objects, like an ageing Walkman or a TV set, move through
domestic spaces and then out into the world, which includes different kinds
of transformations. Sliding down in the hierarchy may mean relocation to
new spaces but also new functions. In a few years time the TV set in the photo
may end up in the bedroom or turned into a platform for video games in the
teenage room, or stored in the basement as a 'back-up set', in case the new one
breaks down (see the discussion in Caron and Caronia 2007, 58ff).

What about the photos in the picture above? To be silver-framed and
included in the kinship altar on the TV set, the piano or the mantelpiece used
to be the aspiration of most family photos. What situations and people get
chosen and how is the collection arranged or changed over time? Forty years
later these photos may have ended up in the basement or in a thrift store.

There are other patterns of upward or downward mobility for photos. They
may for a time be prominently exposed on the desk at work, on the kitchen
fridge or the family home page. Some may have been there so long that they

have become invisible, mere decorative elements. Others may have become part of elaborate albums with captions. When the tradition of album making started to wane, photos might instead end up in shoeboxes in the basement, still waiting patiently after all these years to be pasted into an album. This will, of course, never happen, but it is hard to throw these boxes away. Occasionally people leaf through them and might find a picture that triumphantly makes the journey back to a more prominent position in the home.

The most drastic transformation is of course to be relocated and re-contextualized outside the home. Much media stuff, from TV sets to family photos, ends up in flea markets or, for example, in the local thrift store, which has a shelf for framed photos as well as a section labelled 'everything with cords'. It reads rather like a museum of dead media. Walking along the shelves of electronic appliances in a sense becomes a walk in what Will Straw (1999) has called 'the museum of failures'. Here are evolutionary series of toasters and transistor radios, early computer games and all kinds of forgotten machines. The atmosphere is not much different from that of the local museum nearby.

Will Straw is interested in the analysis of failed cultural commodities – those artefacts that are discarded and residual, found at flea markets, second-hand shops, garage sales and thrift stores. Looking at the growing stacks of old vinyl records, he is reminded that recordings, like other cultural artefacts, do not simply succeed each other in time; they also accumulate in space. Tracing the trajectories of old vinyl records over the world, Straw (1999, 3) notes that:

> the records left unsold at the end of a yard sale are almost never thrown away, because we assume that someone, somewhere will want them and because we have a vaguely moral objection to simply destroying them. No one may want certain kinds of mid 1980s dance singles, or French-language Maoist books of the early 1970s, but there is still a resistance to throwing them out with other kinds of trash. And so we donate them to church rummage sales or charity shops, where they continue to sit, usually unsold, until they are moved along to somewhere else. A whole informal economy has taken shape around this passage, an economy shaped by the trajectories through which certain kinds of cultural commodities move as they seek to find a final resting place.

The final station for most media stuff is not the thrift store, but the local waste disposal station. At our local one you are supposed to sort out your trash in different containers and the crew at the station supervises this process as well as preventing people from reclaiming other visitors' leftovers. This should be the final destination: industrial recycling only! As a concession to the difficulties of performing a final burial, there is, however, a shed in which objects can be placed for reselling by a charity organization.

There is a special recycling space for electronics and in the containers here you can see all kinds of abandoned media, car radios, cassette players, an abundance of loudspeakers in all sizes and shapes, old telephones and a wealth of computer equipment. All this waste is a very visible reminder about the shortening life span of many media products, the result of a successful marketing of the need for continuous upgrading (see Willim 2003).

Watching people throwing away dead media stuff at the recycling station, you can sense the mixed feelings. Seeing the mess of mutilated media stuff in the iron cages, some people hesitate before they let go of their belongings. 'Maybe I should consider giving this phone or computer to the charity thrift shed next door?', but on that door is a scribbled note: 'We don't accept old electronics.' Dead media are not welcome here. The difficulties of recycling media stuff have led to export systems, sometimes thinly disguised as developmental help to Third World countries. In cities like Lagos, Nigeria and Guiyu, China, old computers and TV sets pile up (see Parks 2007).

Such processes of transformative relocations are, of course, even stronger with media stuff that has a highly contextualized past. Down in my basement there are several generations of family albums. Some of them are old enough to contain unidentifiable people and situations. This is a kind of media death that is encountered in both attics and museums. What happens when a text, an image, an object is drained of all its original contextual information? Martha Langford (2001) has explored this theme, by letting people fantasize about family albums of people they do not know and then compare their narratives to insiders who still have the contextual information. What kinds of different afterlives are possible?

Reading Langford, I started to look for abandoned family albums and snapshots adrift outside the home. I found them in antique and thrift stores, in museum exhibitions and nostalgia books. During a break from a conference on tourism I walked the streets of Munich. Outside an antique shop a family album was exhibited. Opening it up I met the name and logo of the proud photographer. 'Photo album of Fritz Johl', it said, but that is the only contextual information found. All the small photos were numbered in an orderly fashion, but there was no text. I leafed through the pages, getting drawn into the intensely private life of a German family some time in the 1950s, feeling more and more voyeuristic. I recognized the classic genre: the album maker and his future wife, the wedding, the honeymoon, parties with friends, holiday travel, the first baby ...

The early parts showed a couple very much in love. The wife was depicted in photo after photo, in the garden, on the beach, clowning in a swimsuit or posing in evening dress, seductively uncovering a leg. There she was, lovingly smiling at her husband, but now she was smiling at me or any customer passing by. I closed the album feeling a bit embarrassed. What would Mr or Mrs Johl have said about the afterlife of their album? And who would buy an album like this? An artist looking for material for a collage, an interior designer in search of 1950s kitsch, or a social historian?

When I returned to the conference I was exposed to another runaway album. A Powerpoint presentation of 1950s German car tourism was built up around a family holiday album the researcher had found. It depicted the first long automobile holiday of another German couple, starting with photos of the new car being washed, getting ready for a trip to Southern Europe. The audience could not help laughing at the family scenes appearing on the screen. They all seem so very much a period piece, full of what now seemed naive 1950-ishness, but all of sudden I remembered my childhood days when I produced multimedia albums exactly like this – the same genre with photos, maps, postcards, ticket stubs, drawings and clippings from our family holidays. Then I didn't realize I was part of a well-established international genre. I can't help thinking where my own albums might end up …

(Un)familiar Images

The family album died as photography became a form of casual mass documentation. New technologies transform old and domesticated ones. Another example is 'the digital revolution' in amateur photography and the marginalization of earlier photo practices (see, for example, Shove et al. 2008), but digitization is of course only one of many historical transformations of photography, from the 1830s onwards (see Henning 2007). Going back to my inventory of dead or hiding media, the oldest piece turned out to be a camera from the 1930s. Why was it still here? Over the years it had been transformed from an outmoded piece of equipment to an heirloom. It is now charged with nostalgic memories of the original owner, my wife's father, and the worn leather casing is no longer a sign of wear and tear but of patina. Slowly it has turned into an antique. This re-aesthetization means that it is no longer possible to discard. ('Maybe it deserves a position in the bookcase in the living room?')

The other cameras I found hidden away in drawers were also pre-digital ones, but not very old. They still looked fine, and had survived simply because they represent investments that are hard to just throw away, although they have no market value and hardly any chance of coming back to use. Next to them I found a couple of disposable cameras, bought for fun, but by now their 'best before date' had expired. They are definitely dead media.

But what about all the photos? There were all kinds, illustrating the changes in formats and style, old studio portraits, the first generation of amateur snapshots, in tiny sizes, the early generations of colour photographs and slides that now are slowly fading away, together with Polaroid shots. What is striking is how conservative the genre of family photography has remained, as Pierre Bourdieu has observed. Although the continuous technological developments of the medium would have made anarchistic, improvised and informal photographic practices possible, this has not – on the whole – happened. Rather, he observes, 'there is nothing more regulated and conventional than photographic practice and amateur photographs' (Bourdieu 1990, 5).

Next to the boxes I found a *Viewmaster* from the 1950s, complete with a disc of 3D tourist views from the USA, a souvenir from my parents' first trip to America. It survived as a piece of kitsch and also as a time machine, evoking childhood enthusiasm for this dreamland. The Viewmaster became one of many tools for mediating and consuming this mythical territory (see Lagerkvist 2006).

Another special case is represented by the wooden boxes of slides I produced in my youth, when I still had ambitions of turning my photography into more artistic forms. These wooden boxes also saved an old slide projector from being thrown away, together they survive as a medium from times when family photography was an evening entertainment show, evoking memories of endless (often boring) performances of holiday slides. The slide show was a media genre that preceded and scripted the platform for the home movie evening and among all the still photos I found reels of film in different shapes and formats and from a later period family videos.

Mediated Families

'Reel families' is a term Zimmerman (1995) has used in her analysis of family performances in home movies. Going through my own media remnants I was reminded of the ways in which families make themselves visible and tangible through all kinds of technologies, each generation constructing its own mediascapes and skills. I come from a media-obsessed family. My grandfather hired a professional photographer in the 1920s to make a documentary 16 mm film of his family life, later his son carried on making amateur home movies, which means that there is nearly a century of home movies scattered around in cupboards and attics, some of them resurrected at family gatherings for yet another show. They depict a family landscape of endless vacations, birthdays and Christmas celebrations. Again, it is the extreme genre conservatism that is striking – from the 1920s to the early 2000s.

My father brought back a cord recorder from the US in the early 1950s and started to record family events, in radio fashion, interviewing us kids about a day at school, having us perform songs or just recording a Christmas Eve and then using it as a soundtrack to the home movie he made of that same evening. I inherited the same obsession in my childhood, trying to become a photographer, a film maker, a diarist, working on all kinds of media to document the family or my own life. In my later teenage years this interest dwindled, but had a short revival when I started a family of my own.

Trying to get an overview of all these kinds of family media leftovers, there are several striking elements. First of all, what survives? Some of these tapes and home movies have been resurrected by the industry that makes its living from giving dead media a second life. Cord recordings transplanted onto tapes and later onto cassettes or CDs, old movies being saved on video cassettes and later on DVDs. Most media stuff is, however, just slowly disintegrating. There are tapes lacking a recorder on which to be listened to, old movies without a working projector, films becoming more and more brittle and pale. The resurrection of such materials often depended on the existence of a family archivist or enthusiast.

Secondly, what are the ups and downs of such documentary activities, the rhythms of years, seasons and situations? Most of my children are heavily documented up till the age of three, then the interest wanes, no more tape recordings or video cassettes. Families are mediated differently over time, as new technologies and changing family ideals interact. How is the family performed in a photo album of the 1920s, as a Polaroid family in the 1970s or a Facebook family from the early 2000s? Styles of representing or doing the family develop in changing mediascapes, as people pose or clown in front of the recorder, the camera or the camcorder. There is a changing choreography of gender, hierarchy, exclusion and inclusion that has been discussed, for example, by Jonas Larsen (2006) in his studies of family photography and by Carina Sjöholm (2008) for home movies.

My family's media obsession probably also had something to do with the urge of trying to be a family, to be able to materialize and perform images of a perfect family life. Both my grandfathers' and my father's media interests were influenced by their journeys of class. They were the patriarchs of upwardly mobile middle-class families, wanting to create a better childhood for their kids and this ambition shaped the ways in which they chose to document, perform and idealize the family.

Almost all families and individuals have some kind of media archives, ranging from the crazy abundance I have described above to just a bunch of letters, a box of mixed photos or a couple of video cassettes. What do these dead or half-dead media products do to people? The silences, the omissions, the gaps are of course just as interesting as the surviving relics.

Haunting

In a short story by E.L. Doctorow (2008) a New Yorker returns to his suburban house after work. Nobody is home yet and he climbs the stairs to the attic above the garage to chase away a racoon family hiding there. In the darkness he moves around among all the old stuff piled up there. So much of it is dead media – college papers, his wife's hope chest, board games, old stereo equipment, photo albums. 'We were a family rich in history though still young', he reflects. At first he is exhilarated by having chased off the racoons, but his mood changes quickly: '... melancholy took over; there was enough of the past stuffed in here to sadden me, as relics of the past, including photographs, always sadden me' (Doctorow 2008, 62). The melancholy mood of the attic mirrors his own depression and leads him to hide away from the family up in the attic for months, surrounded by all the media ruins of a former life.

The collections of old media stuff in attics and basements are never quite dead, they may come back to haunt you in different ways. Going through my own possessions downstairs, I registered mixed emotions as I leafed through photo albums, videos, tapes and letters. I found old diaries from the teenage periods of both my dad and myself and realized I was reluctant to open them. Did I really want to meet these teenagers? I found a bunch of old phone lists that very concretely reminded me how my social network has changed drastically over the years. What happened to all the people listed there?

The topic of haunting has been treated by Rolland Munro and Kevin Hetherington, in their discussions of the ways in which cultural analysis has neglected the process of disposal. Rolland Munro has argued for the need 'to excavate the sump in con-sump-tion' (2001, 109) in his analysis of the disposal of meaning and his perspective has been taken into the realm of objects by Kevin Hetherington (2004). By looking at the trajectories (mental and physical) of possessions he points out that the classic chain of *production – consumption – disposal* does not follow in an inevitable, discrete and linear temporal sequence. He is interested in the power of the presence of the absent, in processes of cultural lingering and haunting. He uses Munro's example of the lingering smell of yesterday's fish in the fridge even after we have eaten it or thrown it into the dustbin. 'The erasure of an object is never complete. There is always a trace effect that is passed on by its absence', he argues (Hetherington 2004, 168). How do objects become transformed, move back and forth and change both value and position? He makes a comparison with the institution of double burials found in some cultures, in which a person is buried in two stages, a preliminary and a final one in order for the bereaved to adjust to the new situation. He recognizes this two-stage process in the ways a discarded object passes several stations before becoming waste. There are the 'first burial places' like the basement or the digital 'wastepaper basket' of the computer ('do you really want to delete this?'), stations that give people a chance to adapt to the transformation or loss. An unread memo will have to rest in the filing cabinet for a suitable period and then be put through the shredder.

While media stuff forgotten up in the attic or in the bottom drawer may have a haunting effect like dark matter, invisible and unknown but still exerting some kind of strange gravity, actual losses may be much more consciously present, sometimes bordering on the obsessive. Asking people about losses they could not get over, one woman answered:

My husband accidentally erased the video of our daughter's baptism. I still wake up at night, crying. I never got to show our daughter her namegiving ceremony! It feels as if I have lost a part of life and I don't know how to get over that loss. Maybe it sounds pathetic to cry over a lost videotape, but it meant much more than that. (Ehn and Löfgren 2007, 196)

Lost and Found

Losses or unfinished disposals may create situations of haunting, a vague feeling of remorse, guilt or longing, but what happens when objects return?

The things. I can't find them. They are gone. They could be there or there … But they aren't. They are somewhere else. The world has moved four millimetres and a crack has emerged. The crack is widening and more stuff falls through it. Now they are in a different place. Unreachable, invisible and bordering on the insane. As I find them again, at the last moment, I wonder: where have they been? Maybe they have been to God for an overhaul. But now they are back, are they blessed or cursed? I don't know. I have to take care. When they return through the four millimetres crack they look shinier than before. With a smell of lemon.

This is the Danish poet Mårten Søndergaard writing for the catalogue of an exhibition on objects from a lost-and-found depot in Copenhagen 2006 (www.thiscouldbeyours.dk). Here the visitors could look at all kinds of items abandoned in buses and trains. But what kinds of strange transformations characterize lost and found media stuff? The thesaurus lists all kinds of forms of strangeness: *weird, funny, quaint, bizarre, outlandish, unusual, uncanny, out-of-the-way, kinky, grotesque.* Many of these synonyms could describe the strangeness of return, rather in the same manner as relocations may give media stuff a new aura. Think about the electronics thrown into the metal cages at the recycling station. Maybe it is this location behind bars that helps to give them an aura of the abandoned and forlorn. A sense of hopelessness is surrounding these media objects, this is definitely the end station and objects now look not only old and worn (although they may be in mint condition), but hopelessly outmoded, clumsy and ugly. How did this happen?

In other situations retrieved media stuff may create pleasant and nostalgic memories. A man talks about finding old boxes of mixed tapes in the basement:

I have put a bit of my soul into these mixes … I rarely play them, but it is fun to know that they still exist somewhere … Unconsciously I feel that the day will come when I am motivated to check out what I have kept … Opening

such a box of tapes you are just caught up. You just have to listen and then
all the memories come back. Directly. (Mildner 2008, B7)

In personal media histories it is striking that while some objects produce feelings
of good memories, others come to symbolize a process of 'tackification'. In
a memoir on teenage life in the 1980s, called 'A Video Box and Two Candy
Bars', the authors remember their childhood enthusiasm at renting their first
videos, complete with the portable video player box from the rental store. In
retrospect that wonderful machine is now taking on the role of something
hopelessly outdated and comical, turning into a key symbol of the naive and
unsophisticated 1980s (Hammar and Wikingsson 2003).

There is something uncanny surrounding the processes of going out of
fashion. One morning a computer game, a piece of music, a camera model
jumps out as unfashionable, tacky or just tired. You suddenly feel ready to stow
it away, discard or delete it, but behind this seemingly rapid transformation is
a long cultural process of wear and tear, a kind of cultural osmosis, in which a
slow trickle eventually will manifest itself as a drastic change. The not-so-new
comes to appear old as it is relocated, seen in a different light or compared to
a new model. In the case of media stuff the industry's traditional interest in
built-in obsolescence is also a help (see Löfgren 2005; Straw 2007).

When encountering a piece of old media, reactions may vary from a giggle,
a shrug, a surprise or a wave of nostalgia and such emotional transformations
are striking when old stuff comes back into fashion, as kitsch or cherished
collectors' items. Media are charged with different potentials for such recycling.
Why is it that jukeboxes, computer games or old cameras rapidly may become
collectors' items and new cults may emerge around Commodore computers or
vinyl records, while nobody sees a similar nostalgic potential in, for example,
answering machines or ink jet printers? Some old technologies manage to keep
up an aura of creativity, like an old battered typewriter, while ageing computer
keyboards usually lack the same aesthetic appeal. The many artists working
with dead or resurrected domestic media are often attracted to the trivial and
ephemeral, doing collages of old Polaroid snapshots or remixing fragments of
fading 1960s home movies.

Moral and Emotional Mediascapes

My discussion has been based on emotional and moral geographies and
economies involved in the life cycles of domestic media, especially focusing
on processes of ageing and transformation. Such trajectories differ. Some
technologies become so mundane that they disappear by just blending into
the surroundings. They become taken for granted and are thus made invisible,
or turned into Thrift's 'epistemic wallpaper'. The radio is a good example of
this, slowly turning into a domestic background noise; like other media it can

become a platform for multi-tasking (Löfgren 2007; Tacchi 1998). Processes of routinization may also transform media technologies into non-technologies. Who thinks about the postcard or the yellow Post-it note as media (or genres)? They are just everyday things.

Other forms of ageing may take the form of remediation, as old media amalgamate into or frame new ones (Bolter and Grusin 1999). They can remain as a more or less ghostly shadow in the new, for example in the ways in which television for many years was dominated by the radio mentality of the pioneer producers. There are other kinds of vanishing, like the one that Elisabeth Shove and Mikael Pantzar (2006) have called *fossilization*. Down in the basement there are media that nobody knows any longer how to use or operate. Looking into a box of old tapes, the labels of *Cr02*, *Metafine*, *Chrome* or *Dolby C* have lost their meaning.

Geography and economy come together as relocations make re-evaluations easier, as Nicky Gregson (2007) has pointed out in her studies of ridding. On the whole, however, media stuff is difficult to get rid off. What is the feeling of tightening your grip or just throwing an object away? There is often a tension there between anxiety, guilt and relief. Cleaning out the attic, the hard disk or the desk drawer sometimes feels like a moral triumph, a fresh start to a new week or a new life. Motion and emotion are intertwined as people get a physical load off their backs and dump some emotional baggage at the same time (see Åkesson 2006). In throwing stuff in the recycling bin or hitting the delete button you are letting go – but of what?

In her interviews with people constantly moving furniture around in their homes, Pauline Garvey (2001, 53) discusses the ways in which feelings are changed by shifting the sofa or cleaning the kitchen. The movement of material possessions holds a dynamic, interactive role in this emotional process, she observes. This seems very true for media stuff. The reshuffling of old records, computer files, boxes of photos or videos is also a way of reshuffling life, in a sense it can be the remaking of both the past and the present. New technologies also reorganize the ways in which people may store or get rid of media stuff. What is the difference between cleaning out a box of video cassettes or a bunch of digital photos on the hard disk?

The moral economy has to do with the cultural charges we give media stuff. How are hierarchies of presence and importance created at home, what is possible to move around or throw away? A jingle or an image may produce strong memories or emotions – like a direct link to the past. Photos, tapes, videos and scribblings may fill that function, but there is also a haptic dimension in emotional reactions. Handling the old Viewmaster, touching the keyboard of an old typewriter or holding the control of a Nintendo games console may feel like a time machine, creating memories through different senses.

A life cycle perspective thus enables us to see media use in constant change, a process that makes the appearances of continuity misleading – is it still the same old media we are talking about? Such a perspective also reminds us that there is no unilinear direction in the life and death of media stuff. It may be retiring or reappearing, turning in and out of fashion and sometimes back again, being routinized into invisibility or having strange comebacks. Who knows what kind of surprises may emerge from the basement the next time?

References

Acland, C.R. 2007. *Residual Media*. Minneapolis: University of Minnesota Press.

Åkesson, L. 2006. 'Wasting', in Löfgren O. and Wilk, R. (eds) *Off the Edge: Experiments in Cultural Analysis*. Copenhagen: Museum Tusculanum Press. (Also a special issue of *Ethnologia Europaea* 2005: 1–2.)

Appadurai, A. 1986. *The Social Life of Things: Commodities in Cultural Perspective*. Cambridge: Cambridge University Press.

Attfield, J. 2000. *Wild Things: The Material Culture of Everyday Life*. Oxford: Berg.

Bolter, J.D. and Grusin, R. 1999. *Remediation: Understanding New Media*. Cambridge, MA: MIT Press.

Bourdieu, P. 1990. *Photography: A Middle-Brow Art*. Cambridge: Polity Press.

Bruno, G. 2002. *Atlas of Emotion. Journeys in Art, Architecture, and Film*. London: Verso.

Caron, A.H. and Caronia, L. 2007. *Moving Cultures: Mobile Communication in Everyday Life*. Montreal: MacGill-Queen's University Press.

Doctorow, E.L. 2008. 'Wakefield', *The New Yorker*, 14 January, 60–74.

Ehn, B. and Löfgren, O. 2007. *När ingenting särskilt händer: Nya kulturanalyser*. Stockholm: Symposion.

Falkheimer J. and Jansson, A. (eds) 2006. *Geographies of Communication: The Spatial Turn in Media Studies*. Gothenburg: Nordicom.

Gitelman, L. and Pingree, G.B. 2003. *New Media 1740–1965*. Cambridge, MA: MIT Press.

Gregson, N. 2007. *Living with Things. Ridding: Accommodation, Dwelling*. Wantage: Sean Kingston.

Hammar, F. and Wikingsson, F. 2003. *Två nötcreme och en moviebox: Hisnande generaliseringar om vår uppväxt i DDR-Sverige*. Stockholm: Bonnier fakta.

Henning, M. 2007. 'New Lamps for Old: Photography, Obsolescence and Social Change', in Acland, Charles (ed.) *Residual Media*. Minneapolis: University of Minnesota Press.

Hetherington, K. 2004. 'Secondhandness, Consumption, Disposal and Absent Presence', *Environment and Planning D: Society and Space*, 22, 157–73.

Hui Kyong Chun, W. and Keenan, T. 2006. *New Media, Old Media: A History and Theory Reader*. New York: Routledge.

Jenkins, H. 2006. *Convergence Culture: Where Old and New Media Collide.* New York: New York University Press.

Kittler, F. 1986/1999. *Gramophone, Film, Typewriter.* Stanford, CA: Stanford University Press.

Lagerkvist, A. 2006. 'Terra (In)Cognita: Mediated America as Thirdspace Experience', in Falkheimer, J. and Jansson, A. (eds) *Geographies of Communication: The Spatial Turn in Media Studies.* Gothenburg: Nordicom.

Landy, M. 2001. *Breakdown.* London: Artangel.

Langford, M. 2001. *Suspended Conversations: The Afterlife of Memories in Photographic Albums.* Montreal: McGill-Queen's University Press.

Larsen, J. 2006. 'Geographies of Tourist Photography', in Falkheimer J. and Jansson, A. (eds) *Geographies of Communication: The Spatial Turn in Media Studies.* Gothenburg: Nordicom.

Löfgren, O. 2005. 'The Production of Newness', in Löfgren, O. and Willim, R. (eds) 2005. *Magic, Culture and the New Economy.* Oxford: Berg.

Löfgren, O. 2006. 'Wear and Tear', *Ethnologia Europaea*, 35:1–2, 53–8.

Löfgren, O. 2007. 'Excessive Living', *Culture and Organization*, 2007:3.

Marvin, C. 1988. *When Old Technologies Were New: Thinking about Electric Communication in the Late Nineteenth Century.* New York: Oxford University Press.

Mildner, A. 2008. 'Spolad kassett', *Sydsvenska Dagbladet*, February 24:B6–7.

Miller, D. (ed.) 2001. *Home Possessions.* Oxford: Berg.

Miller, D. 2008. *The Comfort of Things.* Oxford: Blackwell.

Munro, R. 2001. 'Disposal of the Body: Upending Postmodernism', *Ephemera. Critical Dialogues on Organization*, 1:2, 108–30.

Parks, L. 2007. 'Falling Apart: Electronics Salvaging and the Global Media Economy', in Acland, C. (ed.) *Residual Media.* Minneapolis: University of Minnesota Press.

Shove, E. and Pantzar, M. 2006. 'Fossilization', *Ethnologia Europaea*, 35:1–2, 59–62.

Shove, E., Watson, M., Hand, M. and Ingram, J. 2008. *The Design of Everyday Life.* Oxford: Berg.

Silverstone, R. and Hirsch, E. 1992. *Consuming Technologies: Media and Information in Domestic Space.* London: Routledge.

Sjöholm, C. 2008. 'Smalfilm som semesterminne', in Hedling E. and Jönsson, M. (eds) *Välfärdsbilder: Svensk film utanför biografen.* Stockholm: Mediehistoriskt arkiv.

Straw, W. 1999. 'The Thingishness of Things: Invisible Culture', *An Electronic Journal for Visual Studies*, 2.

Straw, W. 2007. 'Embedded Memories', in Acland, C. (ed.) *Residual Media.* Minneapolis: University of Minnesota Press.

Tacchi, J. 1998. 'Radio Texture: Between Self and Other', in Miller, D. (ed.) *Material Culture: Why Some Things Matter.* London: UCL Press.

Thrift, N. 2004. 'Movement-space: The Changing Domain of Thinking Resulting from the Development of New Kinds of Spatial Awareness', *Economy and Society*, 33:4, 582–604.

Williams, R. 1980. *Problems in Materialism and Culture*. London: Verso.

Willim, R. 2003. 'Claiming the Future: Speed, Business Rhetoric and Computer Practice in Swedish a IT Company', in Garsten, C. and Wulff, H. (eds) *New Technologies at Work: People, Screens and Social Virtuality*. Oxford: Berg.

Zimmermann, P. 1995. *Reel Families: A Social History of Amateur Film*. Bloomington: Indiana University Press.

Chapter 4
The Strange Space of the Body: Two Dialogues[1]

Eva Åhrén and Michael Sappol

... there would be no space at all for me if I had no body.

Merlau-Ponty 1945/1962, 102

First Dialogue

M:

Here's a 'strange space' (Figure 4.1): a book illustration that refers to the 'strange spaces' of the interior of the human body – and asks its readers to subject themselves to a strange process by which they can come to know their strange selves. It was originally printed in a 1926 volume of Fritz Kahn's *Das Leben des Menschen*, a profusely illustrated, encyclopedic work on the science of the human body (Kahn 1923–1931, vol. 3 [1926]: pl. 11).[2] The series was aimed at a mass audience. In different editions, it probably sold more than 70,000 copies, a fairly robust readership for Weimar Germany.

The image addresses the 'reader' – who is a divided self, ambivalently occupying a body in a space outside the page, sitting perhaps on a comfortable chair in a room at home. The reader bears the same relation to the image on the page as the man on the page to the physiological chart (a visual rhyme). On the space of the page, the headline encourages the reader to imagine the divided spaces within his (or her?) body, to adopt the image as his (or her?) anatomical imaginary: 'What goes on in you, when you look at this image?' There is a mirror effect, a regression. In other words: the image represents the gap between embodied experience and representation, and tries to bridge it.

1   This chapter was made possible through support, help and inspiration from the Francine and Sterling Clark Art Institute, the Mütter Museum, the National Library of Medicine (USA), Isabelle Dussauge, André Jansson, Amanda Lagerkvist, Laura Lindgren, Tom McDonough, Michael North and Arne Svenson.
2   Fritz Kahn (1888–1968) was the first great exponent of modernist medical illustration. *Das Leben des Menschen* (5 vols, 1923–1931) deployed over 1,000 illustrations, and was immensely influential.

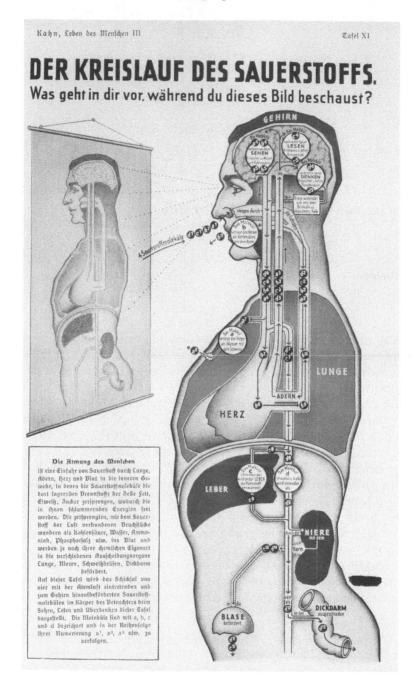

Figure 4.1   Fritz Kahn, 'The oxygen cycle', *Das Leben des Menschen: Volkstümliche Anatomie, Biologie, Physiologie und des Menschen* 3 (Stuttgart: Frankh'sche Verlag, 1926), pl. 11

What the reader sees: a stylized anatomical 'cutaway' profile of a man, divided into parts, who looks at an anatomical chart (turned at a slight angle) that bears the (nearly) identical image of a man divided into parts. The pulsating organs, tissues, cells and fluids are cleaned up and simplified, represented as constituent elements of a plumbing diagram, with intake and outflow and processing chambers, connected by piping, all neatly labelled with arrows to indicate the itinerary of oxygen (*Sauerstoff*) through the production line. The body is turned into a physiological flow chart or an architectural floor plan, a *design* – in a period when graphic design, industrial design and design aesthetics were being professionalized and institutionalized and celebrated as the most modern part of modernity. 'What goes on in you' is industrial production. The brain (and, by extension, the self) is the beneficiary. The corporeal manufacturing process powers up vision, thought processes, and self-consciousness (the end-result), which is visually constituted in the figure of the reader (and by extension the actual reader).

Image and text, together, create an Althusserian 'effect of interpellation,' the sense that 'the image greets or hails ... us, that it takes the beholder into the game, enfolds the observer as object for the 'gaze' of the picture' (Mitchell 1994, 75).[3] Pretty sneaky. But also very bold.

E:

I like the subheading of this illustration: 'What goes on in you, when you look at this image?' You are directly addressed. I am directly addressed. I understand why you chose this image. It is multilayered and seems to embody a modern view of the body – rational, functional, illuminated and enlightened. It represents an embodiment of the mechanized body and tells you: this is You!

The man represented in cross-section, with all his inner tubes and ducts, is shown as watching an image of himself, just as I am watching him. The bodily processes represented are *seeing* and *breathing*. The process of breathing transports oxygen via the lungs, to the blood, and to the different parts of the body. The brain of the viewer (you, me, the man in the image) needs oxygen molecules in order to process and reflect upon the represented image. Quite an ingenious way of addressing me, you, the viewer. See (*sehen*) – read (*lesen*) – think (*denken*). It is a description *of* you, as well as a prescription *for* you. The man in the image, or any viewer of it, is visualized in the 'ambiguous position

---

3   Althusser (1970/1971, 174): 'ideology "acts" or "functions" in such a way that it "recruits" subjects among the individuals (it recruits them all), or "transforms" the individuals into subjects (it transforms them all) by that very precise operation which I have called *interpellation* or hailing, and which can be imagined along the lines of the most commonplace everyday police (or other) hailing: "Hey, you there!"'

as an object of knowledge, and as a subject that knows; enslaved sovereign, observed spectator' (Foucault 1966/2002, 340).[4]

But how can it bridge 'the gap between embodied experience and representation'?

M:

What I mean is this: Embodied experience exceeds representation – and this excess can be quite disorderly, scary in the way it fractures or haunts the self. Kahn offers his image as a technology of self-making, a way of regulating embodied experience. It's a representation of a very orderly self that comes with its own self-referential instruction: Reader study this. Adopt it as your inner political economy.

It's a bit exciting, a bit sexy: this inner space is shiny and efficient, very modern and industrial, and borrows the dynamic power of industrial technology, even perhaps has an industrial rhythm.[5] And comes wrapped in a shiny wrapper, packaged in the outline of an ideal 'you'. In other words, the-self-as-a-site-of-industrial-production is really a consumer product, an object of desire, made for display.

E:

(But I find this idealized, shiny and efficient representation of a body very *un-sexy*, precisely because of its lack of fleshiness, weight, dirt.)

M:

(Maybe you don't find it sexy … But other people might. I can imagine an erotic exchange where the sharply drawn boundaries of the modernist body serve as an object of desire to be annihilated or penetrated … And, in the 1920s, when this picture was published, modernist images and modernist packaging were fashionable, *au courant*, had an aura.)

E:

The image visually argues that the act of seeing and understanding a representation is corporeal, occurs in space outside the printed page. It

---

4   This quote, from *The Order of Things*, refers to the emergence of the 'human sciences' in the nineteenth century, and the painting *Las Meninas* by Velásquez.

5   *Modernity* here is conceptualized as a historical phenomenon, a discursive/material construct that was part of identity formation, social performance, historical experience, and political and cultural agendas in a specific place and time: Germany in the 1920s. See Kern (1978), Berman (1982) and Ward (2001).

insists that decoding representation is a physical, embodied practice, and that representations become (part of) our embodied selves through bodily and intellectual processing. In turn, the experience of being an embodied self is transformed through study and reflection upon representation. Contemplating pedagogical images like this one can therefore make us better human beings, well-informed modern citizens ... That's a grand claim, but somehow I doubt that Kahn had a sophisticated theory of embodiment and representation.

M:

Agreed. Fritz Kahn was a practitioner, not a theorist: he had no sophisticated theory of anything. The close study of images as a method of self-improvement, as a way to modernize the self, was built into his practice. It was *performed* in his texts and images. But not systematically argued there. 'What goes on in you' is as close as Kahn gets to theory, in its depiction of the relation between reader and image, reader and self-representation, reader and embodied experience.

Of course, more 'goes on in you' – and by extension the social experience of the reader – than is shown. Here, the machinery of the body is entirely automated, has no workers, maintenance staff, supervisors, owners or investors. The conditions of its production (sexual intercourse, gestation, birth, growth and development) are entirely occluded. The intake of food, digestion and excretion is suppressed (the subject of other lessons and other images). The body is an isolate, removed from the complex hierarchies and webs of social, political and cultural exchanges that people interact in. The text (headline, labels, arrows and boxed caption) is a disembodied 'voice'. The author (Fritz Kahn) is also an isolate, does not (here) present his credentials or justify how his scientific information was produced and acquired; no authority is cited. The artist, who is uncredited, and the production process by which the image was made, likewise go unrepresented, except perhaps by the chart within the image, which is rendered with a slight sketchiness (and a perspectival tilt) that suggest that the represented 'reality' of the body is precisely that, a representation.

E:

OK, but I would say that the embodied experience not only exceeds representation, but also precedes it, transcends it, maybe eats and digests it, incorporates it, is transformed by it. But there's a problem: If we become aware of our own bodies through different forms of mediation, is it possible to conceive of an embodied experience that is *not* filtered through representation (i.e. mediatized)? Henri Lefebvre puts it this way, but offers no solution: 'Bodily *lived* experience, for its part, may be both highly complex and quite peculiar, because 'culture' intervenes here .... The "heart" as *lived* is strangely different from the heart as *thought* and *perceived* (Lefebvre 1974/1991, 40).

M:

Kahn doesn't provide any solution either. Instead he takes up the proposition from the opposite direction: the heart as thought and perceived is strangely different from the heart as lived. The illustration, an image printed in ink on paper, is there to gaze upon. It asks the reader to accept that the visual representation captures a reality of the body that goes *beyond* embodied experience. This knowledge is not always already present, cannot be acquired intuitively. It must be learned and internalized, taken up by the reader as a blueprint of his inner self. Which will then provide a visual vocabulary that encodes (or recodes) embodied experience – in the space of his own body and in the spaces of the world in which he lives.

In other words, the image is a prop. It shows the reader how to imagine and interpret the body as a textualized, diagrammatic, physiological, and very technological, entity. Words, symbols and image, with all their divisions, combine to present a holistic and universal self – the science of the body is not particular to any location or person, but applies to *all* bodies – a claim that subsumes the multifarious variations among human beings who come with varying features, have varying histories and life trajectories, are female as well as male, young and old, workers and bosses, German and foreign, rich and poor, healthy and diseased.

E:

I think the rhetoric of the image is much more coercive and prescriptive than that. It doesn't modestly *ask* the viewer/reader to accept its premises – it *presupposes* that you will take this representation as fact, scientific fact. The style of the image, the way it makes use of the cross-section, favoured by engineers and scientists, together with the text, makes a rhetorically strong argument – this is the truth about you! Be glad that your brain is fuelled by oxygen, so that you can be enlightened!

At the same time the strong white male body, with his straight profile, his disciplined posture, his narrow waist and expansive chest, tells the rest of us (female, sick, children, weak, elderly, non-white, disfigured, disabled) that maybe we are not really human, but sub-human. This kind of image is more normative than holistic and universalizing. It is, once again, an image made by men, for men. Were (are) they aware of the provocation? Probably not. The Human is almost always represented as Male. White Healthy Male.

M:

Ah well, you're quite right: all universalisms are parochial. The image is visual rhetoric. It tries to persuade the reader – and it is likely that Kahn's readership was predominantly male – to think of himself as a standardized industrial product, which comes off a standardized industrial production line. But it also draws on an older discourse. In nineteenth-century Germany, the ideal of the strong, healthy, and somewhat aggressive, masculine body, governed by science and reason, was linked to the idea of nation-building, national identity formation, the strong nation-state. Politically: in opposition to France, the fragmented polities of the Holy Roman Empire, and the fragmented ethnicities of the Habsburg domains. And socially: in opposition to the decadent aristocrat, the Jew, the homosexual, the undisciplined youth, the alcoholic, the rowdy peasant and the (emotionally, physically) soft woman. So there is an implicit visual argument: use this image as your anatomical/ physiological imaginary, undergo a critical re-education process based on the science of the body, and be healthy and strong and manly. A contributor to science, civilization and nation.

But, like the 'female, sick, children, weak, elderly,' etc., the strong white man also doesn't measure up. The image depicts a utopian industrial body. With no hint of mechanical difficulties, economic competition, labor strife, political disruptions, etc. And the very perfection of the body incites anxiety. Its maintenance requires the reader to cultivate a narcissistic gaze – keep surveying, keep monitoring, keep obsessing – and narcissistic body practices (exercise, diet, hygiene). And this anxious narcissism is in keeping with the methodology of mid-1920s advertising, which was just then developing *Psychotechnik* (applied psychology): visual and textual appeals that activated and recruited the consumer's 'instincts,' 'drives,' psychic needs, insecurities, etc.

So there is coercion and prescription and manipulation.

E:

There's violence here too. This body is cracked open, its secret interior space obscenely exposed. Yet the man doesn't seem to care. We can see into his brain, his lungs, his heart, but he just stands there, looking at himself. That's magic. The perspective is that of a mighty wizard, one who can make the invisible appear before your very eyes, make the hidden visible, unveil nature's secrets. That's not so modern, even if the image has a modern twist on this old theme.

M:

That exposure of the 'interior space' points to an odd metamorphosis that Kahn wants to orchestrate: the interior of the body is converted into a schematic public space, open for inspection, like a model manufacturing plant open for tours, or on exhibition. But are any secrets exposed? If anything, the illustration suppresses the chaotic jumble of spit, sweat, blood, lymph and other fluids, refuses to represent genitalia, or any kind of erotic exchange. The self is so sharply bounded that it almost excludes exposure and penetration, and the possibility of shame. Instead, it is in love with its own image. Which it asks the reader to internalize, to reproduce and exhibit in the private domain of selfhood. Potentially, it's a hall of mirrors. It only seems obscene in its willful denial of the mess that comes with desire, and in its narcissistic abandonment of collective life and social interaction.

E:

This body is a sterile, clean, hygienic machine. It's cool, distanced and denies the fleshiness of being human. Even the refuse is clean; the bladder a neat container and the colon a pipe through which an air stream blows... This body is penetrable only through the eyes (where visual impressions pass on their way to the processing plant in the brain) and respiratory organs (where oxygen enters to travel to the parts where it's needed).

What I meant with obscenity was that the strange and hidden inner parts of the body are removed from the shadows, from the ob-scene backstage to the limelight, from the private to the public (the word *obscene* derives from the ancient Greek term for the place behind the back wall [skene] of the stage).

But at the same time all the secrets, messiness and chaos are still kept behind the scene ....

M:

But I don't see any violence.[6] The body in 'What goes on in you' is not cracked open. It is on exhibition, a museum display. The plane of the page (and of the cutaway body) functions like a shiny pane of glass that separates the viewer from the display object. In 1926 (the year this image was published), the architectural and display use of large window panes was new and modern. People throughout Germany marveled at extravagantly modernist shop window displays. In the same year, widely heralded health exhibitions attracted a large visitorship. For visitors to the GESOLEI and Deutsches Hygiene Museum exhibitions, the contemplation of modernist physiological illustrations, charts

---

6 Industrial manufacturing's violent indifference to organic life is backstage with 'the secrets, messiness, and chaos'.

and vitrine displays was a very charming way to be modern.[7] And unlike alchemists and wizards, who knew and kept secrets, this was public access and public education.

Or maybe just a show of public access: I doubt that most of Kahn's readers had the time, or training, to study and understand the fairly technical explanations of physiological chemistry encoded in the illustration (and explained in even greater detail in the accompanying text). So there is a bit of wizardry here. Like the blustering Wizard of Oz, Kahn wanted to awe his audience, to induce a state of conditioned deference to the authority of modern scientific medicine.

E:

I think a substantial part of the power of medicine, science, and technology in modern society relies on this kind of display of privileged access to that which is hidden, and the possession of the tools to reveal it through representation. Many representations of scientists show them as magicians with the ambivalent power to control the forces of life and the universe. The scientist becomes an object of transference. The lay public projects onto the scientist its fear of modernity's power to destroy as well as its hopes for redemption. Popular medicine and science as a genre is also suspended in this tension between good and evil, between disciplining and potentially liberating its audience. Enlightenment with good intentions can have dark, unintended or unexpected effects. This is a curse of many projects to improve the health and life of people in modern society. The idea of eugenics as a way of controlling the unwanted, degenerative effects of evolution (and modernization) was very widespread in the Western world at precisely the time in history when this image was produced. Technoscientific dreams of total control were not (are not) only dreamt in totalitarian states.

M:

And were not (are not) just mandated from above. There was consumer demand: a market for materials that promised to help control the individual self and the body politic. Which required the suppression of some inconvenient

---

7 'GESOLEI' stands for *Gesundheit* (health)/*Sozialfürsorge* (social welfare)/ *Leibesübungen* (physical exercise). The GESOLEI movement culminated in a series of state-sponsored health exhibitions shown around Germany in 1926. (Kahn served, in a limited way, as a consultant.) In the 1920s the Deutsches Hygiene Museum did not have a permanent space, but sponsored travelling exhibitions. Other commercial and municipally sponsored health exhibitions were also mounted in Germany in the 1920s. See Weindling (1989). For modernist window display and glass surfaces in Weimar Germany, see Ward (2001, 191–240).

truths. Kahn's image suppresses many things that 'go on within you' – hunger, lust, exuberance, pain, exhaustion, anger, boredom, illness. And many things that go on *outside* you: the eviscerating war, 'boom-bust' capitalism, political assassination, street crime, prostitution, domestic violence, unemployment, traffic jams, industrial accidents, etc. All of which made up a large part of the life experience of the author and his readers.

But even if it refuses to specify the conditions of its birth, the image presumes that author, image, text and reader, and especially the reader's body, are located in modernity – a strange space, a strange time, which radically differs from that which went before. And the radical difference is science and technology and continual and accelerating transformation. So 'What goes on in you' offers readers a way to think of themselves as modern: the human body reconceptualized as a technoscientific artifact, a *manufactured* thing made up of standardized, machined parts, but also a *manufacturing* thing: a chemical factory or an industrial engine.

The illustration is of course also a technoscientific artifact, a manufactured thing (which uses modern photo-reproduction print technologies). And a manufacturing thing, a device for mass-producing the self. The subjectivity of the reader, and the mass readership, is assembled on a long assembly line of displayed images (of which the thousands of illustrations which appear in Kahn's work form only a small subset). Technology, science, and the modern world in which such things proliferate – and the world in which the reader and author both live – are thereby 'naturalized', made to seem part of an unchangeable (though constantly changing) order of things.

E:

The illustrated book is part of the process in which the body and the body image is mediatized in modernity. From the late nineteenth century, people of all classes increasingly consumed ever more abundant and accessible media-images of what it meant to be human. And modern. Picture magazines, advertisements, shop windows with their mannequins, wax museums, posters and films – they all contributed to this mediatization, whether crude or refined, popular or high-brow, didactic or commercial. The naturalization of modern technoscience is tied to that mediatization of the life-world, of lived experience. Kahn's image production is a refined contribution to this process. We are encouraged to look at ourselves through a new pair of spectacles: Through these modern lenses we see our bodies; through the representations, which already inhabit our minds we see our selves. We apply a mediatized gaze upon our selves. We are mediatized. Maybe there is no pure bodily experience of self. (Already the mirror in which I see my face is a medium.) So I guess I want to rethink things: representation precedes and transcends, even eats and digests, the lived experience of our bodies.

M:

Exactly so. And this particular mediatization contains a purification ideal. Modernity kept getting ahead of Kahn's readers – 'all that is solid melts into air' (Marx and Engels 1848/2002, 223; Berman 1982).[8] Or modernity kept falling behind them: not enough melting going on to relieve them of the difficulties of life as it had always been lived (need to pay the rent, get a job, find a mate, stay healthy, etc.). They could lessen their anxiety by adopting a self-image that harmonized with, and contributed to, the ride into modernity. For many people, urbanization, industrial technology, science, and the state already seemed to be natural, unchangeable. Kahn's illustrations ratified (reified) what they already knew, and hoped to know.

   And also what they didn't know. 'What goes on in you' had an exotic flavor. It was a newly invented type of image, a deliberate break from the older visual vocabulary of popular medical illustration, which was anatomical and naturalistic (a tradition that dates back to the sixteenth century). It extrapolated and rhapsodized and defamiliarized. It made the anatomical body into something that was novel and diverting. In a world of proliferating advertisement, it advertised the industrial body. And, if the advertisement succeeded, the reader would want to put on such a body and wear it like a shiny new suit, a ready-made body imaginary. Kahn was a commercial author. His livelihood depended on the sale of his books and articles. To get a mass readership, he had to make works that were appealing, entertaining.

E:

The 'older' type of anatomical image still existed in the 1920s and 1930s (Figure 4.2). And still exists today: you can see it in any doctor's or dentist's waiting room, in schools, etc. It has been modernized, and its mapping view of the body is echoed in the latest techniques for visualizing the body, such as PET-scans and magnetic resonance imaging (Dussauge 2008). But you're right – this is different. And the way I as a viewer perceive my own body after seeing this schematic body-design certainly differs from what happens when I contemplate and incorporate the view of the nineteenth-century (or earlier) cartographic anatomical image. Both help me envision the unknown,

---

   8 Marx and Engels, *The Communist Manifesto* (1848: 223): 'Constant revolutionising of production, uninterrupted disturbance of all social conditions, everlasting uncertainty and agitation distinguish the bourgeois epoch from all earlier ones. All fixed, fast-frozen relations, with their train of ancient and venerable prejudices and opinions, are swept away, all new-formed ones become antiquated before they can ossify. All that is solid melts into air, all that is holy is profaned, and man is at last compelled to face with sober senses his real conditions of life, and his relations with his kind.'

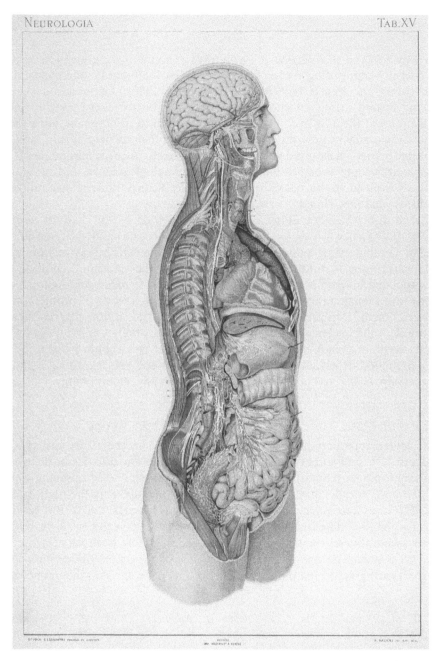

Figure 4.2   **An older style of anatomical illustration. From Laskowski, S. (1894).** *Anatomie Normale du Corps Humain, Atlas Iconographique ...; Dessineés d'après les Préparations de l'Auteur par S. Balicki* **(Geneva: Braun), pl. 15**

opaque inner space of my body. But Kahn's image certainly doesn't help me navigate that strange space; it doesn't give me a map to my kidney or help me understand its structure. It rather makes me feel like I am (in) a building of some sort, with designated spaces for functional furniture, my organs.

M:

Yes, a body architecture. Le Corbusier argued, in 1923, that 'the house is a machine for living in' (Le Corbusier 1946, 10, 89). For Kahn (in this image and many others), the body is a machine for living in. And, not surprisingly, it resembles the modernist architectural spaces being created at the time by Le Corbusier and the Bauhaus designers. Unlike the overstuffed and over-decorated homes and buildings of nineteenth-century Europe – and unlike the overstuffed anatomical illustrations of the Vesalian tradition – unnecessary details are suppressed, replaced by negative space. Which rationalizes the body and makes it crisply legible. And deliberately strange.

E:

I think the image only becomes strange when you start contemplating it. On the surface it is, as you say, 'crisply legible', deceitfully simple, and very un-strange.

M:

'What goes on in you' invites contemplation. And let's also remember that strangeness and familiarity are unstable categories: what is strange becomes familiar, what is familiar becomes strange. For Kahn's readers, the legibility of minimalist functionalism was a novelty and probably seemed a bit freaky.

E:

But it tames the body, makes it a safe space, even while dressing it up in outlandish modern garb.

M:

Maybe it's not only that the image represents the strange modern spaces of the human body, but that subject formation occurs through a process of estrangement. In the hazy 'built environment' of the self (which in other illustrations Kahn represents as a 'head' office or control centre), we host a succession of interior dialogues that perform an auto-interpellation effect:

'Hey M!'
'Yes?'

'Who are you talking to?'
'Me!'

The above example is empty and abstract: a stick-figure named M talking to himself ('me'), a double homunculus. But the structure of that barren exchange echoes how language acquisition, personality development, and identity formation occur cumulatively over time as the internalization of exchanges between M and others, from infancy onward.[9] Meaning is socially produced. The self is dialogical (Voloshinov 1927/1994, 1929/1994) .[10]

The M/'me' entity is an amalgam of 'I' and 'Other', or Self and Others, or Selves and Others. The distinctions are osmotic – Self seeps into Other, Other into Self – and it becomes difficult to discern who is speaking in our inner speech. So the alienation effect is built-in, foundational. And since the cross-talk occurs in the space of a single package of skin, a body, this state of affairs lends itself to, even requires, a certain mystification. Which is shared by all of the other embodied selves that interact in social space, who are arrayed in larger ensembles of estrangements that feed back into the process of self-formation.

E:

I think there's a slippage problem. All this 'Self' talk has a tendency to reify rather than explain. 'Self' keeps moving to the head of the class, claiming priority. Maybe 'Self' is not a useful, or even epistemologically possible, category of analysis.

M:

Yes, 'Self' talk is tricky. But I think it's important to historicize the self and its means of production. Otherwise, we may end up reproducing the self as something outside history, something that's natural and essential, even if we don't dare to speak its name.

---

9 Here, I've bracketed discussion of developmental stages of subject formation. Clearly some developments and events are more decisive than others: some are necessary for further developments to occur, and at some stages language acquisition and personality traits are more acquirable, more malleable than at others. To go further would require a more detailed account than can be given here.

10 Voloshinov: 'meaning is the effect of interaction between speaker and listener' (1929/1994, 35); 'inner speech is the same kind of product and expression of social intercourse as outward speech' (1927/1994, 42). Although published under the name of Voloshinov, many scholars argue that Bakhtin was the principal author of these texts.

'What goes on in you' visually performs and represents the dialogical relation between subject-formation and visualization. Visualizations use metaphor, metonymy, juxtaposition, etc. – derived from a vocabulary of lived and representational experience – to map boundaries between Self and Other. The self forms and re-forms through a succession of gazes into the 'mirror' of social interactions. Kahn's image (and larger image-making practice) is one of those social interactions, a programmatic intervention into the process of subject formation, an auto-geography for modernity.

The Self/Other has historical and spatial particularity. It is made cumulatively, collectively, in exchanges between parent and child, worker and employer, student and teacher, husband and wife, policeman and civilian. In all the spaces where life is lived. And in the 'spaces' of media – newspapers, books, wall charts, motion pictures, computer games, museums, etc. All of which present images that are sources for the visual modeling of the human body. As you suggest, that's mediatization.

'What goes on in you' is a gaze into a looking glass that gazes back, deeply narcissistic and deeply constitutive. The body becomes a machine for inspecting itself, is engineered to construct and consolidate the emergent 'society of the spectacle' – internally as well as externally (Debord 1967/1994). Distinctions between inner self and outer world collapse, one of the signatures of life in modernity which physicist Ernst Mach complained of, or maybe celebrated, around the turn of the century: 'The boundaries between things are disappearing, the subject and the world are no longer separate ...' (Mach, in Virilio 1994, 6).

E:

Isn't the image an attempt to fight that threatening collapse of boundaries, the conflation of self and other, of inner and outer world? It has such clarity; the man is clearly outlined and disconnected from the world around him, of which we see no trace. His inner organs are defined and separated. He is detached from his own representation as if the image wanted to persuade us that we, the viewers, are also detached from it, separate entities in a world of clearly defined subjects. This is a representation that formally states its own ontological status as representation, as well as its epistemological relation to the viewer, at the same time as it instructs you how to use it. It is *not* reality. You see this image and process it in your brain, but the image is *not* you, your 'self', nor 'the Other' – it is an approximation, a schematic illustration to facilitate the understanding of your own body, and all the other bodies inhabiting the world, regardless of their differently formed subjects ...

M:

Maybe we agree. The unified self is a carefully (or sometimes not carefully) cultivated fiction, collectively crafted and enforced by rewards and punishments. It is forged out of embodied social and spatial *practice*: ceaselessly iterated fragments of sensation, dialogue, performance, narrative and representation, all grafted together. But the grafts never entirely take: the amalgam is partial and contingent, and fragile. And this state of affairs can be worrisome, even painful. And so, in the early decades of the twentieth century, there was a growing market for materials to help people hold it together (as there still is today). A proliferation of products and images. Including images that modelled the strange modern space of the human body.

Second Dialogue

E:

We've discussed an image that represents the strange spaces of the human body and its relation to self-formation and embodied experience. Now let's take our thoughts further by exploring a space devoted to the display of strange objects made from actual bodies: the Mütter Museum of the College of Physicians of Philadelphia. This is a site for close encounters with extraordinary objects: bones, flesh, organs, in display cases, filling several rooms from floor to ceiling, on two floors. Things like this usually belong inside a body, living or dead, or in the cemetery. But here they are, on shelves, trapped behind glass, floating in fluid, or supported by wire armatures. Here, we are invited to view and reflect upon the secret insides of human bodies, things that are at once repulsive and attractive: entrails, dead fetuses, brains in jars ... On one hand these things are common, universal: we all consist of this kind of material. On the other hand we see the surfaces and interior of the body in a novel way in the museum, estranged, cut up and displayed in pieces, and, in many cases, deformed and disfigured by virulent disease.

Museums like this used to be found in most North American and European medical schools. Just like the Mütter Museum, they were most often open only to medical professionals and students. In the twentieth century, most of these museums were destroyed or discarded, because they no longer had a vital role to play in medical research and education. The Mütter fortunately survived, and its anatomical and pathological specimens, its models of wax, plaster and papier-mâché are now on public display along with its collection of historical medical instruments and other curiosities. The core of the museum's collection is made up of 1,344 items donated in 1858 by Thomas Dent Mütter, professor of surgery at Jefferson Medical College (Worden 2002, 9 ff). The museum

Figure 4.3   Wax models in the Mütter Museum (late nineteenth or early twentieth century). Photograph by Arne Svenson © 1993

opened in 1863 and moved to its present premises in 1910 (Lindgren 2007, 14).

In a contemporary photograph from the Mütter Museum we see some of the wax moulages: horribly disfigured faces and blistered arms of infants arranged on shelves (Figure 4.3). They are sequestered from the viewer by their glass cylinders and labeled with the names of the diseases that are the reason for their display. These life-size objects, made around the turn of the twentieth century, were skillfully designed and arranged to look like actual specimens. The modellers took extreme care to represent every gruesome aspect of the diseases, and the realistic effect is uncanny.

A photograph from 1909 shows a vitrine with contents from the collection of skeletal pathology (Figure 4.4). It conveys some sense of the three-dimensionality of the museum space: with the wood-framed glass case itself

F.8.T.8S. D.IOM. P.30S.   10AM.2-9-09.

Figure 4.4   Interior of the Mütter Museum, 1909, photographer unknown.
Courtesy of the College of Physicians of Philadelphia

and the skeletons inside; some furniture in front of it and some behind; and on the glass surface some ghostlike reflections of other bones. These old display cases are still there, contributing to the strangeness of the place, its character as a historical anomaly or cultural relic. The old cases, old specimens, and the mainly unaltered architectural space give the visitor a sense of visiting an authentic turn-of-the-twentieth-century museum. This is partly an illusion: the museum (inconsistently) incorporates aspects of late twentieth-century display practices, lighting techniques, and historical interpretations. And the displays have been rearranged. Several layers of historical space coexist in a single temporal dimension.

These photographs evoke something of what there is to experience in the Mütter Museum, but seeing photographs is of course not like being there. Being there, immersed, seeing it for real, surpasses the simulacrum of the printed image. Spatiality itself creates a striking difference between experiencing the museum with its specimens and models, and any type of two-dimensional anatomical image (like the ones we discussed in our first dialogue). Walking among three-dimensional displays of bodily objects is a more intense experience of embodiment than viewing an image of human anatomy on the page of a book. As a visitor to the museum I cannot avoid establishing multilayered (spatial, physical, emotional, referential, indexical) relations to the objects. Seeing the skeletons in the Mütter Museum makes me feel the bones inside my body. I compare the huge thighbone of the museum's 7'6" giant skeleton with my own, or the curved spine of the female skeleton deformed by Pott's disease with my healthy, if slightly achy back. Walking by the display cases with the wax moulages, with their eerily lifelike form and texture, makes me feel my chapping winter skin. My living body establishes relationships with the dead body parts in the museum space, whether I like it or not. The process transforms me and my body image. It estranges me from my body, but also makes me more intensely aware of my corporeal, spatial self.[11]

M:

Yes, the corporeal objects of the Mütter Museum have a special intensity, strongly affect the viewer in a way that Fritz Kahn's 'What goes on in you' does not. (Kahn's images had a special *extensity* – mass publication in a multi-volume set that offered many other images that played variations on his theme.)

---

11   My argument here is based on Maurice Merleau-Ponty's *Phenomenology of Perception* (1945/1962), especially pages 99ff, and 140ff, in which he claims that the body is spatial, always situated, 'in-the-world'. Furthermore the body is not *in* space and time – it belongs to them, 'combines with them, and includes them' (140); we gain access to, and primary knowledge about, the world and the object through movement and our bodily experience of it.

But the anatomical object is also a simulacrum, often produced in tandem with case report narratives published in medical journals, books or museum catalogues. Historically, modes of anatomical representation, display and performance were linked: illustrators observed dissectors making dissections and tried to make their illustrations look like dissections. Dissectors tried to make their dissections look like the illustrations they had seen in books. And specimen- and model-makers did the same. So the 'real' bodies and body parts in the Mütter Museum are in some sense sculptures in the medium of human flesh and bone. The wax and plaster objects cast from living or dead human beings (such as the faces and arms shown in Figure 4.3) are in some sense also 'real', a transcription of bodies or body parts. The models (in wax, plaster and other materials), based on observations of patients and specimens, photographs, drawings, memories, and the modeler's imagination, make no claims to being 'real' body parts. But all of the objects, whether 'real' or not, involve imagination and technical skill, and draw on conventions of presentation.

E:

For sure, all artefacts are real in some sense. All natural objects in museums are also cultural artefacts, transformed by the procedures of collection and exhibition, and by preservation techniques and display strategies. Drying and varnishing, favourite techniques of seventeenth- and eighteenth-century anatomists, made objects look dark and shiny, and were often combined with brightly colored wax, injected into veins and arteries. Preserving organs in fluids, another old technique, became more common when the use of formaldehyde was perfected in the late nineteenth century. And the process continues: the recently developed plastination technique (where polymers replace fats and fluids in the tissues) preserves colors and structures very effectively, and has contributed to the current boom for commercial, touring anatomical exhibitions.

The forms of visualizing the body you mentioned are all part of the same medical, visual culture, and they have always been joined to larger cultural and historical contexts. So, in some sense, the anatomical object is a simulacrum. But, like the relic finger of a saint is a part of the saint's body (if it's not a forgery), the body parts in the anatomical museum are actual flesh or bone. Because of this they have a particularly strong 'reality effect' (Barthes 1968/1989): while representing a certain anatomical or pathological feature the objects also represent their own realness. Plaster and wax casts of faces, bodies and organs are different: their realness derives from retaining the original form of the body they were cast from. (The Mütter Museum, famously, has a plaster cast of the conjoined torsos of the original 'Siamese twins' Chang and Eng Bunker, cast after their autopsy in 1874.) The reality effect of sculpted models of wax, on the other hand, stems from the medium's capacity to mimic the colour, texture and form of human skin and flesh – they can be uncannily lifelike.

M:

But objects derived from the human body are less malleable than objects sculpted to resemble the human body, less amenable to the reshaping hand and eye of the producer. Realness places structural limitations on what the specimen-maker does, acts as an epistemological guarantee that gives the object a representational density, an aura of authenticity – the object doesn't only *refer* to the real, it *is* the real. As such, it has no 'author', no 'artist': The subject contributes his or her body or body part (willingly or unwillingly), which comes with a life-history, an itinerary.[12] The disease entity, congenital condition or developmental accident contingently reworks the body or body part. The anatomist contributes his technical skill, knowledge agenda and aesthetics (if that's the right word). Professional connections to a specific institution and medical/museological network (often a series of exchanges) get the object placed on display. And then there are the weighty and complex webs of signification and neglect (curatorship, the lack of curatorship) that grow around the object, cover it and obscure it. But also explain it, even constitute it. The objects of the museum are 'unrecognized poems in which each body is an element signed by many others' (de Certeau 1984/1999, 128).[13] But these signatures also deface and destroy the object they sign, mark the illiteracy, carelessness, indifference, incompetence, or malicious playfulness of any original designer (if there could possibly ever be one).

So here are complexly layered objects showing obscure surfaces and interiors of the human body – in sickness or health, but always in death, and always in history – gathered by careful planning, and haphazard circumstance, in a complexly layered space. In ensemble, the metonymies signify to the visitor: this is what happens to the human body – death, suffering, disorder, objectification, fetishization, etc. – each bit standing for the whole of human existence. The museum is almost an orchestra of corporeal metonymy. But do the parts fit? Is there harmony in that space? Deliberate atonality? Is the result a symphony?

E:

No, I don't see (hear?) harmony. Nor do I detect any deliberate atonality. The objects all sing their own song in their own tune. There was no mastermind

---

12   Specimens came from near and far, from burial grounds (dug up by medical students or paid bodysnatchers), from hospitals, battlefields, archaeological sites, etc. See Richardson (1988), Sappol (2002) and Åhrén (2002/2009).

13   The phrase comes from a passage on experiencing the totality of Manhattan.

who collected these things and put them together in a coherent pattern or calculated chaos (like, for instance, Sir John Soane's deliciously overwhelming art/architectural museum/shrine in London). There is something that connects all the objects of the Mütter Museum: it is an abstract principle, the scientific utopia of finding and displaying of the order of things. But, in practice, the order is fragmented, incomplete. Where the Kahn picture of our first dialogue conveyed bodily order and discipline, the museum breaks the body into pieces, multiplies it, and shows it as multifarious, wondrous, proliferative. Who to identify with – the dwarf or the giant? Or maybe the museum is not inviting me to identify with them at all, but to assess my normality by relating to their difference (in a hall of mirrors where nothing is normal). The experience of encountering the objects is as disturbing as it is reassuring. The effect is something like watching a horror movie, realizing it is *not* for real, and that life is not so bad after all.

M:

You're right, a visit to the Mütter provides visitors with some degree of reassurance. But that reassurance is spiked with doubt and ambivalence. Like *Alien, Friday the Thirteenth, Halloween* and all those horror films where the monster refuses to die, can never be truly vanquished.

My own close encounter with the Mütter Museum, I think, illustrates that mirror effect. I first visited in 1993, while doing a research stint at the library upstairs from the museum. I was writing my dissertation on the cultural politics of anatomy in nineteenth-century America, and had a chapter on anatomical museums, but had never seen one. I knew something about the Mütter: its reputation (even back then) preceded it. It had a buzz, like a drug that you could get you really high but wasn't good for you. I was reluctant to enter.

When I did, it made an impression. Or rather a certain set of objects did: the wax moulages. These were the casts of faces in colored paraffin showing scars, eroded bone and cartilage, blistered skin, so finely detailed that you could see the pores. They showed the faces of real people, people who had really lived and suffered with these terrible dermatological conditions. On a few pieces you could see a five o'clock shadow, and some showed the thumbs of an assistant helping to hold back the lips while the cast was made, so as to better display the pathological condition. The waxes, some of them wrapped in cotton bandages, the way a patient or cadaver might be, looked almost alive.

They were impossible to gaze upon, impossible not to look at. The pieces displayed something I didn't want to think about, a part of the human condition not usually seen around town: an inner self, *my* inner self (or one of them), an ill-defined homunculus nourished on feelings of intense, unloveable and irredeemable ugliness, an imago that I had worked hard to cover over with my 'normal' face. The other objects receded into the background. Aside from that, the only thing that really registered was a dissected face that floated forlornly in a jar of fluid. (Like many of the wet specimens, it was in disrepair; only one side of the object was attached by string to its supporting framework, so that it was off-kilter.) For me, the Mütter Museum was the museum of the face. And it burned.

There may have been other visitors in the museum that day, but I went unaccompanied, and was essentially alone with my thoughts. Later, my experiences with the museum became more social and professional. But that first visit put me into an altered state, tore open a suture. The museum was a looking glass where I was forced to contemplate the rift. (Divided I stood.) The strange space of the museum activated a painful awareness of the strange space of my own interiority. And vice versa: the museum's strangeness was constituted by my heightened sense of my own strangeness.

There was, I think, one other thing that heightened my experience of the wax faces. They glowed with a kind of antique pathos. They almost called out to me: 'long ago, we were once alive and endured terrible pain and humiliation.' They were ghosts. Their suffering had ended long ago, and yet was sparked to life when I entered their field. I resurrected them with my own imaginative identification, resurrected them with my own ghosts – the selves of my own (not so) distant past, as a child and adolescent, when I had suffered pain and humiliation and (as a result?) awakened into consciousness.

E:

How beautiful and wonderful. Yes, the Mütter is almost a shrine to Otherness. The Others come very close to us when we perceive the traces of their suffering, be it in the form of ancient trepanned skulls or the double liver of the 'Siamese Twins'. We are there together with them; we feel their presence and catch elusive glimpses of their embodied lives. (Maybe they can glimpse us too, like in the movie *The Others*, where the dead and the living inhabit the same space, but only rarely communicate directly.)

M:

The museum puts us into a trance. And reminds us that we – like the people whose body-parts were long ago taken as the raw material out of which the museum objects were manufactured – are suspended in a liminal space between conception and decomposition. The ghosts are all around, eerily curating their own remains, invisibly holding up them up to us so that we can see our own reflection.

E:

At the same time I have a feeling of being detached from those ghosts. To me, the Mütter Museum resembles a ghetto, a prison, or a hospital. The museum is a 'space of otherness', producing alterity by means of segregation (Schick 1999, 44). As visitor I sense this physically and spatially – the specimens are isolated from me in their jars and vitrines. So the bodies of the dead are set apart from the realm of the living, given their own space where they are (unwillingly) confined and controlled.

The museum is, to borrow a phrase from Foucault, 'a negative of that moral city' which the bourgeoisie tried to create.[14] There is no place for decency here; everything is stripped bare, opened up to my immodest gaze. This space stands in opposition to everyday life, in opposition to habitual and unreflective modes of living in, and with, our bodies – it is a *heterotopia*, a 'counter-site, a kind of effectively enacted utopia in which … other real sites … are simultaneously represented, contested and inverted' (Foucault 1967/1998, 239).

Several features of the anatomy/pathology museum fit in with Foucault's *heterotypology* (ibid., 240ff). It is a place of (physical) *deviation*, turning the inside out and displaying corporeal otherness as pathological deformity. It is also *a space which has shifted meaning over time*. It used to be part of the apparatus by which medical science produced and communicated knowledge, and asserted power over the patient and the dead. Now it is a museum of the history of medicine and of death and grotesquery, a place of leisure. It is a space with *special rules for entering*, which defines it, closes it, and opens it: the visitor has to abide by these rules in order to take part of what it has to offer (but there is no controlling what the visitor will feel, think and remember). It is also *heterochronical*, 'linked to slices in time', and, like the cemetery, it is dependent on the loss of life, the ending of time, for its inhabitants. It is *accumulating time* by way of the collecting practice, by establishing a sort of archive, 'oriented toward the eternal'.

M:

Agreed. There is an eerie collapse and convergence of different historical moments, as though all the strata were not just stacked on top of each other, but superimposed, even calling each other out. Like Christian typology where the Biblical narrative is an ongoing prophesy that predicts and structures the present and future.

So the strangeness of the Mütter Museum lies not only in the exteriority of the body interiors and parts, but also in the historicity of the objects. Which are in various states of disrepair and dustiness, and which inhabit wood-framed vitrines of a style no longer used in legitimate museums, and glass jars of a kind that is no longer manufactured (a great difficulty for the museum when they break), and which are haphazardly accompanied by old labels or catalog numbers. The collection is not really organized or displayed as it

---

14   The phrase comes from *The Birth of the Clinic*, Foucault's analysis of the role of the seventeenth-century Hôpital Général (Foucault 1965, 61).

was in the nineteenth and early twentieth century (there have been too many clumsy attempts to update, and make the museum 'relevant'), but its past is ubiquitous.

E:

Yes, the reliquary state of the museum and its objects is an attraction in its own right. There's a romance of the obsolete and dusty. And a kind of nostalgia for the world we have lost.

The scientific anatomical museum had its heyday in the late nineteenth and early twentieth century. Around the same time popular anatomical museums toured the fairgrounds of Europe and sprouted like weeds in the shady entertainment districts of cities in Britain and North America, advertising their sensational and freaky content of wax models, pickled body parts, and live oddities. They were part of a spectrum of popular amusements: natural history museums, Madame Tussaud's, zoos, freak shows, art galleries, peep shows, fun houses, ghost rides, ethnographic collections, operas, shop windows ... All of them were spaces for encounters with the odd as well as the familiar, for thrills and pleasures as well as thoughts and sometimes learning. Proprietors orchestrated a proliferating variety of performances and displays for audience members to gaze upon, and be stimulated.

M:

And that gaze was then returned to the spectator: then, and now, spectators recognize themselves in the display.

Or maybe they don't. The Mütter is a village of the damned, a city of death, disease and suffering. And if the city, as Michel de Certeau argues (1984/1999, 129), creates 'a *universal* and anonymous *subject*' then the museum (and the entire genre of the anatomy museum) creates a universal and anonymous *abject*.[15] Even in the specificity of the pathologies on parade there is a universalizing message: these are the deformations, disfigurements, distortions which human skin and flesh and bone are prey to. The beginning and end, the inside and outside of us.

But this universalizing trope is inevitably thwarted. Even though the directors and curators of the museum have attempted to set up an order of things – here is a class of pathologies, here is a class of a kind of anatomical specimens, here is the structure of human foetal development – the objects resist. They call out to the visitor: *here is the disorder of things*. The display is an anti-geography: the highways, rivers, canals, sewage system, wire networks,

---

15 According to Julia Kristeva (1980/1982, 1), *abjection* is the domain of human experience which is 'neither subject or object': 'The abject has only one quality of the object – that of being opposed to the *I*.'

architectural supports of the body are askew. Even if contained in a jar, the bones and soft tissues of the stillborn congenital twins are impossibly, fatally fused. In other displays, tumors and growths tumble outside of their organs, become embarrassingly visible, a mockery. The orderly avenues of objects suggest that we are entering a planned city with regulated traffic, a realm of law, a world governed by reason. But the objects refuse to comply, seem intent on giving us a lesson in the dire collapse of corporeal existence. They show us how impossible it is to start up a life. And, if started, how impossible it is to maintain. What terrible burdens we human beings are forced to labor under! The displays of human exterior and interior, surface and depth, all inertly, cumulatively, campaign against the fiction of the unified universal subject. They cannot be us, instead they belong to a domain of absolute estrangement. As Kristeva (1980/1982, 1) argues: 'There looms, within abjection, one of those violent, dark revolts of being ... that seems to emanate from an exorbitant outside or inside, ejected beyond the scope of the possible, the tolerable, the thinkable. It lies there, quite close, but it cannot be assimilated.'

And, strangely, this insurrection turns out to be pleasureful. We are relieved of a terrible burden, set free from the obligation to make human or biological sense. For me, like many museum visitors, the Mütter Museum is a special space because it has special reference to the entirely problematic nature of the embodied 'me'. Which disturbs me, but makes me smile, even laugh.

E :

Disturbing or embarrassing things often make people laugh. Kristeva also points out that while the dead body is the ultimate abject, it has the effect of making the living aware of being embodied, alive (Kristeva 1980/1982; Åhrén 2002/2009). Walking among the museum's dead and disfigured, my living body feels alive and well. The presence of death in this strange place does this to me, at the same time as it resonates with my dark inner space, 'a space from below, of mud' (Foucault 1967/1998, 239).

References

Åhrén, E. 2002/2009. *Death, Modernity, and the Body: Sweden 1870–1940*. Rochester: Rochester University Press.

Althusser, L. 1970/1971. 'Ideology and Ideological State Apparatuses', in *Lenin and Philosophy and Other Essays*. New York and London: Monthly Review Press.

Barthes, R. 1968/1989. 'The Reality Effect', in *The Rustle of Language*. Berkeley: University of California Press.

Berman, M. 1982. *All That is Solid Melts into Air: The Experience of Modernity*. New York: Simon and Schuster.

Certeau, M. de 1984/1999. 'Walking in the city', in During, S. (ed.) (1999), *The Cultural Studies Reader*, 2d edn New York: Routledge.

Debord, G. 1967/1994. *The Society of the Spectacle*. New York: Zone Books.

Dussauge, I. 2008. *Technomedical Visions: Magentic Resonance Imaging in 1980s Sweden*. Stockholm: KTH, TRITA-HOT.

Foucault, M. 1965. *Madness and Civilization: A History of Insanity in the Age of Reason*. New York: Pantheon Books.

Foucault, M. 1966/2002. *The Order of Things: An Archaeology of the Human Sciences*. London: Routledge.

Foucault, M. 1967/1998. 'Of Other Spaces', in Mirzoeff, N. (ed.) *The Visual Culture Reader*. London: Routledge.

Kahn, F. 1923–1931. *Das Leben des Menschen: Volkstümliche Anatomie, Biologie, Physiologie und Entwickslungsgeschichte des Menschen*. 5 vols. Stuttgart: Kosmos, Gesellschaft der Naturfreunde/Frankh'sche Verlag.

Kern, S. 1978. *The Culture of Time and Space, 1880–1918*. Cambridge, MA: Harvard University Press.

Kristeva, J. 1980/1982. *Powers of Horror: An Essay on Abjection*. New York: Columbia University Press.

Laskowski, S. 1894. *Anatomie Normale du Corps Humain, Atlas Iconographique …; Dessineés d'après les Préparations de l'Auteur par S. Balicki*. Geneva: Braun.

Le Corbusier. 1923/1946. *Towards a New Architecture*. London: Architectural Press.

Lefebvre, H. 1974/1991. *The Production of Space*. Oxford: Blackwell.

Lindgren, L. (ed.) 2007. *Mütter Museum Historic Medical Photographs*. New York: Blast Books.

Marx, K. and Engels, F. 1848/2002. *The Communist Manifesto*. London: Penguin.

Merleau-Ponty, M. 1945/1962. *Phenomenology of Perception*. London: Routledge.

Mitchell, W.J.T. 1994. *Picture Theory*. London and Chicago: University of Chicago Press.

Richardson, R. 1988. *Death, Dissection and the Destitute*. London: Penguin.

Sappol, M. 2002. *A Traffic of Dead Bodies: Anatomy and Embodied Social Identity in 19th-Century America*. Princeton, NJ: Princeton University Press.

Schick, I.C. 1999. *The Erotic Margin: Sexuality and Spatiality in Alteritist Discourse*. London: Verso.

Virilio, P. 1994. *The Vision Machine*. Bloomington: Indiana University Press, British Film Institute.

Voloshinov, V.N. 1927/1994. Selection from *Freudianism: A Critical Sketch*, in Morris, P. (ed.), *The Bakhtin Reader: Selected Writings of Bakhtin, Medvedev, and Voloshinov*. London: Edward Arnold.

Voloshinov, V.N. 1929/1994. Selection from *Marxism and the Philosophy of Language*, in Morris, P. (ed.) *The Bakhtin Reader: Selected Writings of Bakhtin, Medvedev, and Voloshinov*. London: Edward Arnold.

Ward, J. 2001. *Weimar Surfaces: Urban Visual Culture in Weimar Germany*. Berkeley: University of California Press.

Weindling, P. 1989. *Health, Race and German Politics Between National Unification and Nazism, 1870–1945*. Cambridge: Cambridge University Press.

Worden, G. 2002. *Mütter Museum of the College of Physicians of Philadelphia*. New York: Blast Books.

Chapter 5
Obscure Objects of Media Studies: Echo, Hotbird and Ikonos

Lisa Parks

To be obscure is to be faintly perceptible so as to lack clear definition, to be hidden, out of sight, not readily noticed or seen, inconspicuous, far from centres of the human population. The satellite has been a relatively obscure object of media studies, but it is not alone. As Amelie Hastie points out, many objects – from the ticket stub to the powder puff, from the videocassette case to the antenna tree – that are overlooked in media studies could be used to expand the field in productive ways. As Hastie explains, '… An emphasis on objects and material forms in relation to representational and time-based media might enable a delineation of the social and economic circuits of exchange in which we – and visual culture, in its various forms – participate' (Hastie 2006). Here, rather than adopt an object-oriented disciplinary approach and insist that there should be a field of satellite studies just as we have cinema, radio, television, and cyber-studies, I'll follow the suggestion of Siegfried Zielinski – that media not only have complex relational histories, but can be understood as entr'actes or intervals in a much deeper history of audiovisions. Like film and television, the satellite could be treated as part of an integrated history of media, as a *dispositif* – an arrangement of audio-visions interwoven with other media, architecture, transportation, science and technology, the organization of work and time, philosophical propositions and so on (Zielinski 1999, 18).

While a long-term goal may be to treat the satellite as a *dispositif*, the simple goal of this chapter is to attempt to make the satellite less obscure in media studies by offering descriptive sketches of three satellites – Echo, Hotbird and Ikonos – and discussing possible modes of critical engagement with each of them. In the process I hope to make three points. First, I want to suggest that since the field of media studies is so integrally bound up with, even contingent upon, processes of audiovisual perception it might be useful to explore the opaque and imperceptible so that our historical and critical projects are not totally circumscribed by that which is visible and audible. Is it possible in media studies to develop a critical sensitivity to the obscure as we have to the spectacular? Second, I hope to suggest that by thinking about the processes and exchanges that occur in the space between earth and orbit that we might begin to imagine media morphologies that exceed the nation, the network, the screen and the text – terms that seem to have settled as implicit figurations (and unquestioned conceptual foundations) in the field of media studies. What terms can we use to describe the signal transactions that occur beyond the ionosphere and yet that are fundamental to media cultures on earth? Finally, I hope to suggest that there is a need for materialist histories of satellite technologies and that conducting such work involves taking distribution seriously as a site of media history and criticism. Why is it that we know the names of broadcast networks, major web portals, syndication companies, but we don't know the names of satellites?

Echo

Figure 5.1   The Echo satellite in the NASA lab. Courtesy of NASA

Echo was an experimental satellite launched on 12 August 1960. Developed by Bell Telephone laboratories in association with NASA, it was a passive, mirror or reflector satellite as opposed to an active/repeater. This means that it had the capacity to relay signals but not to store them. When Echo reached orbit, 1,000 miles above the earth, the satellite inflated into a balloon 100 feet in diameter and encircled the earth every two hours. Its reflective mylar surface made it appear brighter in the sky than the north star. On 14 August 1960, during the 21st pass of the satellite, Bell Lab scientists transmitted a recording of the song 'America the Beautiful' to the Goldstone facility in California. The story of the so-called 'transcontinental melody' made front-page headlines and was documented in a Bell Labs film called *The Big Bounce* (1960).[1] During the 22nd pass a 'double bounce' experiment relayed a recorded message from Stump Neck, Maryland to Holmdel, New Jersey and then on to Goldstone, California. On 18 August 1960 Bell Labs sent a message to the Centre National d'Etudes de Telecommunication station at Issy les Maulineaux, France using Echo. And on 19 August 1960 technicians working for Collins Radio in Iowa used Echo to transmit an Associated Press picture of President Eisenhower from Cedar Rapids, Iowa to Richardson, Texas (Elder 1995, 116–21). These early signal exchanges marked the beginning of regularized satellite transmissions.

In addition to being used for inter- and transcontinental audiovisual relays, Echo itself became an international spectacle. Donald Elder reports in his technical history of Echo that during an international trade fair in Damascus, Syria, fair guides stopped their lectures and led audiences to watch Echo in the sky when it passed overhead. In Kabul, Afghanistan King Mohammed Zahir brought an entourage of 700 people with him to view an exhibition about Echo (Elder 1995, 122). The USIA supplied over 200 of its overseas posts with a programme called 'Echo in Space' narrated in over 30 languages. The show appeared in theatres and on television and the USIA deployed mobile field units to distribute the film into villages and remote areas. The USIA also worked out an arrangement to record the messages of foreign diplomats relayed on Echo and then distribute them for broadcast on Voice of America. Radio listeners in 15 nations heard the messages.

When Echo and other satellites such as Telstar and Early Bird were launched during the 1960s there was a concerted effort to visualize and publicize them within popular culture, to celebrate them as spectacles of American ingenuity. Yet as the number of launches increased over the years, the spectacular presentation of satellites diminished. In addition, some satellite programmes, such as Corona, were kept entirely obscure because they were designated as 'classified' and 'top secret'. The history of the satellite thus involves a series of fluctuations and negotiations over the knowledge and visibility of the apparatus, sometimes resulting in its spectacular celebration as a scientific object, sometimes resulting in its banality (due to the high volume of satellite launches), and sometimes resulting in calculated secrecy.

In the process of relaying audiovisual signals, Echo also activated and drew attention to a dynamic field of relations between earth and orbit. The

---

1   The film is available at the Internet Movie Archive, at http://www.archive.org/details/BigBounc1960, accessed 10 January 2007.

very installation of Echo in outer space altered practices taking place on the surface of the earth as people tried to spot it, up and downlink with it, and generate discourse about it. These circuits of exchange, however, are largely imperceptible. Moreover, it is difficult to access recordings of early satellite transmissions since only some have been preserved. In this sense, early satellite relays share more in common with theatrical performance or turn of the century wireless experiments in that they are ephemeral forms of culture that move through and vanish in the air, as opposed to being part of a culture of mechanical reproduction (Parks 2007). What we can attempt to reconstruct, however, are the vectors of signal distribution – the lines, paths and directions of satellite relays.

What I am proposing is an approach to the study of satellites that would embed the obscure processes of signal distribution within cartographic fields of representation as a way of beginning to provide a sense of the material histories of satellites. What places did signals originate from? Which satellites did those signals traverse? Where did those signals end up? It might seem like an impossible task to recover and represent such signal histories, but attempting to do so on even a limited scale may provide a more concrete sense of the dynamic field between earth and orbit. In the case of Echo a successful relay of a telephone conversation from California to New Jersey initiated a series of other such signal excursions and served a catalyzing function so that four decades later the satellite signal economy is so complex it seems beyond mapping or visualization altogether, and yet, it is impossible to fully appreciate what 'media globalization' means unless there is a better way to account for mechanisms of distribution. Perhaps it would be possible to create visualizations of signal traffic so that we imagine audiovisual media not only as narratives, ideologies and programmes/texts, but also as imperceptible phenomena that move across and beyond the earth's surface. We may need new media morphologies to account for such phenomena – for the patterns, vectors, and densities of signal circulation in the world. It might be useful to engage with meteorological and/or aeronautical paradigms to consider the metaphors, practices and forms of signal circulation. If there are signal flows, fogs, storms, flight paths or corridors how might they be specified, visualized, and historicized?

A project called 'Flight Patterns' by artist Aaron Koblin provides a useful model for thinking about how signal traffic might be visualized. This three-minute animation sequence, which appears on YouTube, draws upon Federal Aviation Administration data to generate a time-lapse representation of US flight traffic patterns and density during a period from 19–21 March 2005.[2] The project, which was produced while Koblin was earning his MFA

---

2 Aaron Koblin, 'Flight Patterns', 4 November 2006, available at http://www. youtube.com/watch?v=dPv8psZsvIU, accessed 10 November 2006. High resolution version available at http://www.aaronkoblin.com/work/faa/, accessed 10 January 2007.

at UCLA, shows a running count of the number of aircraft in the sky and animates their movement across a map of the United States inscribed upon a black screen. Aircraft appear as white dots moving along various flight paths and they disappear at their destinations. The project establishes a sense of the relative volume of traffic over different regions of the US. At one point the US map disappears as the flight patterns continue and the animation zooms in to profile planes flying in and out of Los Angeles. We are encouraged to recognize the unique form that takes shape at the site and a frame reads, 'Without geography landmarks and patterns emerge'.

Using this animation as a model to develop representations of local/ national/global signal traffic could lead to a more cartographic and territorial understanding of media globalization. It would allow us to get a specific sense of the patterns of media distribution in the world and the differential flows that make up the global media economy. Such an animation could visualize where and when TV signals depart and arrive, the fluctuating volumes or scales of signal traffic across different locales or regions, and the dynamic field that exists between ground stations on earth and satellites in orbit. Furthermore, such visualizations could help to generate new critical concepts or metaphors, particularly in relation to distribution, derived through a close analysis of signal circulation. Since Echo was one of the first experimental satellites it had relatively limited number of signal exchanges and thus it would be relatively simple to map. A more recent series of satellites, such as Hotbird, which I will discuss in the next section, would be much more challenging to visualize given the high volume of signals this satellite fleet circulates and the wide expanse of its geographic coverage.

Hotbird

While Echo was an early experimental satellite that reflected signals from one point to another, Hotbird is a fleet of commercial communications satellites owned by the French company, Eutelsat. In 2006 Eutelsat transmitted more than 1,500 television channels to 120 million homes (*Eutelsat* 2006, 2). The company uses 23 satellites to serve more than 150 countries, and specializes in coverage over Europe, North Africa and the Middle East. The satellites in the Hotbird fleet are lined up at 13 degrees east which the company claims is 'Europe's most sought after orbital position' (ibid., 3). By lining satellites up in the same orbital location, Eutelsat is able to capitalize upon this position and maximize its capacity by assembling a neighbourhood or cluster of satellites and assert long-term ownership over this address. Though the satellites occupy the same orbital position, they have different launch histories and manufacturers. For instance, Hotbird 1 was launched on 28 March 1995 by the Ariane IV rocket from a launching base in French Guiana and was manufactured by Alcatel Space Industries. Hotbird 2 was launched on 21 November 1996 and was manufactured by Matra Marconi. Hotbird 3, manufactured by a UK company called Matra Marconi, was launched on 2 September 1997 by the Ariane V99 rocket. Hotbird 8, the most powerful broadcast satellite serving Europe, was launched on 8 May 2006 and was manufactured by EADS Astrium. The Hotbird footprints cover Europe and North Africa and extend as far east as Moscow and Dubai.

In addition to mapping signal distribution, it is possible to examine footprints to determine the geographic boundary in which a signal from a given satellite can be received. These footprint maps serve as starting points for understanding how satellite broadcasting turns continents into signal territories. In other words, the satellite footprint is not just an inert technical boundary; it is a transnational trade route, a technological zone, a site of cultural atmospherics, a corporate claim to orbital, spectral and geophysical property (Parks 2005a, 2005b). While these maps hint at broader power relations, they are ultimately insufficient in that they tell us little about material conditions within the footprint boundary. They do not, for instance, provide information about the signals circulated within them. To understand the satellite's role in the global media economy it is important to examine carriage lists, many of which are available at lyngsat.com. This website provides listings of communications satellites, the signals they carry, and the companies that generate them. In 2006 the Hotbird's carriage chart was thirty pages long and identified hundreds of radio and television signals located on different transponders, encrypted with different codes, and beamed into Hotbird footprints using different frequencies. Signals carried by the Hotbird fleet are either downlinked directly by satellite television viewers or by cable providers who bundle and distribute them to subscribers. These carriage lists are important guides for media studies; they provide a slice of the global media landscape by specifying which radio and television signals are circulated where in the world. This varied mediascape include such signals as Iranian Cinema Channel, Baby TV, Viva Polska, Al Hayat, Jolly, Arabesque, Sex Gay TV, Tamil TV Network, Hallmark Channel Turkey, Croatian Music Channel, Pink Plus, or Public TV of Armenia.

The prospect of mapping signal distribution via satellite becomes even more daunting, then, if we consider a contemporary satellite fleet such as Hotbird, which in January 2006 carried 445 radio and 706 television signals from more than 34 countries. What would the map look like? The vectors of signal distribution would be so complex and unwieldy that they might evoke new morphologies that would shift our thinking beyond linear distribution models and toward a theory of cultural atmospherics, media fogs or, as mentioned earlier, signal territories. As Charles Acland suggests in his book *Screen Traffic*, 'When concentrating on the present phase of globalization, the conventional definitions of the borders of social phenomena – nations, populations, ethnicities, countries, and so on – become ever more elusive; new and unforeseen organizations of social categories appear' (Acland 2003, 15).

Given the globalization of signals, it seems clear that we need new categories for media as well. Studying signals transmitted via satellite would require shifting some critical attention beyond the screen, the theatre, the home, the studio, or the nation and beginning to invent categories derived from the patterns of signal distribution. Such a project may share more in common with the flow studies funded by UNESCO during the 1960s and 1970s, which were conducted in the interest of understanding the uneven circulation of television programming on a global scale. As Mimi White explains, 'the number and range of programmes imported and exported around the world were counted by nation-state in an effort to quantitatively map the ways global power and influence are unevenly distributed' (White 2003, 103). Rather than quantify signals to confirm Western hegemony in the global media marketplace (which is something we already know), I am interested in finding ways to visualize signal densities and vanishing points, concentrations and dispersals, and arrivals and departures. What I am proposing is a kind of footprint analysis or a mapping or visualization of distribution that would help to clarify the satellite's relationship to processes of territorialization, geopolitics and world trade. The goal is not to generate a master view of the global signal economy, but rather to generate a series of *partial* visualizations that would provide a sense of how, where, and when signal distribution occurs in the world. Such visualizations could create spaces for the formulation of new research questions and critical concepts, and would draw attention to the dynamic interplays between earth and orbit – the vertical fields of media.

In part to explore this possibility, in 2005 I worked with a cartographer to develop a map of the satellites used by broadcasters from the new states of the former Yugoslavia. Before the break up of Yugoslavia, only one satellite was used by the Yugoslav Television to distribute six hours of programming per day to Europe and North America. After the war ended in 1995, the media sectors of new states were privatized, and by 2005 there were five public and fifteen commercial broadcasters in the region using 15 different satellites to send their signals around the world (Parks 2005b). The 15 satellites (Hotbird 2, 3, 4 and 6 among them) are owned by nine different companies, which have their financial and operational headquarters outside of the region. The map identifies the names of broadcasters, the satellites they used, the location of their footprints and the companies that own the satellites. It was designed to specify the players, ownership and cartographies of the regional satellite economy. The map visualizes the postwar transition from a system of national socialist state broadcasting within Yugoslavia to a system of multiple new national broadcasters and private broadcasters throughout the territory that are integrating with regional and global media economies.

Ikonos

Thus far I have discussed an experimental satellite and a fleet of communications satellites used for television distribution and suggested the need for visualizations of signal distribution and footprint analysis. In this last section I discuss a remote sensing satellite called Ikonos (named after the Greek word for image), which is a privately-owned remote sensing satellite launched on 24 September 1999 by Athena II rocket from Vandenburg Air Force Base in California. Unlike Hotbird, which occupies a geosynchronous orbital position, this satellite moves in a low earth orbit. It was manufactured by Lockheed Martin with investments from companies such as Raytheon, Mitsubishi (Japan), Van Der Horst (Singapore), Hyundai Space and Aircraft (Korea), Europe's Remote Sensing Affiliates, Swedish Space Corp., and Loxley Public Company (Thailand). During its first several years Ikonos was operated by a corporation called Space Imaging, which, in early 2006, merged with Orbimage to form Geoeye, which is now the largest commercial satellite imaging company in the world (Geoeye).

Remote sensing emerged in the US during the 1960s with top-secret spy satellite programmes such as Corona, weather satellites such as the Tiros series, and the Landsat series, which launched during the 1970s (see Mack 1990; Parks 2005a, 78-81). Before the 1990s, remote sensing satellites were used primarily by reconnaissance experts, natural resource specialists, and research scientists. Public access to remote sensing imagery was rather limited until President Clinton's privatization of the industry during the 1990s. By the early 2000s, US companies such as Spaceimaging and Orbimage made high-resolution satellite images more readily available and facilitated their exchange in the global economy. The increased availability of high resolution satellite images, I want to suggest, represents the possibility for making the satellite less obscure both in public culture and media studies.

Since its installation in 1999 Ikonos has been programmed to acquire image data over many parts of the planet. There now exist blockbuster satellite images just like there are blockbuster films and hit television series. The Ikonos image gallery includes annual top ten lists, and special feature sections on a host of disasters including the 911 attacks on the World Trade Center and the Pentagon, Operation Enduring Freedom in Afghanistan, the Tsunami in Southeast Asia, the Pakistan and Kashmir earthquakes, and Hurricanes Katrina and Rita (Spaceimaging). The circulation of such satellite images cannot be separated from a US-dominated global media economy that selects and foregrounds a handful of events as newsworthy, spectacular and profitable and ignores most events in the world. Ikonos images have been used to reinforce already existing psychic and capital investments in particular world events and as such have a way of focalizing the mediated production of world history.

The satellite's involvement in the mediated production of world history can be linked to Doreen Massey's discussion of the 'power-geometry' of time-space compression. Massey suggests that 'different social groups and different individuals are placed in very distinct ways in relation to these flows and interconnections; some are more in charge of it than others; some initiate flows and movement; others don't; some are more on the receiving end of it than others; some are effectively imprisoned by it' (Massey 1993, 61). In the process of developing her argument, Massey writes,

> Imagine for a moment that you are on a satellite, further out and beyond all actual satellites; you can see 'planet earth' from a distance and, rare for someone with only peaceful intentions, you are equipped with the kind of technology that allows you to see the colours of people's eyes and the number on their number-plates. You can see all the movement and tune-in to all the communication that is going on. Furthest out are the satellites, then aeroplanes, the long haul between London and Tokyo and the hop from San Salvador to Guatemala City. Some of this is people moving, some of it is physical trade, some is media broadcasting. There are faxes, e-mail, film-distribution networks, financial flows and transactions. Look in closer and there are ships and trains, steam trains slogging laboriously up hills somewhere in Asia. Look in closer still and there are lorries and cars and buses and on down further and somewhere in sub-Saharan African there's a woman on foot who still spends hours a day collecting water. (Massey 1993, 61)

While the passage makes an uncanny link to the satellite, I want to highlight Massey's use of the orbital view as a place from which to see and delineate power differentials in the world. These 'power geometries' involve various practices and scales of flow and movement, ranging from the broadcast signal transmitted through the ether to the water collector walking on the ground. Instead of assuming satellite omniscience augments a strategic quest for planetary control, Massey treats the orbital view as petri dish in which different scales and vectors of power emerge. Pushing her logic further, we might say that despite the increasing availability of satellite images, some in the world will have access to them and some will not; some will have the power to interpret them and some will not; and some will use them to write world history while others will not.

While satellite images are embedded within and help to expose power-geometries, they are also implicated in the changing configuration of media studies. If there is any doubt about the intersection of satellite imaging and film and media studies, we could consider some more literal intersections. For instance a 2001 animation featured in the NASA Scientific Visualization Studio made out of 'mosaic-ed' Ikonos, Terra and Landsat images takes us directly from an orbital position to a one hovering over the Hollywood sign (NASA).[3] The sequence implies that the remote sensing satellite 423

---

3   The sequence is available at http://svs.gsfc.nasa.gov/vis/a000000/a002100/a002108/, accessed 30 January 2006.

miles above the earth offers a meta-perspective that exceeds the visualizing potential of even Hollywood's. Ikonos images were also used to expose the location of CBS facilities during the production of *Survivor: Africa* in Kenya's Shaba National Reserve in August 2001 (Spaceimaging).[4] Dan Bollinger, a fan from Layfayette, Indiana, sent a request to Space Imaging to acquire the image data over Kenya, and after spotting the CBS facilities, Space Imaging executives waived their 'special mission' fee of $3500 because they felt this would be good publicity for Ikonos (Vergano 2001). Rather that simply reveal a Hollywood sign or production site, satellite images could be used to examine the environmental and economic effects of Hollywood's on-location productions around the world, whether with *The Beach* in Thailand, *Titanic* in Mexico or *Survivor* in Kenya. Maybe the meta, or panoptic perspective of Ikonos isn't as daunting when it exposes what Hollywood studios have been up to in other countries' coastlines or national parks.

Figure 5.2   Screen capture of the KFC Face From Space video

---

4   These images are included in the Space Imaging online gallery, available at http://www.spaceimaging.com/gallery/survivor/default.htm, accessed 30 January 2006.

It is more likely that Ikonos images will be used for corporate landscaping than eco-friendly film production. One of the most bizarre uses of the satellite in recent times involved Kentucky Fried Chicken's development of an 87,5000 ft^2 Colonel Sander's logo that could be seen from orbit. More than 50 designers, engineers and scientists worked for 3,000 hours to create the 'Face from Space' near Area 51 in Rachel, Nevada, also known as the 'UFO Capital of the World' (*Kentucky Fried Chicken* 2006). As a KFC press release explained, 'The event marks the official debut of a global re-image campaign that will contemporize 14,000-plus KFC restaurants in over 80 countries over the next few years' (ibid.). KFC purchased an Ikonos image of the site and distributed it through global media circuits to address its 4.5 billion customers worldwide including those in emerging markets of India, Russia and Brazil (ibid.). While satellites have historically passed over the earth to observe 'naturally unfolding' phenomena, now events are staged precisely so they can be viewed from an orbital perspective. Whether we call this corporate landscaping or orbital marketing, remote sensing satellites are now being used to pitch products and address global consumers just as other media such as commercial television or the world wide web.

Our sense of world history is increasingly regulated and produced by satellite technologies, whether through the circulation of signal traffic or the staging of events for satellite capture. More and more, world historical activities are oriented in relation to orbit. While the orbital perspective is highly subject to manipulation, its distance and distinction from the earth nevertheless infuses it with a powerful *globalizing optic*, where the event in frame suddenly seems to take on a global scale and significance simply by virtue of being visualized from an orbital platform. The globalizing optic of satellite imagery produces a sense of the worldly and the historic as a visual trick. Indeed, on multiple occasions, states, NGOs and news agencies have strategically exploited the globalizing optic of satellite imagery to draw international attention to and confer 'world-historical' status upon events as well as their discovery and/or coverage of them (Fair and Parks 2001). The implication of such a condition is that the states, corporations and agencies that own and operate satellites, and the individuals that have the knowledge and expertise to interpret and use satellite imagery, also have the power to govern the hyper-real production of world history.

If there is any recent development, however, that makes us reflect upon such power, as well as the obscurity or presence of satellites in our media culture, it may be Google Earth. This web application, which builds upon past initiatives such as Al Gore's Digital Earth Project of the 1990s (Parks 2003), emerged in 2005 and has since generated a vibrant community of users from geographers to businesspeople to artists. Scholars across fields are also using and writing about this application from an array of perspectives. What I want to address here, however, is the fact that Google Earth tends to efface information about satellites at its interface. When the basic version of Google Earth emerged, there was no information available at its interface about the satellites that acquire the image data used to construct the world as a navigable digital domain. Google Earth announces on its website that the information comes from a variety of sources and is 'mosaic'ed' together and 'a single city may have imagery taken from different months' (*Google Earth*, 'Images and Dates'). On another page in response to the question 'When were these pictures taken?' the company explains, 'Our photographs are taken by satellites or aircraft sometime in the last three years' (*Google Earth*, 'Common Questions'). With such vague responses, unfortunately a web application with great potential to inform large numbers of users about satellites such as Ikonos ends up keeping them in the dark. If satellites are increasingly relevant to the production of world history, then public literacy about these technologies and the images they generate is more important than ever. Furthermore, if citizen-users around the world now have opportunities to participate in the collective production of Google Earth, it is vital that they understand the material conditions and technologies with which such digital communities are made.

We can get a better sense of the difference it makes to have satellite image details by examining the layer of Digital Globe, the only company that provides date information for satellite images that are part of Google Earth. When the user activates the Digital Globe layer, colour-coded squares and 'I' icons appear in the visual field. Clicking on an 'I' opens a frame with data about the image including the acquisition date, cloud cover, and an environmental quality rating. The satellite name is not provided, but it was likely acquired by Digital Globe's Quickbird satellite. If the user clicks on 'preview', she enters a meta-browser featuring the single satellite image captioned with information about how to purchase it or others from Digital Globe.

Clearly, Digital Globe is providing date information (when other companies are not) as part of a marketing strategy. When the user activates the layer, Digital Globe and its product stand out in bold relief amongst a sea of unidentified and undifferentiated satellite image data. In a sense, there is scant difference between this strategy and the KFC ad in that both corporations use satellites to turn plots of earth into giant billboards. Still, I find the Digital Globe layer useful because the inscription of date information and colour-coded squares helps us to understand which parts of the earth's surface a satellite has scanned and when. The colour-coded squares (sometimes called scene footprints) function as traces of a satellite's pass over a specific part of the earth. When composited like this they form a historical record of satellite image data acquisitions, as well as a slice of Digital Globe's inventory. Just as visualizations of signal distribution and footprints can help to emphasize the different morphologies and materialities of satellite media, this view provides a series of tracks or traces of the obscured technology in use.

Conclusion

If we accept that experimental satellite relays, the distribution of hundreds of radio and television channels, and constantly snapped satellite images of earth are relevant to the field of media studies, then we either need a theory of obscure media or we need ways of making the satellite less obscure. In this chapter I have suggested that the uses of Echo, Hotbird and Ikonos have generated a set of interplays between earth and orbit that demand different media morphologies and mappings. This call for maps of signal distribution should not be understood as a positivist gesture to see, know and master, but rather an attempt to generate critical spaces for exploring other obscure objects of media studies, whether other satellites, distribution systems, or objects that linger in a peculiar twilight between visibility and invisibility. I have only discussed three satellites and there are about 3,500 functioning satellites in orbit. Each satellite is a symptom of a complex institutional history and imperceptible signal traffic. While imagining and specifying the contours and vectors of this traffic, it is also important to explore in greater detail the financial, temporal, regulatory and intermediale dimensions of the satellite economy and to recognize that this part of the culture industry synthesizes and alternates between scientific, military, entertainment and educational modalities. There is a potential, in other words, to refine the materialist approach to global media and world history by considering the specific roles that satellites have played in their production.

I would like to close by making three points. First, studying the signal territories and cultures that form in relation to satellites can help make them less obscure in media studies. In his provocative essay 'Popular Secrecy and Occultural Studies', Jack Bratich makes a case for a tactical secrecy, suggesting that publicity is a 'a truth-telling strategy' often aligned with the Enlightenment project and is swept up in the fickle dynamics of concealment and revelation that shape our public culture. He asks, 'In an age where secrecy is virtually everywhere as a strategy of domination, can we begin to experiment with an insurgent secrecy, a minor secrecy or a popular secrecy?' (Bratich 2007, 48). While Bratich offers a sharp analysis of the nuances of power and knowledge in the current political moment, I find it difficult to apply this logic to satellites since their secrecy (or, as I have termed it in this chapter, obscurity) is so bound up with institutions that require more public oversight and scrutiny, whether NASA, the US military, or the telecommunications and media conglomerates. In short, the stakes are too high for keeping the satellite secret, in both media studies and public culture.

Second, studying satellites may enable media scholars to develop new ways of conceptualizing and visualizing the dynamic field of signal distribution that has taken shape across continents and between earth and orbit for nearly 50 years. The year 2007 marks the 50 year anniversary of the first earth satellite, Sputnik, and while there are plenty of images and models of satellites, we still do not have adequate visualizations of signal territories and traffic. Generating visualizations, then, can help us imagine signals as material phenomena and allow us to invent new conceptual categories and metaphors for media theory and history. The vectors of signals moving through the world may share more in common with patterns of weather systems or other creatures of flight, whether birds, hurricanes, or aeroplanes. Developing ways of conveying the imperceptible or obscure aspects of satellite media, may also allow us to recognize our field's overlap with environmental studies.

Finally, although one could conceive of an object-oriented field of satellite studies, such a project, it seems to me, ultimately privileges institutional legitimation over passionate investigation. Put in the historical perspective of what Zielinski calls the 'deep history of the media', the satellite is but a passing interlude of audiovisions (Zielinski 2006). Thus rather than fetishize the object and call for another field of study, I am more interested in using satellites and signal exchanges as objects to think with in a way that may expand possibilities for historical and critical research in media studies. At the very least, Echo, Hotbird and Ikonos spark questions about the definition of signal territories, paths of signal distribution, and new economies of global imaging. Perhaps the best we can hope for in an object is for it to spin us into the orbit of new curiosities.

Author's Note

I would like to thank Amanda Lagerkvist, André Jansson, Jim Schwoch, Charles Acland, Amelie Hastie, Anna McCarthy, and members of the UC Santa Cruz Department of Film and Digital Media Studies for their encouraging comments and helpful feedback as I was writing this chapter. An earlier version appeared in the online journal *Mediascape* with a full set of illustrations. See http://www.tft.ucla.edu/media scape/Spring2007/archive/volume01/number03/columns/parks.htm.

References

Acland, C. 2003. *Screen Traffic: Movies, Multiplexes, and Global Culture*. Durham, NC: Duke University Press.

Bratich, J. 2007. 'Popular Secrecy and Occultural Studies', *Cultural Studies*, 21:1, 42–58.

Elder, D.C. 1995. *Out from Behind the Eight-Ball: A History of Project Echo*. Springfield, VA: American Astronautical Society.

Eutelsat. 2006. 'Hotbird' publicity brochure, Eutelsat website, available at http://www.eutelsat.com/news/media_library/brochures/hotbirds.pdf, accessed 30 January 2006.

Fair, J.E. and Parks, L. 2001. 'Africa on Camera: Televised Video Footage and Satellite and Aerial Imaging of the Rwandan Refugee Crisis', *Africa Today*, 48.

Geoeye. Undated. Corporate website, http://www.geoeye.com/, accessed 10 January 2007.

Google Earth. Undated. 'Common Questions about Google Earth', available at earth.google.com/faq.html#1, accessed 11 January 2007.

Google Earth. Undated. 'Images and Dates', available at earth.google.com/images_dates.html, accessed 11 January 2007.

Hastie, A. 2006. 'Curator's Statement for "The Object of Media Studies"', *Vectors: Journal of Culture and Technology in a Dynamic Vernacular*, Spring 2006, available at http://vectors.iml.annenberg.edu/index.php?page=7&project Id=65, accessed 3 January 2007.

Kentucky Fried Chicken. 2006. Press Release, 14 November, available at http://www.kfc.com/about/pressreleases/111406.asp, accessed 7 January 2007.

Mack, P. 1990. *Viewing the Earth: The Social Construction of the Landsat Satellite System*. Cambridge, MA: MIT Press.

Massey, D. 1993. 'Power-Geometry and a Progressive Sense of Place', in Bird, J., Curtis, B., Putnam, T., Robertson, G. and Tickner, L. (eds) *Mapping the Futures: Local Cultures, Global Change*. London: Routledge.

NASA. Undated. Scientific Visualization Studio, available at http://svs.gsfc.nasa.gov/vis/a000000/a002100/a002108/, accessed 30 January 2006.

Parks, L. 2003. 'Satellite and Cyber Visualities: Analyzing "Digital Earth"', in Mirzoeff, N. (ed.) *The Visual Culture Reader 2.0*. New York and London: Routledge.

Parks, L. 2005a. *Cultures in Orbit: Satellites and the Televisual*. Durham, NC: Duke University Press.

Parks, L. 2005b. 'Postwar Footprints: Satellite and Wireless Stories in Slovenia and Croatia', in Franke, A. (ed.) *B-Zone: Becoming Europe and Beyond*. Barcelona: Actar Press.

Parks, L. 2007. 'Orbital Performers and Satellite Translators: Art in the Age of Ionospheric Exchange', *Quarterly Review of Film and Video*, 24, 207–16.

Spaceimaging. Undated. Ikonos Image Gallery http://www.spaceimaging.com/gallery/, accessed 1 March 2006.

Vergano, D. 2001. 'Scouting the New "Survivor" Location', *USA Today*, 29 August, available at http://www.usatoday.com/life/television/2001-08-29-survivor-location.htm, accessed 1 March 2006.

White, M. 2003. 'Flow and Other Close Encounters with Television', in Parks, L. and Kumar, S. (eds) *Planet TV: A Global Television Reader*. New York: New York University Press.

Zielinski, S. 1999. *Audio-Visions: Cinema and Television as Entr'actes in History*. Amsterdam: Amsterdam University Press.

Zielinski, S. 2006. *Deep Time of the Media: Toward an Archaeology of Hearing and Seeing by Technical Means*. Cambridge, MA: MIT Press.

PART 2
Dislocation, Disruption, Disobedience

Introduction to Part 2

André Jansson and Amanda Lagerkvist

This part of the book aims to reveal how the estranging sense of dislocation, evoked in certain cities or landscapes, is tied to disobedient or disruptive spatial/symbolic practices or events; that is, practices and events that escape, re-articulate, or destroy dominant spatial scripts and myths. Through a broad variety of analyses ranging from addressing representations of space and aesthetic styles, to material cultures of architecture, performance and practice, strange spaces are depicted as produced through *dislocation,* as well as *disruption* and *disobedience.* Strangeness is thus always predicated upon moments of transformation or interruption. You may thus, as did Jean Baudrillard (1986), who tried to mundanely *se promener* in the city of Los Angeles, end up feeling utterly dislocated there: 'If you get out of the car in this centrifugal metropolis, you immediately become a delinquent; as soon as you start walking you are a threat to public order, like the dog wandering in the road. Only immigrants from the Third World are allowed to walk' (1986, 58). But the top storey affords dislocations – even dizziness – as well, at least for this one Frenchman in LA in the 1980s:

> The Top of the Bonaventura Hotel. Its metal structure and its plate glass windows rotate slowly around the cocktail bar. The movements of the skyscrapers outside is almost imperceptible. Then you realize that it is the platform of the bar that is moving, while the rest of the building remains still. In the end I get to see the whole building revolve around the top of the hotel. A dizzy feeling that continues inside the hotel as a result of its labyrinthine convolution. (ibid., 59)

Strange and mediated geographies may thus belong both to the street level and the top level of an edifice, to exteriors and interiors. When the urban script is interrupted, when codes are lost, and cognitive mappings are confounded social subjects are left to emotionally fare as they might and to grapple with *strangeness.* Such a situation is portrayed in 'Beside Myself with Looking: The Provincial, Female Spectator as Out of Place at the Stockholm Exhibition 1897' where Ylva Habel discusses the journalistic burlesque representations of a pastoral woman, Lovisa Petterkvist, who 'detourns' space by plunging herself into the modernist space of the Stockholm exposition of 1897. Alienation is the upshot for someone who does not possess the proper socio-spatial codes and distinctions, the proper training to observe and look, but equally important, space is also rendered strange through Petterkvist's injudicious strategies.

Strangeness may thus be generated through disobedient, carnivalesque, rebellious, or criminal *action* and occurs when spatial codes are disrupted, sometimes as the outcome of *rapid change* – both in an orderly and disorderly fashion. As we encounter cities, they also entail within their baffling spatial organization, within their often inherent heterotopian constitution, spectres of both past and future haunting their here-and-now-ness. The coexistence of temporalities in cities (often as a result of dramatic change and modernization) – layerings of history, legend and memory that display complex '*juxtastructures*' *of timespace* (cf. Shields 1991) – render them utterly strange. Lefebvre describes struggles over space in terms of the 're-appropriation' of space occuring when an original dominating function has been abandoned, and where new codes and rhythms are emerging while layers of history and memory still remain (1974/1991, 167). In 'La Villa Rouge: Replaying Decadence in Shanghai', Amanda Lagerkvist shows how such rhythms of strangeness may be tuned into, through the locus of a rearticulated space: a restaurant building/ museum that used to be a recording studio for jazz in the interwar era – a house simultaneously encompassing the material and mediatized imaginary of the 1930s metropolis, traces of the revolution and the spaces of globalism. Lagerkvist explores this strange space for global elites, which invites as sense of belonging through estrangement, and communicates a tradition of novelty. Restaging a previous era of sin, violence and decadence by reenacting an urban script that appeal to the disobedient undersides of human behaviour, potentially re-enforces ideologies of, for example, gender and of Western cultural superiority, in the new city.

Replaying decadence in new Shanghai is one way to render space alien, but perhaps also simultaneously to *magnify* its meanings. In 'Sin Cities, Back Roads of Crime' Will Straw proposes that this is the effect of the curious crime space imaginaries of US print cultures of the 1930s and 1950s, detailing the many faces of vice within these visual paradigms of criminalized spaces. The first, from the 1930s, draws on colonial rhetorics of exploration and entails 'documents' and/or semi-illicit pornography of sinful cities of the world that visually depicts a labyrinthian glocality of shadows and networks of sin. The other is the 'True Crime Magazines' of the 1950s – a mode of visualization on par with certain strands of film aesthetics – that portray crime through the use of white bodies on brightly lit rural back roads. In demonstrating that strange crime spaces are thoroughly relative and belong to shadows as well as light, to cities as well as the countryside, Straw suggests that visualizations of crime actually invite us to think about how crime and secrecy serve to extend space, rendering it elastic.

Another crime space imaginary is invoked by Chris Wilbert and Rikke Hansen in 'Walks in Spectral Space: East London Crime Scene Tourism', where they interrogate those strange inclinations to murder displayed when tourists play with the dark sides of human history and imagination, through walking in horror in the ghostly spaces of the city. Wilbert and Hansen stress the performative aspects of tourism and accentuate the theatricalization of experience within these touristic practices, exploring this phenomenon in the light of the overlayering of urban space with different modern and postmodern narratives told by the guides, narratives that are in dialogue and contest with the vast range of recurring mediations of the Jack the Ripper murders.

As dark tourism exemplifies, whether represented or performed, space is always the outcome of struggles over meaning. Our Lefebvrian points of departure direct us to the processes of spatial production that are the outcome of social struggle and power relations. Communicative space is extremely powerful; it prescribes and prohibits, it speaks but eschews the basis of its own story, eagerly concealing the order behind its (omni)presence (Lefebvre 1974/1991, 142). Struggles over the meaning of urban spaces – in academic discourses or planning etc. – correspond to the struggles over space itself, social struggles that involve not only the 'grammar' or code of a particular place, but also, and most importantly, material preconditions. This dimension is highly present in the final chapter of this section, which is the only one that deals with outright disobedience. In 'The Soul of the City: Heritage Architecture, Vandalism and the New Bath Spa' Cynthia Imogen Hammond discusses the strangeness that the repeated incursions of vandalism by unknown perpetrators on a glass structure renders the order of a place. The Georgian city of Bath in the UK becomes contested as struggles over space abound. Heritage and preservationist activism and vandalism here confront the urge to modernize and the need within a global economy to compete in order to attract cosmopolitan tourists. But this new spa building, obscure as it stands in the location in Bath itself, carries estranging communicative effects also in its architectural features: the glass is self-effacing and simultaneously mirroring, while virtualizing, the ideal image of the city, thus pointing to the strangeness of that ideal of clean Georgian space in the first place. Hammond's discussion illustrates something we wish to emphasize in this section more broadly, namely that as much as strangeness is the result of systemic struggles, grasping its *scope* is not reducible to them. Hence, we retain the need to gauge the strangeness of space without reductionist gestures or simplifications, be they modernist, Marxist or postmodernist. The purpose is ultimately to delineate the heterogeneity of both spaces and the ways in which we make or do not make sense of them.

References

Baudrillard, J. 1986. *America*. London: Vintage.
Lefebvre, H. 1974/1991. *The Production of Space*. Oxford: Blackwell.
Shields, R. 1991. *Places on the Margin: Alternative Geographies of Modernity*. London: Routledge.

Chapter 6

Beside Myself with Looking: The Provincial, Female Spectator as Out of Place at the Stockholm Exhibition 1897

Ylva Habel

Much has been written about the turn-of-the-century World's Fairs and their phantasmagorical landscapes, which offered their wide-eyed visitors encyclopedic representations of the world's latest technological, demographical and territorial wonders (Çelik 1992; Gunning 1994; Tobing Rony 1996). In several ways, the great exhibitions constituted 'total' media situations by virtue of their capacity to accommodate and integrate a variety of other media phenomena (Habel 2006). Discursively important sites for staging and showcasing Western modernity to the public, the great expositions furthermore assembled a variety of spatio-temporal planes to set off the latest achievements of technology, architecture, and culture against each other (Ekström 1994, 12f, 20, 170; Jülich 2006). I would argue that the compression of time and space presented in these contexts contributed to defamiliarizing visitors from their everyday sense of orientation, requiring them to learn new ways of looking at, and interacting with the exhibition architecture. Borrowing Janet Ward's formulation, spatiality was rediscovered (Ward 2001).

In accordance with Michel Foucault's discussion of heterotopias in 'Of Other Spaces', the exposition was 'capable of juxtaposing in a single real place several spaces, several sites that are in themselves incompatible' (Foucault 1986, 24). Heterotopias, he claims, are linked to specific temporal structures that diverge from traditional time, offering heterochronies that slice or aggregate time in different ways. The museum, according to this logic, represents endlessly accumulated time, while the fairgrounds are constituted by transitory time, which is 'absolutely temporal' (ibid.). Expositions, which, in various forms incorporate the spaces, objects, and cultural practices of both these places, fuse transitory and aggregated temporal structures into an intensified representation of an ideal, futurological present.

In much the same way as did the World's Fairs, the 1897 Stockholm Exhibition constituted a privileged space for staging and enacting the utopian promises of modernity. The great exhibitions in Sweden were often regarded as historical points of orientation, one of the most visible discourses around them being the patriotically high-toned appreciations of their historical value

– often written long before the opening date (Habel 2006, 342). At the same time the exhibition and the surrounding discourse constituted an arena for negotiations about cultural values in a wide sense, providing a collective, scrutinizing overview of current times (Ekström 1994, 12).

In accordance with its national and international predecessors, the Stockholm Exhibition offered visitors a pedagogical space in which to look at and interact with the new; the exhibition experience also entailed learning new modes for navigating and spectating (Gunning 1994; Ekström 1994). The most critical competence required of the audiences was that they should be able to negotiate a proper balance between immersion and detachment in their ocular and embodied encounters with the various media situations in the exhibition grounds, and, moreover, to distinguish different scopic regimes and representations of time frames from one another. The exhibition's condensed microcosm offered visitors manifold spectacles of spatio-temporality that could be entered and interacted with through various illusionary techniques, and in several ways, the audience could assume spectatorial positions from which to see and compare past, present, and coming ages. Hereby, the represented temporal distance between exhibition visitors and the objects and sceneries they observed were framed as simultaneously accentuated and eliminated (Habel 2006, 339). In important ways, the exhibition could be likened to a history machine, producing a scripted, collective experience of the ideal present, which was mediated through this elliptic, or multilayered representation of history and futurity.

The exhibition visit required not only complex navigatory skills in space and time, but a process in which the cultural capital of the individual visitor became condensed in unobtrusive, embodied visual competence. In this chapter, I would like to visualize how these spatio-temporal negotiations become problematic when represented through the outlook of the fictive character Lovisa Petterkvist, a full-bodied, provincial exhibition visitor from northern Jokkmokk, who stands out as a remarkable figure in the Swedish women's journal *Idun*'s (1888–1963) exhibition coverage on women's participation. In interesting ways, *Idun*'s material on this topic spans from serious intertextual discussions about the representation of women at the Stockholm Exhibition, to Petterkvist's more carnivalesque accounts of her attempts to negotiate her experience of the heterotopic grounds. As I will argue, the insertion of her gendered and class-marked body into exhibition space helps to visualize contemporary norms about visual education.

Women's Mobility in Local Material Culture

Even if the great exposition was a demarcated space, its heterotopian qualities seeped into, and dynamized the surrounding city, sensitizing the landscape and its inhabitants, visualizing it anew. This process, I would argue, also prepared ground for renegotiating women's mobility in the capital, socially as well as literally. Before discussing Lovisa Petterkvist's cultural and physical disorientation in exhibition space, something needs to be said about the problematics of women's mobility in this historical context.

During the last decade or so, the research interest in urban female flânerie around the turn of the century has moved from a general discussion about the problematics of women's mobility and agency (Buck-Morss 1986; Gleber 1997, 1999) towards more locally based case studies, which consider what concretely material cultures and technologies have brought to bear on women's conditions of possibility in various urban contexts. Shelley Stamp's study on 1910s Chicago cinema culture, *Movie-Struck Girls*, is one example; others are Erika D. Rappaport's *Shopping for Pleasure: Women and the Making of London's West End* (2000), Janet Staiger's *Bad Women: Regulating Sexuality in Early American Cinema* (1995), and Lauren Rabinovitz's *For the Love of Pleasure: Women, Movies, and Culture in Turn-of-the-Century Chicago* (1998).

In her book *The Art of Taking a Walk*, Anke Gleber has claimed that Western forms of consumerist flânerie around the 1900s was not much more than a circumscribed ersatz for true mobility (Gleber 1999, 174). In several ways, the case-oriented, locally defined studies mentioned above, gainsay Gleber's somewhat generalized view of a domesticated female mobility, first and foremost by exemplifying women's contested agency, what routes they have taken through the urban landscapes to get access to pleasure, and showcasing the particular mass media contexts that have expanded their range of action.

What interests me in this historical context is what particular situations and material objects may bring to bear on women's mobility. As cultural historian Anders Ekström has shown in his essay 'The Lady with the Bicycle: The Cultural History of Things', a good share of fin-de-siècle social interest in woman's increased mobility in Sweden and its capital was focused upon her public appearance and carriage, her way of moving, her dress, and her way of observing the *umwelt*. Even if the bicycle gave her increased access to the urban sphere, and new behavioural patterns, her movements were pervaded by a continued internalized, individual monitorization of the body, its posture and dress. Contemporary bicycle-related writing offered detailed descriptions of the appropriate feminine saddle-posture, subjecting women's bicycle clothing to close scrutiny (Ekström 1997, 19–29, 38).

Significantly, gender-oriented studies departing from specific cultural objects and their immediate, spatial-social situations, such as bicycles, cinemas and visual culture, exemplify the ways in which women's mobility was simultaneously enabled and constrained by modernity's material culture. As yet, there is not much discussion about the specific conditions offered by the fantasmatic and pedagogical space of the great exhibition. Much like the early bicycle, which enabled disciplined and temporary female excursions into the urban landscape, the Stockholm Exhibition offered an expansion of legitimate public space to explore. Furthermore, this event entailed the possibility for unconcealed, mobilized spectating in a way that was not allowed for respectable women traversing the everyday urban landscape. A quote from the Swedish etiquette book *The Lexicon on Social Life* openly addresses the perceived problematics between female looking, mobility and sexuality in public space: 'A lady should keep her eyes in check both on the street and in company, so that they do not fly around friskily, betraying the owner's eagerness for conquest, or something even worse' (*Umgängeslifvets lexicon/The Lexicon of Social Life*, in Lennartsson 2001, 87; see also Gleber 1999, 175). Doreen Massey reminds us of the fact that 'spaces and places are not only themselves gendered but, in their being so, they both reflect and affect the ways in which gender is constructed and understood'. She writes:

> The limitation of women's mobility, in terms both of identity and space, has been in some cultural contexts a crucial means of subordination. Moreover, the two things – the limitation on mobility in space, the attempted consignment/confinement to particular places on the one hand, and the limitation on identity on the other – have been crucially related. (Massey 1999, 179)

During the exhibition season in 1897 in Stockholm, the site rose as a city within the city, a space where everyday gender norms could be renegotiated, where identities could be formed or polished, and where new forms of behaviour could be adopted. I would argue that the event enabled renegotiations around 'women's place' on the public arena. If women's range of action around the Stockholm Exhibition was certainly as visible here as elsewhere, there, the exhibition gave women an 'entitlement to space' (Skeggs 2004, 15).

Moreover, in accordance with Massey's claim that social understandings of gender and spatial relations constitute each other, the placement of the Stockholm Exhibition contributed to a disjunction – however slight – of the moral geography of the city. The exhibition was raised at Djurgården, a green belt in the eastern outskirts of Stockholm. Both before and after its season, Djurgården was partially considered a territory of entertainment, and thus, multifarious licence, where social classes mixed on restaurants and cafés, and where drinking, courtship and sexual activities could be hidden in the mottes (Lennartsson 2001, 275). In conjunction with the building of the Stockholm Exhibition, the nearest areas were tidied up to be presentable: fairgrounds were moved and 'tasteless' buildings reconstructed and re-purposed (Snickars 2006, 138). However, if the exposition and its surroundings offered women an educational, safe space, it simultaneously lent this guise of respectable mobility to the many women of 'easy virtue' who came to the capital during the exhibition season (Stockholm City archive, The Police, Prostitution section 24, DVa1 1859–1918).

Women at the Exhibition – Contentious or Irenical?

In accordance with other international expositions, the coming into being of the Stockholm Exhibition 1897 was surrounded by a collectively mediated and repeated inscription of the event as historical breakpoint. Considered from the perspective of gender, this is one of its most visible traits, even if it cannot be understood univocally. The Stockholm Exhibition and its framework was an ambiguous arena for heavily symbolic gender performances, on the one hand offering a set of conventional positions for women and men to assume – as committee members, congress participants, exhibitors, or visitors. Specifically for women, on the other hand, the exhibition was an official site for international renegotiations of women's social positions.

For male profiles operating within and around exhibition committees, their participation entailed important career steps (Sachs 1946; Ribbing 1998). Their work was surrounded by great publicity, ascribing them and the event an historical aura already at the preparatory stage. Every issue of *Nordens Expositionstidning*, the official Stockholm Exhibition paper, was adorned by the photo of a leading male profile and in one case, a female one (*Nordens Expositionstidning*, No. 40: 1897). In many ways, the Stockholm Exhibition and its framework was a sphere where performative and lofty manifestations of heroic masculinity abounded, in ceremonies, and congress festivities (Lundell 2006, 56f). But depending on from which perspective the exhibition's media framework is seen, this gender relationship is not uniform in all areas of representation. One the one hand, preconceived masculine and feminine domains were visualized very clearly during the season; women were invited to participate as exhibitors at a late stage in the planning process, and were mainly included to represent handicraft sections. On the other, women enjoyed a relatively broad representation in one of the most prestigious sections, the art hall (Hasselgren 1897, 467–95).

In a US context, the representational gender imbalance on the World's Fairs was a hotly contested issue from the 1870s until the First World War. From the perspective of technological innovation, Ruth Oldenziel shows that women did not quietly accept to be marginalized by masculinist definitions of what belonged inside or outside of the industrial world. She argues:

> Sensing the importance of these classificatory frameworks, middle-class women began to challenge them; they invested enormous financial, organizational, and intellectual resources in several world's fairs' in the United States and beyond, in an effort to counter the fairs' growing importance as new sources of male authority. Both moderate and radical women's activists petitioned for congressional funds, staged protests, planned sessions, organized alternative programs, and mounted exhibits to display women's skills, products, and inventions at a number of world's fairs. (Oldenziel 2001, 132)

Internationally, moreover, the World's Fairs had become arenas for feminist congresses, where women's societal roles and interests were discussed among national and international organizations (Ekström 1994, 44; Grever and Waaldijk 2004). Around the turn of the century, these congresses had become a standing ingredient in the exhibition programme, strengthening the international dialogue between various national feminist associations. During the Women's Congress in Washington in 1880, the first step had been taken to constitute 'The International Council of Women', an association that was later consolidated on the international meeting at the World Columbian Exposition in Chicago, 1893 (*Idun*, No. 38, 1897, 301). The Stockholm Exhibition's Nordic Women's Congress was held by the newly inaugurated Swedish Women's National Association, which had come into being as a consequence of the international contacts established at the Chicago congress (ibid.; Ekström, 159f).

If women's range of action around the Stockholm Exhibition were often more firmly associated with norms of housewifely femininity than that of their US sisters, the intertextual production evidence that these official encounters opened up a sphere for articulating a variety of feminine positions. One case in point was a pamphlet published by a number of Swedish women's organizations in connection to the Chicago event, *Overview of Swedish Women's Social Standing: Issued on Account of the Chicago Exposition 1893*. The dual-language (English/Swedish) publication offered a chronological account over Swedish women's social and judicial standing through the times, including a presentational record over the country's women's associations. The introduction impressively characterizes the Swedish woman by referring to olden times, when she shouldered a big responsibility for a large-scale household, and did not hesitate to participate in fierce battle as shield-maiden. If she was crossed, enemies should fear her vengeance. In this mythical-historical account, housewifely qualities merged with those of the Amazon (ibid., 3). Considering that World exhibitions were launched as sites for peaceful international contestation, the Swedish presentation vis-à-vis their sister organizations carried somewhat bellicose overtones.

In contrast, when the Stockholm Exhibition's Nordic Women's Conference was notified in *Idun*'s columns in 1897, the idiom was subdued and peaceful, quietly distancing itself from the rhetoric of the Chicago pamphlet. The journal hailed the congress participants as 'peacemakers of the North', wishing them good luck in their quest for progress, humanity and compassion (*Idun*, No. 38, 1897, 303).

A recurrent question appearing in *Idun*'s coverage of the 'Nordic Women's Conference' was how different women's issues should be advocated, and with what level of arrogation. Several speakers expressed some caution against endangering the reputation of already existing organizations by expansive, radically emancipatory aspirations. This, they argued, would mean that women transgressed the boundaries of their natural calling, moving into the men's domain (*Idun*, No. 39, 1897, 312f).

In this discursive context, the bodily performance of normative femininity became an important instrument in warding off critique against feminist interests. In her study *Moviestruck Girls* on women's mobility in the film and media culture of 1910s Chicago, Shelley Stamp shows how the contemporary suffragette movement had to deal with fiercely biologistic counter-discourses that caricatured them as mannish. In apocalyptic or comical depictions in film and press, they were represented as physically masculinized, coarsened and vulgarized by aggressive agitating. Suffragette leaders countered this propaganda by responding in kind; Emmeline Pankhurst appeared in a film sequence, directly addressing film spectators, showing through her worthy, moderated and feminine gestique how very wrong the movements' adversaries were (Stamp 2000, 173). Even if there was no similar controversy related to the Nordic Women's Conference, the cautionary arguments ventilated there suggest that the perceived social risks with feminist work were understood according to a similar, embodied logic. Several of the reprinted speeches make pledges that women's feminine grace would not be harmed as long as their engagement in feminism was balanced and sobrietous. Turning back to Ekström's discussion about the hyper-visualized, self-monitoring female bicycle rider, his context also elucidates the intricate gender balance acts constituting the body movements and gestures of the New Woman. The well-balanced and regulated movements of exemplary feminine cycling could be regarded as a material concretization of the imperative discourse on women's continued graceful and abstemious emancipation (Ekström 1997, 19–29, 38).

Similarly, a certain sense of internalized gender performance monitoring runs as subtext in the *Idun* coverage of the 'Nordic Women's Congress', coming to the surface in the writers reviewing approach vis-à-vis the appearance of different speakers; their feminine grace and style is consistently visualized and commented upon, treated as both the result of uncorrupted nature and competent gender performance (*Idun*, No. 39, 1897, 312f and No. 40, 321). As I will attempt show in the following, the interest in feminine appearances and behaviours acquires a heightened visibility during the Stockholm Exhibition's season.

'To the Exhibition! To the Exhibition!': Ladies and Communication

As argued by several of the writers in the anthology *1897: Media Histories around the Stockholm Exhibition* (2006), the 1897 event opened up a space for reflective civic spectatorship – a space addressing and soliciting an active and curious public. In contemporaneous Swedish culture, visuality constituted the point of departure for a range of pedagogical initiatives directed to laymen and experts (Ekström 1994, 2000, 2006; Habel 2006; Ekström et al.

2000; Lundren 2006). Both before and during the exhibition season 1897, several efforts were made to educate the public in object lesson pedagogy, which, if successfully mastered, allowed the trained viewer to appraise and analyze the world of objects 'at a glance'. Ensuring civic learning through properly calibrated and embodied forms of spectatorship was one of the most noticeable areas of public concern ventilated around the opening of the 1897 Stockholm Exhibition.

In accordance with a big concert, the event was preceded by a series of mediatized rehearsals, of which the preliminary stagings of exhibition walks in publicly circulating texts were one of the most visible phenomena. The Stockholm Exhibition opened in mid-May, but the reading public got the opportunity of making recurrent virtual guided tours on the yet unfinished, repeatedly photographed site at least nine months beforehand. *Nordens Expositionstidning* provided the possibility to strike up an acquaintance with the rising buildings on site in September 1896. In the rich publicity around the exhibition, the round-trip narrative is a textual genre that merits specific attention; its apparent logic of virtual, or 'armchair travelling' can be linked to other contemporary media that offered viewers a sense of presence in distant places: actuality films, sciopticon images and stereoscopes. In accordance with them, the textual guided tour invited Stockholm readers to assume a touristic relationship towards their well-known city in a more general sense (Snickars 2001) by making it enticingly strange. The imperative to see the rising exhibition was accordingly followed by the Stockholm citizens, who were not satisfied by observing it at a distance however; in March 1897, the site had to be closed on account of all the curious strollers and picnickers that hindered the construction work with their presence (Ekström 1994, 119).

In the context of the Stockholm Exhibition's opening, something needs to be said about the educational discourse to which both serious and comical accounts made frequent reference – the round-trip narrative. This was a textual genre that fostered the touristic gaze, and constituted one of the most important instruments for educating visitors beforehand in how best to navigate the exhibition landscape. The round-trip narrative, which reached wide circulation in the press and in handbooks, became critical in familiarizing both Stockholmers and rural visitors with the exhibition in image and text. By describing a couple of *Idun*'s round-trip scenarios, I will exemplify the ways in which female readers-visitors were addressed, and how their perceived experiences of the exhibition and its outskirts were inscribed by the reporters. Textually staged according to contemporary object lesson pedagogy (aiming to develop the faculty of observation), these round-trip narratives offered the public concrete manuals for ideal, educational spectatorship.

Before and during the exhibition season, two round-trip narratives dominated *Idun*'s columns. The first, 'Women at the Exhibition' was a series of articles with an informative, guiding purpose, penned by Adolf Hellander. After the opening date, this and other ideal 'exhibition walks' were followed by more carnivalesque equivalents about fictional, provincial visitors, and their comically disoriented misconceptions and misrecognitions of what they saw and experienced on site. The second round-trip narrative, 'In Stockholm', was a comical feuilleton written in this style by the fictive exhibition visitor Lovisa Petterkvist from Jokkmokk in the far North of Sweden. Before accounting for her spectatorially deviant and spatially delirious adventures, however, I will first give a few examples of how *Idun*'s female constituency was addressed in the orientating narratives.

In his article series, Hellander did not only act as an official guide to areas of expected female interest, but also took upon himself to incite women readers in gentle but encouraging words to see the exhibition. About five weeks before the opening, he started a thematically structured mapping of the sections in detailed, guided tours. All in all, readers could walk in the footsteps of this 'invisible' cicerone through 13 articles. One of the important objectives in Hellander's writing was to provide an intermediate link between the exhibition and an imagined timid, but curious female readership in Stockholm and other parts of the country. Without directly addressing the question about women's problematic relationship to the public sphere, he tried to anticipate their misgivings about venturing to the exhibition by assuring them that they had an important role to play in lending festivity and glory to the coming event. When summer neared, Hellander vividly imagined *Idun*'s women readers as throbbing with expectation for the exhibition to open:

> Under usual circumstances, *Idun*'s honorable readers have probably started dreaming about life at the bathing resorts, the tranquilizing serenity at the peaceful summer cottage [...] But a new and burning longing has come over you, revered readers, a longing which has to be satisfied, if you are not to regard the summer 1897 as totally missed – which is to behold, to be a part of the surging life of the great Nordic exhibition in Stockholm. And therefore, I think, the same slogan rings out in *Idun*'s wide readership in all parts of the land: 'To the Exhibition! – To the Exhibition!' And surely you will come. Moreover, what would become of the striking exhibition if the ladies did not turn up, if you left it all to us men, the 'non-fair' sex? It is you, my ladies that will bring glory, life and the proper color to this national enterprise. It is your warm interest, your alluring enthusiasm, your acclaim, which significantly paves the way for the exhibition's success; and finally it is you, that will give the exhibition sojourn out there, its tone of nobility and happy concord, a spirit which should characterize the peaceful, great festivities of the peoples. (*Idun*, No. 8, 1897, 67)

Those of Hellander's 'walks' leading to the halls and pavilions specifically for women are some of the most suggestively written narratives. In the Nordic Museum, the journal had furnished a section of its own, the *Idun* Lounge, aimed to provide one of several oases for women visitors. Presenting the lounge, he metaphorically paused on the doorstep, carefully describing its position and interior for the readers:

> And now, my charming ladies ... let us imagine that the exhibition is open and that we all find ourselves out there in the midst of the bustling, exultant life. Well, we glide through the crowd, and hurry over the big, beautiful, flowered open space in front of the Industrial hall, and make our way towards the Nordic Museum's stately, palace-like building. We enter through the big middle entrance, just where the temporary annex begins, and walk straight ahead. We do not need to *search for* the Idun Lounge, since it rises right there, slightly to the right, dominating the immediate interior surroundings. First we need to climb a couple of steps, because the floor is slightly raised, and then we find ourselves in the midst of the airy, light, cheerful, and high-arched summer lounge. Two of its sides are entirely open, only protected by a white balustrade, from which slender, decorative pillars stretch towards the ceiling. From the open arcades one has the most splendid view all over the great hall, where visitors teem, and where the largest part of the women's handicrafts is displayed, by the way. Straight ahead from the lounge entrance there is a big alcove, accommodating a comfortable panel sofa, over which hangs a beautifully chiseled shelf with a library. Here we can thus find all *Idun*'s annual volumes, bound in the well-known red covers, all the guides [published around them], and the entire library of *Idun*'s novels, and finally, miscellaneous reference books. There are certainly a lot of things to look through, while we rest there for a while. (Hellander in *Idun*, No. 14, 1897, 108)

In another context, I have exemplified how a significant draw in the exhibitions' display strategy was a smooth integration of visitors as temporary inhabitants in a given hall's scenery. In this way, they were allowed to break through the more or less invisible representational membrane of the display, having its staged space at their disposal while they stayed there to rest, read or look at objects (Habel 2002, 208). *Idun*'s Lounge offered its female visitors a space of retreat, display and overview, where they could socialize with other visitors.

From a more general perspective on the attractions included in the exhibition's mediating total framework, it constantly incited intensified communication among visitors through encounters, cards, letters, graffiti and telephones (Ekström 1994, 141); in order to gain confirmation and substance, the experience needed to be communicated to others. During the Stockholm Exhibition, specifically women visitors were invited to pavilions offering the simultaneous possibility for intensified communication with the surrounding world and privacy from it. Adolf Hellander encouraged his female readers to come to the exhibition site and boldly 'throw themselves into the world crowd', but also assured them that they would find a sheltered resting place away from

Figure 6.1   The *Idun* Lounge at the Stockholm Exhibition, 1897. Image from
**   *Idun*, No. 35, 1897**

the nerve-wrecking bustle in the *Idun* Lounge. Women visiting 'The Ladies
Pavilion' could retreat there to freshen up their appearance, write a letter, or
telephone gratis. 'My dear ladies, please enter and stretch out comfortably',
Hellander wrote. 'Let your gazes glide over the entire lounge, look at the
beautiful, arched ceiling with its daintily painted frieze, let your eyes linger at
Idun's fair picture', he continued gently, 'and finally let your beauty sense be
caressed by the exquisite perfume flowing from all the beautiful flowers. And
hear the music, which seeps in here, subdued, melodious and sweet, calming
your nerves, instead of agitating them! [...] Everything has been devised and
arranged to provide for your comfort, and to bring you into the happiest and

most charming mood!' (*Idun*, No. 14, 1897, 108–9). After having called forth this sensory composition of colour, sound and fragrance, Hellander guided the visitor to a more secluded part of the lounge:

> One thing I had almost forgotten to mention, my ladies – but let me whisper it in your ear as a secret. In the upper left-hand corner you see an elegantly painted screen. You can suddenly disappear behind it, going through the door into a stylish little ladies room, where nothing that you would expect is missing. Perhaps the summer heat has spoiled your carefully curled hair! Do not worry, in there you will find both a spirit lamp and handy little tongs, which your nimble fingers can skillfully manage. Your coiffure will soon regain its charming look. (ibid.)

This part of the lyrical description leads us back to the earlier observation that a woman's presence in public was a performance, involving high demands on feminine perfection – which in this case needed to be maintained over the course of a whole day, away from home. This issue is mentioned by *Nordens Expositionstidning*, specifically referring to the Stockholm ladies' elegant clothing as a problem. Many of them, the paper reported, ended up stranded at the exhibition, since they had often chosen too sensitive outfits, and shoes with much too thin soles to be able to manage long walks, sudden weather changes and crowds. On his latest visit, the writer had spotted a young lady with an unhappy countenance, waiting for her friend outside the exhibition entrance; her dress was already stained by the moist weather, and the frilled bangs on her forehead straightened (*Nordens Expositionstidning*, No. 14, 1897, 3).

With this problem in mind, we might return one last time to Hellander's description, since it partly aimed to prevent such little disasters. Also in this passage, he offers a manual for the female readers' future visits:

> Perhaps it is unnecessary in our times to mention that there is a telephone. Furthermore, we know that you ladies love the telephone – it is so heartily pleasant to confabulate with a lady friend about something that you have forgotten, or want to remind her of. You may, for example, have gone to the exhibition ahead of your friend, who was not yet 'ready'. You want to tell her where you are, and where you could meet. This is almost impossible in the teeming exhibition crowd. You can telephone here pretty safely: 'Come to the Idun lounge, where you can meet me at 12 noon.' This is a pretty important thing as it is always good to have a set meeting place, which cannot be mixed up with others. When you, my dear ladies have rested for half and hour or so, enjoyed a little of what *Idun* has been able to give you … you can prepare yourself for mingling into the large crowd, heartily contented with *Idun*'s hospitality. […] Well, my dear readers, now I have ended my invisible ciceroneship … In *Idun*'s name, be heartily welcome! (*Idun*, No. 14, 108–9).

'In Stockholm': Lovisa Petterkvist's Exhibition Letters

In accordance with other great exhibitions, the Stockholm event 1897 saturated its visitors' experiences with a sense of the absolutely new and modern. Whether technological or cultural novelty actually constituted the greatest share of what was represented in the grounds was less important than foregrounding it through strategic and striking juxtapositions. When time and space was thus condensed with modernity as determinant principle, visitors were required to navigate in both physical and intellectual senses among the temporal cutouts, cultural microcosms and technologies that were presented at the exhibition. Simultaneously, they needed to interpret correctly the different rhetorical accentuations and juxtapositions of objects that had been staged to highlight modernity. One case in point discussed by Solveig Jülich in the article 'Media seen as modern magic' was the new media technologies X-ray and film, whose kindred magic was accentuated by being demonstrated in the medieval buildings of 'Old Stockholm' (Jülich 2006, 229).

In addition to the already mentioned textual round-trips, the Stockholm Exhibition was enframed by normative articulations guiding the visitor on how best to benefit from the pedagogical purpose embedded in displays and halls. In his article 'The vertical archive: Synoptical media and historical vertiginous sensations', Ekström takes his point of departure in observational pedagogy, whose philosophy was that the eye could be turned into the noblest sense for learning if trained by structured and carefully positioned acts of looking (Ekström 2006, 291ff). Frans Lundgren's discussion similarly draws attention to critical reviews questioning whether the exhibition displays were organized well enough to answer to the norms of object lesson pedagogy (Lundgren 2006, 309f). But if the Stockholm Exhibition and its official publicity framework were structured to choreograph and educate civic spectatorship, respectively, there was a parallel production of comical, often richly illustrated narratives, which in significant ways turned tables on the applied logic of visuality, offering an alternative, often vertiginous version of exhibition space, and its relations of looking. These carnivalesque narratives followed the principles of object lesson pedagogy as an unacknowledged, but recognizable point of reference, their humour built around the many deviant behaviours and viewing acts that the fictive country cousin engaged in. Recurrently the rural experience – as well as the accompanying, clumsy body – is described as a clogged sieve, through which the exhibition experience was filtered in insufficient, uneven and unpredictable portions. In accordance with contemporary, apocalyptic depictions of urban modernity, the vulnerability of a 'dense' provincial subject is represented in spatial terms (Singer 2001, 59–99).

Figure 6.2   *Uncle Hybinette,* **in the humour magazine** *Strix***, No. 11, 20 May 1897**

One of these fictive provincials was Lovisa Petterkvist, who had come with her husband Ruffen to write about her exhibition experiences for *Idun* in a series of letters. The writer behind the pseudonym was Alfhild Agrell, a young author belonging to the generation of left-oriented modernists debuting in the 1880s (Manns 1997, 57–9, 268). From the early 1890s and onwards, her epistolary novel *In Stockholm: Also a Travelogue* (1892) became so popular that it was reprinted several times, and translated into Danish. One of the reasons for the narrative's exceeding popularity may have been Petterkvist's peculiar touristic gaze on the capital. During the early 1890s, the Swedish Tourist Association had established Jokkmokk as one of the popular northern tourist destinations (*STF Yearbook* 1891–1893; Atmer 1998, 26). Through Petterkvist's character, Jokkmokk instead came to sightsee in Stockholm, which resulted in a series of comical episodes.

Mrs. Petterkvist writes a series of 'exhibition letters' for the journal *Idun*, describing her many disorienting visual experiences. Her attitude towards her task as exhibition reporter alternates between deep rural humbleness and unbending, ridiculous pride. She is both anxious to blend in with the unfamiliar urban environment and vociferously talkative. The letters, vividly coloured by her provincial colloquialisms, take their point of departure in a series of burlesquely depicted exhibition episodes. In contrast to the timid ladies that Adolf Hellander addressed in his round-trip narratives, Petterkvist needs no gentle prodding to throw herself into the 'world crowd'. On the contrary, she pushes her way through these crowds in order to get the best possible viewing position, and when it is taken, she immediately gives voice to her impressions, heeding no complaints. One example is given in the second letter, where she describes the problems she experienced trying to see and hear the inauguration address made by the king.

> Along came another four-horse team, and inside was the most stately lady and a uniform, which – and then the crowd made up a *wall*. 'Excuse me', I said jovially, 'I'm here to see something!'
>
> Apparently all the others were there with the same intent, but to a much higher degree …
>
> 'What do they do now?' I asked one of them from my position behind a wall of backs.
>
> 'An old man is making an inaudible speech', said Ruffen.
>
> 'Then he could refrain from taking up the time', I said.
>
> 'What do they do *now*?
>
> 'They are singing the cantata'.
>
> 'Who wrote that? Is it —?'
>
> 'Hush, hush, hush!', it sounded around me like a nest of vipers.
>
> 'I'm sorry', I said mildly, 'but *could* you gentlefolk tell me how I'm supposed to know *anything*, if I'm allowed neither to *see*, nor *hear* nor *ask*?'
>
> 'Not when people are singing!'
>
> 'Oh', I said just as mildly, 'that' means nothing to *me*'.
>
> The cantata was incredibly beautiful … naturally I followed the lyrics in an undertone, and then they hushed me *again*! [...] But when our own beloved king stepped up to make his address, *nothing* could hold me back! I stepped up on the bench. The very same moment the sun broke through the clouds, making the yellow feathers in his majesty's shimmering helmet shine like fire flames against the deep blue velvet. His mighty voice pierced through the wind and into my heartstrings – Oh! They could shout both down and hush as much as they liked. I had seen a sight that I will *never* forget … (*Idun*, No. 22, 1897, 174–5)

Petterkvist does not understand the logic of being part of a crowd, but is preoccupied by her own desires.

Put in relation to the earlier discussion about the discourse on feminine perfection in public spaces, her heavily embodied, and excessively gendered presence constitutes a recurrent disturbing moment on the exhibition site as well as in the surrounding city. In some respects her adventures in Stockholm is 'comedy of manners', in which the class habitus of Petterkvist and her spouse frequently run up against the alien city and its population. Their encounters with locals give rise to a babel of tongues regarding questions of taste, courtesy, and meaningful practices. Speaking with Bourdieu, Petterkvist voices a grass-root class disposition, which is communicated to others without hesitation (Bourdieu 1984, 170). She is almost likened to a barge, with great difficulties in changing an already-taken direction. Even if she is eager to adapt her comportment and observations according to what she perceives is suitable for the upwardly mobile exhibition visitor, this ambition collides with her many whims and unconsidered, loud remarks, as in the above example. Her attempts to wield symbolic capital by feigning to belong to a higher, more urban social class fail, and she instead makes a spectacle of herself, or offends other people. Bourdieu argues that: 'the social rank and specific power which agents are assigned in a particular field depend firstly on the specific capital they can mobilize, whatever their additional wealth in other types of capital' (Bourdieu 1984, 113). In Petterkvist's case, her newly acquired status as a reporter makes her overestimate the significance of being a representative of the 'fourth estate'. Comical episodes result from her misguided adoption of high society dress details, as when she wears a pince-nez, or 'imposing' plumes on her head for a press congress dinner.

In her discussion on Victorian discourses of obesity, Joyce L. Huff argues that heavy bodies were considered as 'out of place' in modernity's standardized physical environment: 'The fat body was singled out and stigmatized in an environment tailor-made for a hypothetically average body' (Huff 2001, 45). The corpulent body, she argues, was furthermore seen as 'resistant to normalization', appearing as visibly and disturbingly individuated. This corporeal waywardness is certainly represented in Petterkvist's failure to perform feminine manners, and her loud personality stands in metonymical relationship to 'excessive' embodiment.

Delirious Visions through an Ungraceful, Feminine Body

Lovisa Petterkvist's exhibition visits are narrated through personal and dramatized tableaux: each experience is noticeably filtered through her robust persona. In contrast with the visually acculturated visitor, she lacks the cultural capital to appreciate and navigate the exhibition grounds according to their spatial script, and her letters evoke an almost delirious adventure into an unpredictably fantastic landscape. For her, properly channelled visuality becomes impossible, since it is constantly impeded by, corrupted or short-circuited by her interfering body. Unintended synaesthetic experiences interfere with her intention of 'learning how to look', causing a kaleidoscopic vision of the grounds. The fleshy corporeality of Petterkvist's provincial body hereby prevents her from acquiring a 'the survey habit of mind' prescribed for turn-of-the-century educational spectating (Yeo 1991). The prescribed, pedagogically reflexive act of looking, with which the reader is assumed to be familiar, is constantly impeded by her failure to direct and concentrate her gaze in a structured manner. In contrast to modernity's astonished spectator, who enjoys the relatively controlled pleasure of vacillating between belief and disbelief (Gunning 1996), Petterkvist constantly gets destabilized by oscillating between these poles; she gets caught up in false perceptions of spaces and objects, and misunderstands their intended meanings and interrelationships.

During the first week of the Stockholm Exhibition, *Nordens Exposition-stidning* recurrently quoted S.A. Andrée's pamphlet, *The Art Of Studying Exhibitions*, aiming to teach curious visitors how to sharpen their faculties of observation to view the exhibits in a more instructive and exhaustive manner. Andrée claimed that the viewing experience needed to be based in a good sense of spatial orientation. Thus, the act of looking should be organized according to three stages: first the visitor's gaze should take in the overall perspective of the hall in its entirety, thereafter moving to the various sections, and finally zooming in on a detailed study of the different objects. This strategy was modelled on Johann Amos Comenius pedagogy for 'successful seeing' (Comenius in Alpers 1983, 95). Andrée cautioned visitors against looking aimlessly, since there would be a risk that they became disoriented, deluded into 'unknowingly and needlessly go back and forth' in 'the multifarious drama' that the exhibition offered, and

> unexpectedly encounter the same halls that you have already visited, surprising yourself by seeing objects that you have already seen. Your impressions thereby become unclear and tangled, visual images more or less overlap each other, and finally you experience the same confusion as if you were in a labyrinthic space. (*Nordens Expositionstidning*, No. 14, 1897, 6)

Petterkvist constantly fails to acquire these observational skills. She exerts the quickly browsing, appraising gaze of the economizing housewife in a small-town shop, and becomes dizzy when she cannot take in and interpret the visually striking disposition of halls and objects at once. Her fragmented concentration span stands in the way of the exemplary spectatorship Andrée describes and her ability to navigate the eclectic milieus of consumerist spaces is not a useful competence in this here; her lack of cultural capital makes her unable to appreciate the value of the objects she sees, or decode the context of their presentation. In more senses than one, exhibition space becomes unreadable to her and therefore disturbing. The Eskilstuna section, displaying coffee pots in accordance with exhibition mass aesthetics, strikes her as a boastful waste of shop space. When Ruffen points to it, remarking that Eskilstuna has first-rate products, she counters fretfully: 'Well, yes ... they certainly *do*, but there's no need to over-run the whole place with an entire *line* of kettles, when they're only displaying *one* model!' (*Idun*, No. 23, 1897, 180).

The *Idun* reader who had already been preliminarily guided through the exhibition grounds by Hellander's narratives, was given a very different story through Lovisa Petterkvist's exhibition letters – to put it mildly. She cannot participate in the plays with perspectives, and many of the sights become indecipherable through her many faulty associations and her wayward juxtapositions of what she sees. On her exhibition visits, she looks energetically with her whole face ('Ruffen and I went around gaping'), or rather, with her entire body. A series of uncontrolled optic disjunctions seems to take place in front of her confused gaze. Thereby the exhibition landscape is alternately transformed into kaleidoscopic, almost delirious views, alternately it disintegrates into incomprehensible fragments. She wades in unsorted impressions that she cannot make sense of. The impressive totality of the exhibition presented in the round-trip narratives stands in striking contrast to the disenchantingly jumbled and disjuncted landscape she traverses – arbitrarily strewn with halls and objects.

Many a time, aims and contexts in the exhibition space are therefore lost on Petterkvist, and she often misses to get an overall impression, instead getting stuck on irrelevant details. She cannot make the three-stage calibration of her vision from overview to detail, and so gets dizzy and confused when entering one of the largest buildings, the Industrial hall. The ceiling beams immediately distract her gaze:

> My first impression of the hall was that I found myself in a colossal crow's nest; I got so giddy by all the snow-white spars, crosses and bars in the ceiling ...
>
> 'Look at the floor, Lovisa, he-he', said Ruffen, '*that's* where the exhibition is!'
>
> 'Sure, Hubby, sure', I said, blinking. But I'm getting frightfully giddy, my head is swimming'.
>
> 'Don't look at *everything*, but try getting an overall impression, that's what I do'.
>
> 'Alright', I said obediently, and blinked even worse.
>
> 'We-ell – how is it going?'
>
> 'I think that I could work it out, if I only knew if it is the showcases that display the objects or the objects that display the showcases'...
>
> 'Oh, how I wish that there was one single gap that I could look "straight through"; I become so frightfully dizzy – –! (*Idun*, No. 23, 1897, 180)

In the above passage it also becomes clear that Petterkvist does not perceive the required adaptation of her gaze as conditioned by inner reflection, but regards it as a purely physical manoeuvre. However, when they are about to finish their visit to the Industrial Hall, she cannot let go of the question whether it is the showcases that exhibit the objects or vice versa, and says: '"Now I know, Ruffen" ..., I said thoughtfully, "It is the things that display the showcases – !"' (ibid.). To be sure, she returns on her own:

Figure 6.3   The Industrial Hall

I could not get peace of mind, everything in my head was a jumble – and that *that* was not as it should be, I *felt*. After an hour, I came back again, warm, red and glowing. 'Well, old girl, what about the overall impression now?', Ruffen said, who was also warm, red and glowing [from beer – at the nearest bar where he has been waiting for her]. 'Now I get it', I said. 'It was in the wings!' *There*, the balustrades were painted and covered with tapestries, and there one did not become dizzy by colours; there I did not 'run up against' things wherever I went, and there I really *saw* that it was the showcases that displayed the things and not the other way around. (ibid.)

In his article 'The Vertical Archive', Ekström accounts for the air balloonists lofty perspective, discussing the dis/embodied premise of observing the exhibition from above. Exemplifying descriptions of balloon trips, he argues that this perspective became conceptualized as a distanced, objective and clarified viewing position, from which the world below was given its 'true proportions' (Ekström 2006, 286). Reciprocally, the aerial gaze was linked to an almost ethereal, disembodied observer. The contrast between the balloonist's allegedly transparent and Petterkvist's toilsome acts of looking could not be greater. Her viewing experience is firmly anchored in a heavy, refractory body, which is steadfastly earthbound, yet constantly destabilized by her visual adventures. She is flooded and saturated with impressions, getting adrift; she runs up against things and views with her gaze, and collides with people and objects in both literal and metaphorical senses.

Juxtaposed with each other, the balloonist's and Petterkvist's examples show that perceptions of space are unavoidably linked to specific forms of embodiment that are not accommodated equally well. Furthermore, this discursive interrelation between vision, embodiment, gender and class speak to what Michael Warner calls 'the rhetorical strategy of personal abstraction', i.e. the possibility to represent, and speak from a bodily transparent power position of non-identity (Warner 1992, 382). In the case of the balloonist, this position is not only enabled by the masculine and scientific context in which his heroic figure appears, but significantly by the literally uplifting vehicle, which is also a media technology. In her article 'Telegraphy's Corporeal Fictions,' Katherine Stubbs' summarizes the logic of 'the rhetorical strategy of personal abstraction', arguing that:

> [T]he public subject's right to speak was predicated on his disembodiment and anonymity; he could assume the speaking position of the disinterested public citizen only because he inhabited a body perceived as so normative (white, male, upper class) that it did not count as a body at all. [...] For those persons whose bodies deviated from the white male standard, corporeal difference was equivalent to hyper-embodiment; unable to abstract themselves, they had no grounds from which to speak in the public sphere. To attempt to speak would only call attention to their bodies, to their lack of authority; for women, this meant that appearing in too public a fashion was a form of dangerous promiscuity. (Stubbs 2003, 102)

In accordance with Stubbs and Warner's argument, Lovisa Petterkvist's sturdiness bars her from assuming the position of the educated spectator. The metaphor of her and other provincial visitors' experience as a clogged sieve, could be understood almost literally – impressions are not taken in, and cannot be communicated. In her dissertation *Malaria Urbana*, the ethnologist Rebecka Lennartsson writes about the stratification of urban space in turn-of-century Stockholm, arguing that it was layered according a classed and gendered economy of looks, granting the upper strata entitlement to visual scrutiny of the lower. The masses were considered to lack powers of observations, at least from the perspective of the well-to-do flâneur:

> Perhaps they are not blind, but nonetheless lack the ability to really see. They gape with the same empty, vacant and unintelligent stare ascribed to other subjected categories: the savage, the child, the woman, and the poor. They do not belong to the select few who are entitled to look, to read, and

savor with their gaze. Instead they are counted as the genealogical troops of displayed objects. (Lennartsson 2001, 74)

Even if Petterkvist's friction-filled spectatorship is clearly related to this discursive construction of the ability to see and orientate oneself, the comedy of episodes she narrates do not only gain their comical twist through failures. If she lacks observational training and ideal embodiment, her spectatorship is still reflexive in a fleshy sense. In a prosaic way she constantly puts herself in relations to what she sees by interfoliating her narratives with descriptions of her troubles with her dress, chafed feet, sweating, homesickness and hunger. In these situations, unintended synaesthetic experiences occur, and are discursively symptomatic of a 'disturbed' viewing experience; she fails to canalize all her concentration to the eye. In contrast to the balloonist, whose body 'forgets itself' from its elevated position (Ekström 2006, 285), her body constantly interferes.

The above examples of Petterkvist's visit to the Industrial Hall shows that she can neither create metaphorical nor literal distance towards the exhibition's halls and objects; every experience has such a strong impact on her that she gets increasingly exhausted and confused – beside herself with looking. By and by, she grows more cautious about where she directs her gaze. When she makes a visit to the Art Hall together with Lycko ('he who takes care of the women at *Idun*'), she initially screws up her eyes, observing everything with anxiously narrow slits.

Lycko and I entered the 'International' me with my eyes as pinched together as I possibly could without risking falling on my face. 'That is a somewhat original way of viewing art', Lycko said, 'Not to close your eyes to it, but to *show* that you are doing it'. 'I do not want to see anything until I reach old Sweden', I said. [...] Then we finally entered the Swedish section. 'Well, I do say, that if I had not already prepared myself for what a difficult moment it would be for me, poor ignorant, to look at 'modern art', I would have perished already from the start. But now I closed my eyes energetically to everything that yelled too horribly, and really saturated myself with all the beauty I could find instead. [...] To think that it should be so horribly exhausting to study art! (*Idun* No. 26, 1897, 204–5)

Despite all the difficulties, Petterkvist really wants to learn, and therefore makes a new visit to the art hall in her next letter. This time, too, she has problems seeing correctly and constantly gets caught up in an unexpected economy of looks that will not let her enjoy her spectating in peace; either she is disturbed by being looked at herself by the exhibition janitors, or – by the artworks themselves:

It sure is awful when people are laughing behind your back – at least when you are alone! I *felt* how one of the janitors sniggered when I passed by. Maybe my hat was – or maybe there was a gap in my dress – or, heavens! Imagine if the bodice of my dress was coming apart at the back – it *was* a bit tight! To study art with such feelings in your head is not easy, I can assure you gentlefolk, but I made an effort to do my duty. Now I gazed upon a young Water sprite, sitting in a stream and practicing his song. He is plucking at a dirt-yellow violin and screaming himself blue in the face, probably to take over the stream [...] As I was walking about there, I thought – that it was no longer *me* who looked at the paintings, but *they* who were looking at me. I sat down – on the lap of a *gentleman*! I jumped up again, bumped into some other people, turned red, apologized, and carried on so foolishly that I was ashamed. But I did not give up. (*Idun*, No. 27, 1897, 212)

Petterkvist's encounter with the exhibition assumes a surprising reciprocity, where the power balance between the seeing subject and seen object is momentarily unpredictable. The media situations – in this case the paintings – are not in any given way benevolent, but 'bite back' by looking back at the confused Petterkvist (Gitelman and Pingree 2003, xviii).

Conclusion

Both before and during the Stockholm Exhibition, the round-trip narrative was a textual genre that was instrumental in several respects. By way of authoritative descriptions it guided and directed public interest, simultaneously as it inscribed an official discourse on the significance of the event. Moreover, it functioned as an intertextual node for multifarious advertising: for the journal to which it was linked, for the described pavilions, for the exhibition, and finally for the surrounding city landscape, which was made dynamic by touristic energies. As Adolf Hellander's presentations in *Idun* show, the round-trip narratives targeting the female readership were intended to incite boldness, and to assure them of safety.

As exemplified here, the round-trip narratives also had a double relationship to the popularized discourse of object lesson pedagogy. The comical round-trip narrative hereby functioned as the constituting outside of the serious, official ditto. The reading public's familiarity with the 'art of studying exhibitions' was the premise for the humour in Lovisa Petterkvist's carnivalesque exhibition letters, in which all the recognizable spectator competences where inverted. Her kaleidoscopic vision of the exhibition grounds speaks of a prevalent discourse where gendered and classed embodiment, vision and space constitute each other. All these aspects are condensed and coded into the power of vision, or the lack thereof. 'It is great fun to se the exhibition again', Lovisa Petterkvist ends her last letter, 'but a bit disheartening, too, because now I really notice how poor I have been in my "exhibition pieces". You have to learn to look, as well as you have to learn everything else, and I believe I have learned something' (*Idun*, No. 35, 1897, 275).

References

Agrell, A. (Lovisa Petterkvist) 1892/1895. *I Stockholm: Också en resebeskrifning* [*In Stockholm: Also a Travelogue*]. Stockholm: Fr. Skoglunds Förlag.

Alpers, S. (1983). *The Art of Describing: Dutch Art in the Seventeenth Century*. Chicago: University of Chicago Press.

Atmer, P. and Katrin, A. (1998). *Livet som leves där måste smaka vildmark: Sportstugor och friluftsliv 1900–1945*. Stockholm: Stockholmia.

Bourdieu, P. 1979/1984. *Distinction: A Social Critique of the Judgement of Taste.* Cambridge, MA: Harvard University Press.

Buck-Morss, S. 1986. 'The Flaneur, the Sandwichman and the Whore: The Politics of Loitering', *New German Critique*, 39 (Autumn): 99–140.

Çelik, Z. 1992. *Displaying the Orient: Architecture of Islam at Nineteenth-Century World's Fairs.* Berkeley: University of California Press.

Ekström, A. 1994. *Den utställda världen: Stockholmsutställningen 1897 och 1800-talets världsutställningar*. Stockholm: Nordiska Museet.

Ekström, A. 1997. 'Damen med velocipeden: till tingens kulturhistoria', working paper from the research programme *Stella: Modern vetenskapshistoria 1850–2000*. Uppsala: Uppsala University.

Ekström, A. 2000. 'Konsten att se ett landskapspanorama: Om åskådnings-pedagogik och exemplarisk realism under 1800-talet', in Ekström, A., Lundgren, F. and Bergström, M. (eds) *Publika kulturer. Att tilltala allmänheten, 1700–1900*. Uppsala: Uppsala University.

Ekström, A. 2006. 'Det vertikala arkivet', in Ekström, A., Jülich, S. and Snickars, P. (eds) *1897: Mediehistorier kring Stockholmsutställningen*. Stockholm: Mediehistoriskt Arkiv, Statens Ljud- och Bildarkiv.

Foucault, M. and Miskoviec, J. (1986), 'Of Other Spaces', *Diacritics*, 16:1, Spring, 22–7.

Gitelman, L. and Pingree, G. (eds) *New Media: 1740–1915*. Cambridge, MA: MIT Press.

Gleber, A. 1997. 'Female Flanerie and the Symphony of the City', in Ankum, K. von (ed.) *Women in the Metropolis: Gender and Modernity in Weimar Germany*. Berkeley: University of California Press.

Gleber, A. 1999. *The Art of Taking a Walk: Flanerie, Literature and Film in Weimar Culture*. Princeton, NJ: Princeton University Press.

Grever, M. and Waaldijk, B. 2004. *Transforming the Public Sphere: The Dutch National Exhibition of Women's Labor in 1898*. Durham, NC: Duke University Press.

Gunning, T. 1994. 'The World as Object Lesson: Cinema Audiences, Visual Culture and the the St Louis World's Fair, 1904', *Film History*, 6: 4.

Habel, Y. 2006. '"Kvinnorna på utställningen"': Dygder och odygder på Stockholmsutställningen 1897', in Ekström, A., Jülich, S. and Snickars, P. (eds) *1897: Mediehistorier kring Stockholmsutställningen*. Stockholm: Mediehistoriskt Arkiv, Statens Ljud- och Bildarkiv.

Habel, Y. 2002. *Modern Media, Modern Audiences: Mass Media and Social Engineering in the 1930s Swedish Welfare State*. Stockholm: Aura Förlag.

Hasselgren, A. 1897. *Utställningen i Stockholm 1897: Beskrifning i ord och bild öfver Allmänna Konst- och Industriutställningen*. Stockholm: Frölén & Compani.

Huff, J.L. 2001. 'A "Horror of Corpulence" Interrogating Bantingism and Mid-Nineteenth-Century Fat-Phobia', in Braziel, J.E. (ed.) *Bodies Out of Bounds: Fatness and Transgression*. Ewing, NJ: University of California Press.

Idun, 1896–1897.

Jülich, S. 2006. 'Medier som modern magi: Tidiga röntgenbilder och film', in Ekström, A., Jülich, S. and Snickars, P. (eds), *1897: Mediehistorier kring Stockholmsutställningen*, Mediehistoriskt Arkiv: 1, Statens Ljud- och Bildarkiv, pp. 230–71.

Lennartsson, R. 2001. *Malaria Urbana Om byråflickan Anna Johannesdotter och prostitutionen i Stockholm kring 1900*. Stockholm/Stehag: Brutus Östlings Bokförlag Symposion.

Lundell, P. 2006. 'Pressen är budskapet: Journalistkongressen och den svenska pressens legitimitetssträvanden', in Ekström, A., Jülich, S. and Snickars, P. (eds) *1897: Mediehistorier kring Stockholmsutställningen*. Stockholm: Mediehistoriskt Arkiv, Statens Ljud- och Bildarkiv.

Lundgren, F. 2006. 'Social samling: Att ställa ut samhället kring 1900', in Ekström, A., Jülich, S. and Snickars, P. (eds) *1897: Mediehistorier kring Stockholmsutställningen*. Stockholm: Mediehistoriskt Arkiv, Statens Ljud- och Bildarkiv.

Manns, U. 1997. *Den sanna frigörelsen: Fredrika-Bremer-förbundet 1884–1921*. Stockholm/Stehag: Brutus Östling Bokförlag Symposion.

Massey, D.B. 1999. *Space, Place, and Gender*. Minneapolis: University of Minnesota Press.

Nordens Expositionstidning, 1896–1897.

Öfversikt af den svenska kvinnans sociala ställning: Utgifven i anledning af verldsutställningen i Chicago år 1893 [*Overview of Swedish Women's Social Standing: Issued on Account of the Chicago Exposition 1893*]. 1893. Stockholm.

Oldenziel, R. 2001. 'Man the Maker, Woman the Consumer: The Consumption Junction Revisited', in Creager, A.N.H., Lunbeck, E. and Scheibinger, L. (eds) *Feminism in Twentieth-Century Science, Technology, and Medicine*. Chicago: University of Chicago Press.

Rabinovitz, L. 1998. *For the Love of Pleasure: Women, Movies, and Culture in Turn-of-the-Century Chicago.* New Brunswick, NJ: Rutgers University Press.

Rappaport, E.D. 2000. *Shopping for Pleasure: Women in the Making of London's West End.* Princeton, NJ: Princeton University Press.

Singer, B. 2001. *Melodrama and Modernity: Early Sensational Cinema and Its Contexts.* New York: Columbia University Press.

Skeggs, B. 2004. *Class, Self, Culture.* London: Routledge.

Snickars, P. 2006. 'Mediearkeologi: Om utställningen som mediearkiv', in Ekström, A., Jülich, S. and Snickars, P. (eds) *1897: Mediehistorier kring Stockholmsutställningen.* Stockholm: Mediehistoriskt Arkiv, Statens Ljud- och Bildarkiv.

Snickars, P. 2001. *Svensk film och visuell masskultur 1900.* Stockholm: Aura Förlag.

Svenska Turistföreningens årsbok, 1891–1893.

Staiger, J. 1995. *Bad Women: Regulating Sexuality in Early American Cinema.* Minneapolis: University of Minnesota Press.

Stamp, S. 2000. *Movie-Struck Girls: Women and Motion Picture Culture after the Nickelodeon*, Princeton, NJ: Princeton University Press.

Stubbs, K. 2003. 'Telegraphy's Corporeal Fictions', in Gitelman, L. and Pingree, G. (eds) *New Media: 1740–1915.* Cambridge, MA: MIT Press.

Tobing Rony, F. 1996. *The Third Eye: Race, Cinema, and Ethnographic Spectacle.* Durham, NC: Duke University Press.

Ward, J. 2001. *Weimar Surfaces: Urban Visual Culture in Weimar Germany.* Berkeley: University of California Press.

Warner, M. 1992. 'The Mass Public and the Mass Subject', in Calhoun, C. (ed.) *Habermas and the Public Sphere.* Cambridge, MA: MIT Press.

Yeo, E.J. 1991. 'The Social Survey in Social Perspective, 1830–1930', in Bulmer, M., Bales, K. and Kish Sklar, K. (eds) *The Social Survey in Historical Perspective, 1880–1940.* Cambridge: Cambridge University Press.

Chapter 7
La Villa Rouge:
Replaying Decadence in Shanghai

Amanda Lagerkvist

... rhythm (linked on one hand to logical categories and mathematical calculations – and on the other to the visceral and vital body) would hold the secrets and the answer to *strange questions*.

Lefebvre 1992/2004, 14, italics added

Rhythms of Shanghai – Another 'Jazzy Opus'

In the collection *Shanghai Lounge Divas*, rhythms from the past become encroached by the rhythms of today, in a remix of jazz originally recorded by Pathé in their Shanghai studio in the 1930s. Listening to these old songs – now encrusted by the sleek, groovy sounds of *the global lounge* – is one way of reading, or perhaps rather, hearing out the sensuous and obscure imaginary of today's Shanghai. This CD seems illustrative of that specific experience of layers of time, those laminations of modernities and urban ideologies (cf. Qingyung 2006) that stand out as the quintessence of this hodgepodge place, where 'yesterday' always meets 'tomorrow'. The unorthodox blending of styles within these remarkable tunes discloses those temporal co-existences and 'juxtastructures' of past, present and future (Crang and Travlou 2001), that are distinctive of this globalizing Asian mega city, a place where the 1930s as a mediated mythology and memory of the city (see Boyer 2002), are constantly materialized and replayed. While it seems valid to concede in line with – for example – Mike Crang that 'places have always had different temporalities orchestrated through them' (2001, 191), this, as I will attempt to show, has a particular pertinence in the context of contemporary Shanghai. While the city seems to exemplify broader cultural tendencies of modernization, that are patterned on global traits of late modern capitalism, the co-existences of temporalities are here radically heightened. Hence, in this chapter, I will gravitate to the singularities and specificities of Shanghai on this note. Drawing on Andreas Huyssen, I will discuss Shanghai as a locus of accumulated non-synchronicities where 'a very hybrid structure of temporality seems to be emerging' (1995, 8). Hence, I will suggest that rather than simply recalling the past or moving together, if at different paces, into the future the rhythms of the city also exemplify new emerging forms of modernization, pervaded by extraordinary frenzies of temporal overlap.

Shanghai is now experiencing a regeneration that seems unprecedented in history. The speed with which the new architectonic structures are erected, and the advanced visions for infrastructure, digitalization and sustainable development, surely make this a space of futurity: a space where the future seems to have arrived already. Tourists and expatriates flock to this city of consumption and opportunity. Within the master plan covering 1999–2020, the goal is to make Shanghai the world centre of trade, finance, economy and shipping. But such planned futures of control, harmony and order do not exhaust what Shanghai is about. Despite the grand efforts to organize this space of the future, on behalf of the municipal government in Shanghai, nothing seems to be in the 'proper place' there: architectonic styles, urban ideologies and different temporalities clash and coexist in a perfectly (dis)harmonious yet fully intertwined medley of rhythms. Shanghai is a maelstrom, which shamelessly copies and repeats themes, images and ideals and puts them together, in a feverish movement toward 'the future'. Stylistic ensembles in design, visual codes, commodities, fashion, media forms and representations (especially but not exclusively from the 1930s), are mishmashed and offer, instead of a fixed Jamesonian pastiche or Baudrillardian simulacrum, a sense of *movement* through this jumble. To be in today's Shanghai may be described as experiencing an urban rhythmicity which situates (and unsettles) the visitor in a temporal spiral that seems to be rotating backwards and forwards at the same time. This is a place of strange and multiple rhythms of modernity: a modernizing Asian city reinventing itself materially and symbolically by every means possible.

To grasp such a strange space, provided that it is possible, will require attending to the 'contrapuntal rhythms' that make up the experience of the city (Highmore 2005, 150). Rhythm in Henri Lefebvre (1992/2004) holds an almost mystic quality, unifying the objective world and the subjective embodied experience while simultaneously aligning space and time, showing that they are impossible to tell apart (see also Massey 2005). Turning to rhythms is an attempt to move ahead in analyzing the restructurings of Chinese cities and an effort to qualify the plea for paying more attention to time in these new spaces (Ma and Wu 2005). Rhythmanalysis is, as Ben Highmore argues, a social and cultural phenomenology, which requires 'a predilection for complex orchestrations of time and space, a desire to pick out the different beats and pulses of urban experience and find ways of registering their syncopated arrangements' (2005, 9). Rhythms are in Lefebvre's writings to be understood as inseparable from an understanding of time in terms of repetition. They allow us to understand the everyday in – for example – towns, cities and urban life. But that is not all there is to it. As Lefebvre argues, 'differences induced or produced by repetitions constitute the thread of time' (1992/2004, 8) and, consequently social times disclose diverse, contradictory possibilities: delays and early arrivals, reappearances (repetitions) of an (apparently) rich past, and revolutions that brusquely introduce a new *content* and sometimes change the form of society. (ibid., 14, italics in original)

Shanghai is in transition. Changes are so swift they may often be perceived as 'revolutionary'. It is also a place where routines, historical phenomena, enduring patterns and slow paces remain. Through rhythm we may detect this wider cultural system we call the city of Shanghai. Although rhythms are sought out by Lefebvre as the folding of space and time into one another in the *everyday*, I will use them to understand another (perhaps even opposite) dimension with its own innate procedures and workings: a leisure space of a restaurant building. An analysis of rhythms seems to be a particularly apt way of teasing out the specific urban imaginary pervaded both by an obsession with time (cf. Zhang 2001) and an emergence of hybrid time that Huyssen posited above, which is at work in Shanghai of today.

Lefebvre spells out the need for interdisciplinary approaches to capture rhythms and asserts the primacy of time over space. Without omitting the spatial and places, he writes, the rhythmanalyst makes herself more attentive to times than to spaces. And further, she 'will come to listen to a house, a street, a town as an audience listens to a symphony' (1992/2004, 22). Rhythm may therefore make possible the analysis of those obscure spatio-temporal settings and stagings in which the futurity and historicity of the city are produced and enacted – spaces in which the rhythms of Shanghai are (re)played and may be perceived. This way one may also listen to rhythms to tease out the sensory-emotive politics involved in cities, as they are places not only of official discourse, planning, physical structures and circulation of people, but strange spaces of affect, of cryings and prayings, of emotions and embodied interactions (Thrift 2004, 74; see also Jansson and Lagerkvist 2009). Peter Brooker puts this eloquently as he, in a discussion on New York, argues that the urban imaginary is not only a result of our cognitive grasp of the urban experience. Refinement of such approaches is needed:

so as to recognize how the urban imaginary comprises an internally differentiated set of representations and ideological shadings and how these are joined by a range of psychic and affective tones, lived in the mind and the body, across both space and time. We need to recognize too that the urban imaginary will be particular to individual cities and how New York's physical and visual regime has given a distinctive base to its cognitive and affective rhythms, producing a kind of jazzy opus of regularity and extravagance in which the real and not so real are played off against each other. (Brooker 2005/06, 24)

Shanghai is today, and was in the interwar era, another jazzy opus, or perhaps *that other jazzy opus* of the world. It is similarly a mythscape where the mediated past is replayed over again. It is also a place of concentrated pulses and energies, of bodies, sounds, smells, noises, traffic, and of people swarming, spooning, eating, dancing, talking, moving and dressing up. Rhythmanalysis, conceived in this way, seems to corroborate Simmel's and Benjamin's famous depictions of the metropolitan experience as an overload of impressions on the nerve system. Some credence to such an argument may be drawn from the words of one Chinese travel narrator Li, who represents Shanghai in terms of the multi-sensory thickness of the street level, as on the bustling pavement of Nanjing Road East

> All around were even more brightly lit apartment buildings, plazas, offices, restaurants and beauty salons, and the incessant rumbling of motors, the squealing of children, the Babel-like chattering of the crowd, the trembling of the stores' rock-and-roll. All of it was packed into a moist night air, shaking your spirit. (Li 2005)

In addition, paying attention to the rhythmicity of Shanghai also involves acknowledging the thrust of the performative for a cultural analysis of the city. Urban scripts (Lindner 2006) depend upon what I call 'urban performers', stressing the potential for agency and invention even in and through the repetition of theses discursive formations. Urban performers are the products and producers of urban rhythms. As I have discussed at length elsewhere, the Westerner – who in the past used to be very much at home here – has returned to observe, relish or partake in the rigorously organized and planned modernization of Shanghai. But more than this; the city's cultural or mediatized memory can only be articulated through collective remembering: through acts of recapitulating and even inventing the past through performance (Lagerkvist 2007 and forthcoming). The need to pay attention to the bodily, performative and emotive aspects of how these spaces are appropriated and produced through the movements of visiting subjects notwithstanding (cf. Crang 2001), I will in this piece primarily focus on the material, discursive and communicative aspects of the strangeness of space itself: its mediatized settings, its stagings, and mediations – and their potential meanings.

A Space of Colonial Nostalgia and Globalist Venture

> In the center of Shanghai's busy bar street, this charmingly tiny cafe/ pub gives the visitor a feeling of having traveled back in time to colonial Shanghai, circa 1930s – a decadent time when Shanghai was known as Paris of the East. The walls are covered with nostalgic advertisements and photographs. A beautiful gramophone and other 1930s memorabilia of old

Shanghai are scattered around. While quiet and empty in the afternoons, the place gets crowded in the evenings. Shanghainese dishes and Western cakes range between CNY30-50. Alcoholic beverages, coffees, and tea are professionally made or mixed for your pleasure from CNY30-50. (http:// travel.yahoo.com/p-travelguide)

An ad on a Yahoo.com travel guide, for the café and bar 1931 on 112 Mao Ming South Road exemplifies how the Golden Age of the 1920s and 30s is ever *present* in Shanghai. The tourism and nostalgia industries have concertedly made efforts to bring this era to life and this occurs here, as I have shown elsewhere (Lagerkvist 2007), through encouraging *time travel* among the visitors. During the interwar period the international glitterati infamously enjoyed a decadent, adventurous and cosmopolitan lifestyle in Paris of the East, a place of conspicuous consumption, mass culture, sinfulness, and violence. Today and ever since the resurrection of Shanghai began in the early 1990s, the 1930s has been exploited by different interests profiting from the legend of adventurous Shanghai, including its entrepreneurial spirit, mass and media culture. This era of violence, prostitution, consumption, mass culture, mafia gangs, opium trade as well as of cultural pluralism and tolerance has materialized in a memorabilia craze for the 1930s in posters, calendars, reissues of jazz music, old films, photographs, books, clothes, restaurants, hotels and cafés (Lu 2002). Art Deco architecture itself tells stories about colonial Shanghai. *The Broadway Mansions*, a 1934 hotel located to the North of the Huangpu River or *Peace Hotel* on the famous waterfront promenade, the Bund, are traces from a colonial era now deceased, an era when different independent jurisdictions within French, international and Japanese settlements, and the Chinese jurisdiction, existed side by side in the city. *Legendary Shanghai* was produced already during this very era:

> In the twenties and thirties, Shanghai became a legend. No world cruise was complete without a stop in the city. Its name evoked mystery, adventure and license of every form. In ships sailing to the Far East, residents enthralled passengers with stories of the 'Whore of the Orient' They described Chinese gangsters, nightclubs that never closed and hotels which supplied heroin on room service. They talked familiarly of war lords, spy rings, international arms dealers and the peculiar delights on offer in Shanghai's brothels. Long before landing, wives dreamed of the fabulous shops: husbands of half an hour in the exquisite grip of an Eurasian girl. (Sergeant, quoted in Ou-fan Lee 1999, 76, note 2)

In cherishing and celebrating pleasure and excess the city condemned by the central government for decades, due to its sinful and capitalist history, is now encouraged to relive its notorious past. In today's phase of globalization, when Shanghai reappears as a modern metropolis before the world, there are settings in time and space in the city, that I elsewhere describe as the *chronotopes of nostalgic dwelling*, that invite certain performances among visitors and in particular the cosmopolitans, expatriates and other recurring visitors (Lagerkvist 2007) but also among the nouveau riche Shanghainese. One such place is the restaurant and museum, La Villa Rouge on 811 Hengshan Road (Figure 7.1).

La Villa Rouge has a slogan which gives the visitor an idea about what to expect in here: 'Where History Happens'. This is both, as mentioned above, a place for performing the past but the slogan also reveals that time is of utmost importance here: no less than three eras are encapsulated by the very building itself. The house was constructed in 1921, the same year as the Chinese Communist Party was founded in Shanghai, by the American record company EMI. The building was a recording studio in this mansion in the French Concession and it was later taken over by the French recording company and record label Pathé, which also made gramophones, phonographs and records. The French entertainment consortium Pathé et Frères (which was involved in the music as well as in the film industry[1]) was founded in 1896, although the label Pathé existed already in 1894 when Pathéphone was established. Pathé recorded jazz in New York and Paris, and it had a branch in China in the 1920s and 1930s: Pathé Orient Ltd (which was bought by Columbia international Ltd in 1928, and also became part of EMI in 1931). Jazz-tinged mandarin tunes were recorded by and released on Pathé through out the 1930s in China. Pathé was the largest trading and manufacturing enterprise of records in the country and recorded those jazz items that have appeared again today – on, for example, *Shanghai Lounge Divas*. In 1952 the company changed into the Shanghai Record Factory. In 2001 the building was transformed into XuJiaHui park and in 2004 it was publicized as cultural relic of the Xuhui district. In September of 2007 La Villa Rouge was put on the district's list for heritage architecture.

La Villa Rouge becomes an apt illustration of how social time, as Jon May and Nigel Thrift have argued, may be conceived as radically *uneven* and *heterogenous* (2001). In the following I will explore the rhythmanalytic approach for tuning into the urban material imaginary of Shanghai and for seeking out those complex orchestrations of time and space that comprise the

---

1   Prior to World War I, Pathé was an emblem of international enterprise, making films and motion picture projectors. By 1908 it was a major international film empire. The company had branches all over the world in the early twentieth century, in New York, London, Brussels, Moscow, Kiev, Calcutta, and Singapore. The company expanded rapidly and began manufacturing raw film, and monopolized the business as they built studios, laboratories and motion picture theatres. It controlled the entire film industry from ownerships of cinemas to cameras and film (Mattelart 2000, 29; *The Macmillan International Film Encyclopedia* 1994, 1063; see also Gunning 1993; Abel 1999; Dahlqvist 2005).

contemporary city. Doing rhythmanalysis, argues Highmore, is as much about listening as seeing, is as much about immersion as gaining epistemological distance. Its analytical potential is not provided by a methodology of distance (Highmore 2005, 145) but to allow oneself to be taken in by the city. In Lefebvre's words: 'to capture the rhythm, one needs to have been *captured* by it. One has to *let go*, give and abandon oneself to its duration' (Lefebvre 1996, 219, italics in original).

So let's go in, let go and listen.

Strange Times – Strange Spaces

Figure 7.1   La Villa Rouge. Photograph by Johan Lagerkvist, 2008

We enter through XuJiaHui park a couple of minutes from Metro City – one of the most lively areas of Shanghai, where there is a constant buzzing noise from billboards, traffic and audiovisual ads, and where thousands of people swarm the junction where huge department stores face each other, including the giant dome of glass, hosting Starbucks Coffee and a countless number of shops. People in circulation with shopping bags, headphones, iPods and cell phones, cameras and digital displays appear at night as shimmering

Figure 7.2   Featuring Pathé Orient in a space of colonial nostalgia. Photograph
              by the author, 2007

points of light making up Globalizing Shanghai in constant movement and hypermediation.

Inside, silence settles around you (Figure 7.2). The orchestra performing in the bar every evening, has not yet started to play, and the wooden walls seem to muffle the busy city. Once, this house was never quiet. It was a recording studio for jazz music, during Shanghai's heyday of modernization in the 1920s and 30s. When the communists came to power in 1949 the place continued to host recordings of for example the national anthem of the People's Republic of China, and other revolutionary songs. This is so strange – and yet, it is immensely appealing! The response is immediate and visceral. The move from the business of the junction to the tranquility of this brick building is estranging enough. As an analytical orientation, Ben Highmore argues, rhythmanalysis allows us to recognize such *multiple* rhythms of modernity. It invites us to 'consider the speeding-up and slowing-down of social life' (2005, 9). As the move from the frenzy city to its quiet waterholes displays, Shanghai is a place of both speed and duration: of traffic flow and traffic jams, of energetic mediatization and meditative silence, of speedings-up of futuristic visions and slowings-down of historical nostalgia and inert lingering. This emphasis may grasp the gradual change in the social geology of the city and display the laminations of different eras and urban ideologies as well as social rhythms co-existing within the urban experience. For Henri Lefebvre, paying attention to rhythms will overcome dualisms of time and space and provide us with a third element – an active ingredient which may be used in thinking through a dialectical relationship. 'So rhythm overcomes the relationship between time and space – rhythm is on the side of spacing, on the side of durational aspects of space and place and the spatial arrangements of tempo' (Highmore 2005, 9). Lefebvre acknowledged such multiple rhythms of modernity and he tried to describe the various speeds of circulation, the different spacings of movement and the varied directions of flows (ibid., 11).

Today in globalizing Shanghai, the building is a first class restaurant – La Villa Rouge – and a museum of the history of Pathé and other labels which later continued to record in the studio. In La Villa Rouge 1930s colonial Shanghai is staged through a simulated setting of dark French Art Deco furniture of mahogany, walnut and fruit- or rosewood, but other styles connoting 'colonial times', of for example chestnut hardwood also produce an environment of serene elegance with tables, chairs, breakfronts and cabinets with subtly carved details, typically with some Asian ornamentation. In this place one encounters waiters from a world apart, and consumes delicious food in rooms filled with giant gramophones and black and white portraits of female jazz stars – both reminiscent of the evaporated golden past. In enumerating but a few of these luminaries, they include Bai Kwong, Gong Qiu Xia and Woo Ing Ing. These divas were central to the success of Pathé, and were active in both music and filmmaking in Shanghai, until their exodus to Hong Kong after 1949. On the walls posters of the singers and the Pathé brand also appear.

In the grasp of affectual moves – from subliminal estrangement to a strong ironic stance, to a feeling that you have a stake or even a habitat here (and that you will never want to leave) – visitors roam behind the walls of this beautiful place, which offers an obscure spatial experience. And as you have a seat, in the bar or by the dining table, myth and reality, symbols and materiality, past and present are thoroughly mated. This even more so, since the brand is on offer gastronomically on a 'Pathé menu', with an exquisite selection of dishes. Infused by the brand, the name, the image, you devour it as you digest its taste.

Cosmopolitan Shanghai was a metropolis displaying a large number of material emblems of its modernity as well as media forms and technologies: cars, electric light and fans, radios and foreign style mansions, guns, cigars, perfume, 1930 Parisian summer dresses, high-heeled shoes, beauty parlors, jazz and films, and Japanese and Swedish matches (Ou-fan Lee 1999, 76). This memory of this material and mediated urban culture is exploited and profited from in the production of decadent space in La Villa Rouge today. But in La Villa Rouge records from the revolutionary era, cultural relics of the revolution, are also displayed in showcases (Figure 7.3). The glass separates them from the more tangible textures of massive or fake oak furniture, golden stripped tapestry and heavy velvet curtains that comprise the scenography for a replay of the disreputable decadence of this metropolis. Forty years of Communism – the Long March, the Great Proletarian Cultural Revolution, New China and the thoughts of Mao – are now contained behind glass.

One house – three ideologies. Once a studio in the modern capitalist metropolis, then a studio of communistic ideology production, and now a museum of all that was and at once a restaurant for global elites debauching the pleasures of the city: La Villa Rouge encompasses spatial and temporal discrepancies and incommensurabilities. It is seductive! It is strange! But why is such an obscure space produced, and to what purpose? And what is it that makes it so strange and yet familiar at the same time? Marie-Claire Bergère has provided an insightful analysis of how legendary Shanghai has once again been invoked in the city. The past is celebrated for commercial purposes and the revived attraction of this era is instrumental for attracting foreigners and foreign direct investment. The former image of Shanghai as a glamorous and adventurous, international place makes it more easy for Westerners to feel at home there. The past is here used as precedent to validate bold policy moves. In a general way Shanghai nostalgia plays a very important role, as it represents a symbolic capital summoned up to convey a sense of historical continuity: to project an image of Shanghai as the worlds' best potential entrepreneur city and a place where the spirit of entrepreneurship penetrated and continues to penetrate city life with a stereotypical notion of the Shanghainese as shrewd capitalists (2004, 50; see also Pan 2005).

In order to further the understanding of how the complex and strange rhythms of Shanghai are composed, we need however to pursue a spatiotemporal

Figure 7.3   **The revolution contained behind glass. Photograph by the author, 2007**

analysis. La Villa Rouge is a re-appropriated space (Lefebvre 1974/1991, 164-5). Intended for a different time and purpose, it is now a restaurant and museum, exemplifying a kind of remediation of urban space (Graham 2004) and materializing the spatio-temporalities of the global; those strange times and strange spaces that seem to evolve from globalization. This heterotopian space, defined by Michel Foucault as a space capable of juxtaposing in one real place several different yet converging sites, entangles several dimensions that are in themselves incompatible or foreign to one-another (1967/1998). It could be delineated as a *heterochrony* typically linked to *slices of time*, which functions at full capactiy in a break with traditional experience of time and temporality. This space offers a kind of temporal co-existence which makes the past into something more than 'history'. It is, as mentioned above, 'where history happens'. La Villa Rouge is a socially produced space where time – as made, lived and experienced (May and Thrift 2001; Crang 2001) – never stops building up and where it accumulates indefinitely (see Soja 1996, 160). But what is actually there in the heterotopian house, apart from food, wine, prosperous Chinese, Westerners and Japanese chefs? And can we discern through the set-up, the rhythmicity of Shanghai as it is played out in this space?

Sexualized Bodies and Media Artifacts: Nostalgia for Futures Past

Let us shift our attention to another space for a moment. By the entrance to the Xintiandi Museum, which displays the ways and lives of modern Chinese people in the interwar era living in the typical Shikumen housing, the sign reads: 'Before, SMS, MSN or ICQ – Come discover how Shanghai used to be.' This museum is part of the nostalgia surrounding the era of the 1920s and 30s, but also shows the importance of media forms for the city imaginary – today and yesterday. In here portraits of modern women, posters, photographs, film clips from the Shanghai film industry and the interior textures and home settings for these modern creatures, including phonographs, records and magazines from the 1930s, are emphasized as the embodiment of that past era. La Villa Rouge, similarly, is an invented environment where modern women and modern machines from 80 years ago seem to both 'abolish and preserve time and culture, that appear somehow to be both temporary and permanent' (Soja 1996, 161). The centrality of media forms of today's China as well as their importance in the past, is demonstrated in the space of La Villa Rouge. The nostalgia is here about a past futurity and the memory of the modern. The distinctive aspect of the memory craze for 1930s Shanghai which is prevalent at La Villa Rouge, celebrates the role of media artifacts, media culture and media forms that were common during that time. This seems to somewhat contradict Huyssen's analysis of the chaotic, fragmentary and free-floating mnemonic convulsions of our culture. These, he argues, do not seem to have a clear political or territorial focus, but they do express our society's need for temporal anchoring when in the wake of the information revolution, the relationship between past, present and future is being transformed. Temporal anchoring becomes even more important as the territorial and spatial coordinates of our late twentieth century lives are blurred or even dissolved by increased mobility around the globe. (Huyssen 1995, 7)

Huyssen analyzes the memory craze of our era in the light of the informational city, but in Shanghai we actually need to address the exploitation of another media age, and its mobility and newness, as *a form of temporal anchoring*. Here we may begin to understand the role of old signs of modernity and futures past – of media apparatuses and representations, of mass culture in the bygone – for the futurity and digitalization of the contemporary city. There is a cycle of emotions involved. Tom Gunning (2003) discusses what people *feel* in encountering a new medium or technology. We move from astonishment, to second nature – an accustomed familiarity – and into an uncanny feeling. Newness entails many things for human beings and 'a discourse of wonder draws our attention to new technology, not simply as a tool, but precisely as a spectacle, less as something that performs a useful task than as something that astounds us by performing in a way that seemed unlikely or magical before' (ibid., 45). But more importantly and more interestingly: the cycle from wonder to habit need not run only one way, he claims. Re-enchantment is possible through aesthetic de-familiarization and the uncanny re-emergence of earlier stages of magical thinking. New technologies evoke astonishment because of their prophetic nature and because, Gunning concludes, of 'their address to a previously unimagined future' (ibid., 56). This seems to me to explain why media forms take on such an importance within the city imaginary. Media from the past represent a kind of interrupted future, they may now be re-enchanted as Shanghai rises as if by a stroke of magic. And a space – La Villa Rouge – may bestow meaning to these objects and render them a renewed symbolic capital. This, I argue, occurs within a dialogue between 'new media' as cultural forms, that is between our cultures of connectivity, and older media forms through which is spelled out that Shanghai was always the modern place – a place of new beginnings and of envisioned futures.

But something escapes the analysis. This strange space of a replayed mass and media culture, and an invoked future past, is above all *gendered* and not least *sexualized*. Women's bodies themselves, on display on huge photographic reproductions on the fancy restaurant walls, are here monuments of Old Shanghai and its delights and pleasures. Notwithstanding their prudence in costume according to today's standards, they however connote the 'sin city' and an era where their silk-dresses, hairstyles and postures dramatically broke away from the signs of traditional femininity in China (Figure 7.4). As Marina Warner has discussed, the allegory of female form can easily be mobilized to embody certain ideals and values and to make an alluring argument for them: 'To lure, to delight, to appetise, to please, these confer the power to persuade: as the spur to desire, as the excitement of the senses, as a weapon of delight, the female appears down the years to convince us of the message she conveys' (Warner 1985/1996, xx, cf. 11). The female form is perceived as generic and universal, and can therefore be thought to embody values such as liberty, justice and victory. Shanghai had another generic reputation however, with much less praiseworthy connotations, and the city was a *she*: the 'Whore of the Orient'. The female figures – the lounge divas – simultaneously represent a legendary era, and an unveiled male fantasy about female sexuality saturating city space.

Figure 7.4   Female bodies and media artifacts. Photograph by Johan Lagerkvist, 2008

This reverberates with analyses of the commodification of the female body and the simultaneous sexualization of artifacts and goods in consumer culture. Eroticized women are in a commodity culture, argues Abigail Solomon-Goudeau (1996), a supplementary emblem to the commodities themselves. Both in the modern metropolis of the interwar era – where women were used in advertising for beer, soap, cigarettes etc. – and today, this has credence in Shanghai. The female form thus implies an allowing atmosphere of sexual liberty, libidinal indulgence, pleasure and consumption. Representations of women's bodies inscribed into the material setting, mediate the spirit of Shanghai – as an allowing place where nothing is prohibited and everything is possible. There is also a connection between women's bodies and the launch of for example automobiles, new ultramodern technologies of war and media technologies. Despina Kakoudaki (2004) interprets the role of the pinup in initiating interest in new media, art forms and new technologies as dependent upon its wide range of cultural uses and its ability to both 'pass' for mainstream images and to simultaneously retain the excitement of their pornographic and explicit sexual component. Calendars, posters, advertisements, magazine covers, in the past, she argues, only featured beautiful girls and the 'presence of the smiling young woman is often what actually initiates and animates these various spaces, media or print forms before anything else can occupy them' (ibid., 344).

There are thus several conjunctions in the spaces considered between mass culture, media technologies and femininity at work, and they engender an allowing atmosphere of sexual liberty, consumption and profligacy. Woman animates the room where we are spending this evening and the spirit of Shanghai becomes possible when women offer themselves – while the whole of Shanghai is on offer simultaneously (and we may note that there are thousands of prostitutes from the countryside in the city!). Hedonism and a libidinal impulse are on hand for visitors.

What is more, however, within the mythology of the thirties, materialized within the city imaginary, the past is a setting in which modern women with modern garbs, attires and hairdos – Shanghai lounge divas – become a kind of *naturalization* of the city's inclination toward the new, the future, the modern and towards the ever-changing and at all times developing space of New Shanghai. As Marina Warner also states, a web of symbols in cultural history ensconce that the female form represents both *original matter* and *made artifact*. The two become interchangeable (1985/1996: 224–5). In their natural artificiality, modern women of the past represent the natural yet by all means construed futurity of this place. The strange rhythms of the city may be discerned, as I have shown, both in its sexual politics and in the memory of media modernities but foremost, I believe, in their interrelationship. Nostalgia for sinful cosmopolitan Shanghai and 'her' modern Women and sexualized bodies that appear as part of the media culture of the past, communicate this space efficiently, saying to the world: Look we have always been modern – see, this long ago we were already modern. Enjoy our offerings, consume our spirit!

Conclusion

La Villa Rouge is something of a mystery. The house stands there through almost 100 years of music and political drama with strong sensory and emotive experiences on offer for the customer, gastronomically as well as identity-wise.

The purpose of the nostalgia industry is not only to bring back the past, but to create a sense of *retroactive futurity*. It uses old buildings, filled with old media forms and place-based nostalgia, and women's sexuality unleashed, to stage and replay the adventurous city, to say to the world that Shanghai Modern – its natural element – is back and here to stay.

I have propounded in this chapter that laminated spaces of temporal coexistence may be conceived in terms of their capacity of engendering estranging experiences. The strangeness of space here comes about through three enmeshed facets. One place becomes strange through a temporal and rhythmic overlap which is achieved by the inter-stagings of old media forms and 'modern women' to invoke a future of the past, a modernity that once was. These markers of modernity, evoked in a geo-social order of the globalizing city, are however accompanied by another utopian vision now abandoned – signs of the revolution. All this summoned in a leisure space for today's mobile global tribes.

This strange space has many things to offer. The city's cultural and mediated memory of entrepreneurial spirit and decadence may here be brought back to life and in effect replayed, through the movements of visiting subjects and their 'collective remembering' in this chronotope of nostalgic dwelling. But La Villa Rouge is not just a dwelling place beyond the fast and furious commerce of the city. It represents a place where the futuristic dreams of today's Shanghai are spawned, as an invented tradition where Shanghai becomes a locus of ever-developing new forms (Lagerkvist 2006). That sinful capitalism, previously condemned, is now encouraged to become the very essence of the city: its spirit of decadence is encouraged to be relived and replayed, and constitutes an instrumental and productive part of the city's futuristic and utopian vision. This corroborates very closely with Andreas Huyssen's (1995) observation about the role of nostalgia and the importance of those excluded eras and blind spots in history for envisioning futures and inventing a new utopia:

> In the search for history, the exploration of the no-places, the exclusions, the blind spots on the maps of the past is often invested with utopian energies very much oriented toward the future ... Nostalgia itself however, is not the opposite of utopia, but, as a form of memory, always implicated, even productive in it. (ibid., 88)

Nostalgia for media forms and objects become part of a layering of space, resulting for customers in a thrilling sense of being out-of-place and out-of-bounds – a utopian impulse of being *decadent* – and yet remaining fully in control. Nostalgia as utopia is thus linked to this heterotopian space, and following Foucault we may acknowledge how La Villa Rouge in all its exceptional composition underscores a given yet evolving social order. And one *connects* in here. The sense of belonging in this space of sophisticated extravagance is established through the replay of cultural memories of that previous era of western presence, reign and plenty for Shanghai. Portraits of female jazz stars representing that other era seem to comprise a tradition which contributes to a natural sense of newness. La Villa Rouge offers an impossible fusion between newness and solid tradition. In fact, the place affords a sense of historical continuity of newness – of novelty and futurity as tradition. (Therefore, in here and in this context, the obsolete long playing records from the revolutionary years – that remind us of communist visions and the overthrowing of what went before – may make all the sense in the world.)

Inspired by Henri Lefebvre I have attempted to pay attention to rhythms in order to overcome dualisms of time and space and to analyze the city in terms of its complex asynchronic synchronicity. Rhythms provide us with a third element that may disclose that time and space are always co-implicated. But even more important: rhythm is not equal to speed. Rhythmanalysis is thus 'the measure of dynamic relationships and it insists on the plural rhythmicity of the city' (Highmore 2005, 9). The difficult and pressing task is to describe rhythms in terms of time as *repetition and change*. Rhythms that are manifold, changing, contrapuntal and often disharmonious, yet amounting to an urban space of strangeness that sweeps you off your feet, with – despite all its disheveling and contradictory aspects – a tangible, substantial and unwavering rhythmic experience in stock.

A visit to La Villa Rouge enacts an obscure script on a strange stage that alienates and invites us at the same time. The rumbling rhythms of this heterotopian building offer a befuddling sense of both estrangement and belonging – or of belonging through estrangement. Visitors turn into strangers in the Shanghai night in search for the memory of Shanghai. We step into globalist and nostalgic rooms, settings where there seems to be a license for almost everything. When rummaging such houses and our own 'pockets' of repressed desires, we need to listen to the strange remixes of the city's futures past: rhythms through which we may be made aware of the vast geopolitical, socio-economical and gendered repercussions of venturing into them.

Acknowledgement

I wish to thank Anja Hirdman for advising me on important literature on the allegory of female form.

References

Abel, R. 1999. *The Red Rooster Scare: Making Cinema American, 1900–1910.* Berkeley: University of California Press.

Bergère, M.-C. 2004. 'Shanghai's Urban Development: A Remake?', in Kuan, S. and Rowe, P.G. (eds) *Shanghai: Architecture and Urbanism for Modern China*. Munich: Prestel.

Boyer, C.M. 2002. 'Approaching the Memory of Shanghai: the Case of Zhang Yimou and *Shanghai Triad* (1995)', in Gandelsonas, M. (ed.) *Shanghai Reflections: Architecture, Urbanism and the Search for an Alternative Modernity*. Princeton, NJ: Princeton University Press.

Brooker, P. 2005/2006. 'Terrorism and Counternarratives: Don De Lillo and the New York Imaginary', *New Formations*, 57, Winter, 10–25.

Crang, M. 2001. 'Rhythms of the City: Temporalised Space and Motion', in May, J. and Thrift, N. (eds) *Timespace: Geograhies of Temporality*. London: Routledge.

Crang, M. and Travlou, P. 2001. 'The City and Topologies of Memory', *Environment and Planning D: Society and Space*, 19, 161–77.

Dahlquist, M. 2005. 'Global versus Local: The Case of Pathé', *Film History: An International Journal*, 17:1, 29–38.

Foucault, M. 1967/1998. 'Of Other Spaces', in Mirzoeff, N. (ed.) *The Visual Culture Reader*. London: Routledge.

Graham, S. 2004. 'Introduction: From Dreams of Transcendence to the Remediation of Urban Life', Graham, S. (ed.) *The Cyber Cities Reader*. London: Routledge.

Gunning, T. 1993. *Pathé 1900: Fragments d'une filmographie analytique du cinema des premiers temps* (ed. André Gaudreault). Quebec: Presses de la Sorbonne Nouvelle, les Presses de l'Université Laval.

Gunning, T. 2003. 'Re-Newing Old Technologies: Astonishment, Second Nature and the Uncanny in Technology From the Previous Turn-of the-Century', in Thorburn, D. and Jenkins, H. (eds) *Rethinking Media Change: The Aesthetics of Transition*. Cambridge, MA: MIT Press.

Highmore, B. 2005. *Cityscapes: Cultural Readings in the Material and Symbolic City*. London: Palgrave.

Huyssen, A. 1995. *Twilight Memories: Marking Time in a Culture of Amnesia*. London: Routledge.

Jansson, A. and Lagerkvist, A. 2009. 'The Future Gaze: City Panoramas as Politico-Emotive Geographies', *Journal of Visual Culture*, 8:1, April, 25–53.

Kakoudaki, D. 2004. 'The Pinup: the American Secret Weapon of World War II', in Linda Williams (ed.) *Porn Studies*. Durham, NC: Duke University Press.

Lagerkvist, A. 2006. 'Future Lost and Resumed: Media and the Spatialisation of Time in Shanghai', paper from the European Science Foundation – Linköping University Conference *Cities and Media: Cultural Perspectives on Urban Identities in a Mediatized World*, Vadstena, 25–29 October. LiU Press Electronic publication: http://www.ep.liu.se/ecp/020/.

Lagerkvist, A. 2007. 'Gazing at Pudong – "With a Drink in your Hand": Time Travel, Mediation and Multisensuous Immersion in the Future City of Shanghai', *Senses and Society*, 3:2, July, 155–72.

Lagerkvist, A. forthcoming. 'Velvet and Violence: Performing the Mediatized Memory of Shanghai's Futurity' in Frassinelli, P. P., Frenkel, R. and Watson, D. (eds) *Traversing Transnationalism*. London/New York: Continuum.

Lefebvre, H. 1974/1991. *The Production of Space*, London: Verso.

Lefebvre, H. 1992/2004. *Rhythmanalysis: Space, Time and Everyday Life.* London: Continuum.

Lefebvre, H. 1996. *Writings on Cities*. Cambridge, MA: Blackwell.

Li, L. 2005. 'China, Society, Feature Series, On the Road to Shanghai ... The Beginning', 31 May, http://josephbosco.com/wow2004/wow.html.

Lindner, R. 2006. 'The Cultural Texture of the City', paper from the European Science Foundation – Linköping University Conference *Cities and Media: Cultural Perspectives on Urban Identities in a Mediatized World*, Vadstena, 25–29 October. LiU Press Electronic publication: http://www.ep.liu.se/ecp/020/.

Lu, H. 2002. 'Nostalgia for the Future: The Resurgence of an Alienated Culture in China', *Pacific Affairs*, 75:2, 169–86.

Ma, L.J.C. and Wu, F. (eds) 2005. *Restructuring the Chinese City: Changing Society, Economy and Space*. London: Routledge.

The Macmillan International Film Encyclopedia, 1994. New York: HarperCollins Publications.

Massey, D. 2005. *For Space*. London: Routledge.

Mattelart, A. 2000. *Networking the World, 1794–2000*. Minneapolis: University of Minnesota Press.

May J. and Thrift, N. (eds) 2001. *Timespace: Geograhies of Temporality,* London: Routledge.

Ou-fan Lee, L. 1999. 'Shanghai Modern: Reflections on Urban Culture in China in the 1930s', *Public Culture*, 11:1, 75–107.

Pan, T. 2005. 'Historical Memory, Community Building and Place-making in Contemporary Shanghai', in Ma, L.J.C. and Wu, F. (eds) *Restructuring the Chinese City: Changing Society, Economy and Space*. London: Routledge.

Qingyun, M. 2006. 'Shanghai, People's Republic of China' *Cities: Architecture and Society*, Venice: Marsilio Editori.

Shanghai Lounge Divas. 2006. EMI Music Hong Kong.

Soja, E. 1996. *Thirdspace: Journeys to Los Angeles and Other Real-and-Imagined Places*. Malden: Blackwell.

Solomon-Goudeau, A. 1996. 'The Other Side of Venus: The Visual Economy of Feminine Display', in Grazia, V. de and Furlough, E (eds) *The Sex of Things.* Berkeley: University of California Press.

Thrift, N. 2004. 'Intensities of Feeling: Towards a Spatial Politics of Affect', *Geografiska annaler,* 86, 57–78.

Warner, M. 1985/1996. *Monuments and Maidens: The Allegory of Female Form*. London: Vintage.

Zhang, Z. 2001. 'Mediating Time: The "Rice Bowl of Youth" in Fin de Siècle China', in Appadurai, A. (ed.) *Globalization*. Durham, NC: Duke University Press.

Chapter 8
Cities of Sin, Backroads of Crime

Will Straw

This chapter examines two ways of representing criminalized space within popular U.S. print culture. One of these styles is represented by the cover illustrations for two books by Hendrik de Leeuw, a once popular and largely forgotten author of travel narratives and adventure literature (Figures 8.1, 8.2 and 8.3). De Leeuw's *Cities of Sin* and *Sinful Cities of the Western World*, first published in the 1930s, moved in and out of the markets for semi-illicit pornography through multiple editions over a 30 year period. Sometimes the covers to these books highlighted their claim to documentary truth and usefulness. On other occasions, as with the editions shown here, cover illustrations promised access to shadowy worlds of sexual commerce described in titillating detail. These unusual images present spaces which are oneiric, exotic and sexualized, marked (most notably in Figure 8.2) by a highly stylized play with light and shadow.

The other sort of space emerges later, in the mid-1950s, on the covers of US true crime magazines (Figures 8.4, 8.5 and 8.6). The three magazines shown here each display a compositional figure which quickly became a cliché of magazine cover illustration during this period: on an isolated patch of earth or road, a body lies still, while other human figures stand near it. Virtually unseen in true crime magazines before 1955, this figure would be repeated across dozens of cover photographs over the next decade. These images are all staged tableaux, shot by commercial photographers and posed by models. Nonetheless, each strives for a matter-of-fact, journalistic sense of *actualité*, of an action caught in the midst of its unfolding.

The two kinds of spaces to be discussed here did not succeed each other in any strict genealogical sense, as if one neatly took over the cultural function or space of the other. The styles and compositional figures of each represent, nevertheless, distinct ways of endowing criminalized spaces with a sense of the peculiar. In the covers to de Leeuw's books, 'strangeness' is a function of the openings enacted by each image, of the lines of association which link each locale to a set of other places criss-crossed by sexual traffic and criminal intrigue. A sense of each image's incompleteness is conveyed by the elaborate play with light and shadows which obscures key details of each space, producing an aura of secrecy and concealment. In the magazine covers of the second set, in contrast, strangeness resides in the peculiarly disconnected quality of the space and the clarity with which it is shown. All three of these

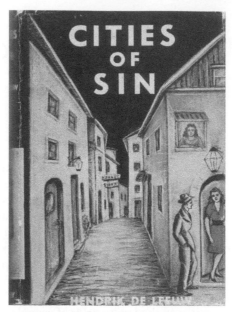

Figure 8.1   *Cities of Sin*, Hendrik de Leuw. Front cover. Artist unknown. New
               York: Willey Company, 1943. Author's collection

Figure 8.2   *Cities of Sin*, Hendrik de Leuw. Front cover. Artist unknown. New
               York: Gold Label Books, 1943. Author's collection

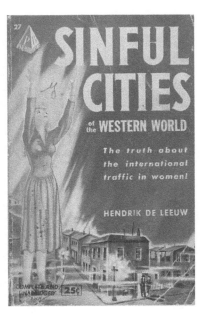

Figure 8.3   *Sinful Cities of the Western World,* Hendrik de Leuw. Front cover. Artist unknown. New York: Pyramid Books, 1951. Author's collection

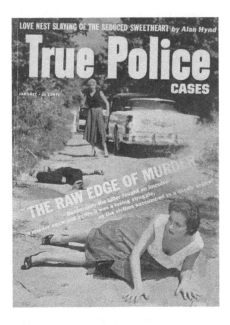

Figure 8.4   *True Police Cases.* Front cover. Vol. 9, No. 95, January 1957. New York: Fawcett Publications. Artist unknown. Author's collection

Figure 8.5   *Homicide Squad.* Front cover. Vol. 1, No. 1, October 1957. Artist unknown. Author's collection. Sparta, IL: Precinct Publications

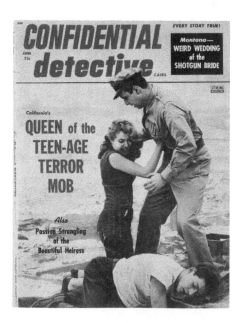

Figure 8.6   *Confidential Detective Cases.* Front cover. Vol. 7, No. 1, June 1956. New York: A Sterling Magazine. Artist unknown. Author's collection

true crime photographs are characterized by high levels of illumination, by an almost total absence of the shadows which conventionally serve to render crime scenes atmospheric. These are images lacking the familiar stylistic gestures which link crime to hidden, secret forces.

In the analysis which follows, I will use each set of images to gather up resources with which we might think about the relationship between criminality, place and visual style. The covers of de Leeuw's books invite us to reflect upon the ways in which crime and secrecy serve to magnify space, rendering it elastic and extendable. This magnification, I will argue, underscores the preoccupation of so many interwar popular cultural texts (like pulp magazines and vice exposé books) with globalized networks of vice and sexual traffic. The staged photos of the true crime magazine cover, on the other hand, suggest the restriction of criminalized space, its confinement within places of bright illumination. In the move from our first set of images to the second, a sense of menace is no longer conveyed through a shadowy, nocturnal obscurity. It will come to reside, more and more, in the isolation of empty spaces subject to the heat and light of day.

De Leeuw's Sinful Cities

The covers of *Cities of Sin* which accompany this chapter are from two different editions of that book, both published in the US in 1943. The Dutch-born author of *Cities of Sin*, Hendrik de Leeuw (1891–1977) had moved to the United States as a young man, then written a series of travel narratives while serving as Far Eastern representative for the Firestone Tire and Rubber Company. The first publication of *Cities of Sin*, in 1933, coincided with (and clearly sought to exploit) the release of the League of Nation's *Report on Traffic in women and children in the East*, a document whose themes are the same as those of De Leeuw's book. *Cities of Sin* offers itself as a study of 'prostitution, white slavery and trade in women and children', and is organized around chapters devoted to individual cities of the East (Yokohama, Hongkong, Shanghai, Macao, Port Said, and Singapore). De Leeuw's 1935 follow-up, *Sinful Cities of the Western World*, contained chapters on Paris, Amsterdam, Berlin and New York, but was taken up, for the most part, with journeys through the vice districts of cities of the middle East. In 1939, de Leeuw pursued adjacent themes in his *Flower of Joy*, on the production and consumption of narcotics in Asia.

Mail-order advertisements for *Cities of Sin* could be found in the back of 'spicy' periodicals of the 1930s and 1940s, even as the book was listed or reviewed in academic journals of the time, such as *Social Forces* and the *American Journal of Sociology*. In one of the few scholarly references to *Cities of Sin*, Yingjin Zhang notes how quickly de Leeuw's persistent moralism and documentary zeal give way to hallucinatory, highly visual fantasies. Following its businesslike preface, *Cities of Sin* opens with 'Dreams of Lost Woman', a chapter set entirely in dreamy italics that recounts, in Zhang's efficient summary, de Leeuw's 'final private vision of a stream of naked, languishing, and hysterical women careening toward the cavern of lust' (Zhang 1999, 179). 'Cities fade', de Leeuw writes, in an apocalyptic vision that collapses the cities of Asia into a generalized image of sex, opium and human traffic (1933/1943, 20).

The unsigned covers for *Cities of Sin* shown in Figures 8.1 and 8.2 convey something of this oneiric strangeness. Both are deliberately vague as to the precise geographical locales covered within these books. So, too, is the cover for the 1954 paperback edition of *Sinful Cities of the Western World* (Figure 8.3), whose lower third, curiously, seems to be an expanded view of the scene contained in Figure 8.1. Like the covers of most vice exposé publications from mid-century, these simultaneously convey the book's promise to travel far, in search of sexual exoticism, and the titillating possibility that vice may be found close to the reader's own world. Just as each cover shows a very specific context of sexual encounter, it offers lines of vision and thematic flight which suggest a myriad of spaces beyond it, within a larger world of sin and vice. As we shall see, this vision of interconnected global circuits of vice and criminality shaped the spatial imaginary of a great deal of Western popular culture in the first half of the twentieth century.

Secrecy and the Magnification of Criminalized Spaces

It is common to think of crime as something that simply contaminates a space, infusing it with a sense of menace or insecurity. We might reflect more broadly, however, on the ways in which criminality acts upon spaces to expand their volume and meaning. In his well-known article on the secret, Georg Simmel noted that secrecy posed 'the possibility of a second world alongside the obvious world', creating, in Michael Taussig's summary, 'a world split between a visible exterior and an invisible depth that determines the exterior' (Simmel 1906, 462; Taussig 1999, 56). This sense of invisible depth is rendered in literal fashion in crime fictions which propose secret criminal conspiracies operating through hidden networks or venturing out from spaces of concealment. Crime fictions typically work on the basis of elaborate metonymies and displacements which function to expand the radius of criminal action within narrative worlds. As Dennis Porter has observed, the hard-boiled detective novel is organized around the structure of the 'spreading stain', in which the social and geographical reach of crime is magnified as a narrative unfolds (Porter 1983, 333). The shadows and obscurity which typify popular crime imagery (in films and magazine illustration, for example) magnify space by blurring its limits, suggesting its potential for unlimited extension.

Recent work on the representations of crime in pre-World War II France has described the expansion of space characteristic of cultural texts which linked the local mystery to the global conspiracy, the urban *fait divers* to the worldwide criminal network. In the Fantomas serials of print and cinema, Dominique Kalifa suggests, we glimpse 'the shift from a monocentric world toward a noticeably polycentric one. Telegraph, telephone, radio broadcasting, ocean liners, and transcontinental trains make Fantômas a great traveller who visits such places as Scotland, Russia, Mexico, Columbia, and Natal' (Kalifa 2004, 187). In his book *Shanghai on the Métro*, on the prevalence of spy figures in French popular culture of the 1930s, Michael B. Miller shows how the localized spaces of Parisian criminality were seen, inevitably, to be caught up within global intrigues and conspiratorial networks that ran through exotic, faraway locales (Miller 1995). Metonymic lines tied each urban mystery to broader struggles unfolding across the distant arenas of colonial power. And, finally, Adrian Rifkin, writing of the French crime newspaper *Détective*, speaks in similar fashion of the ways in which, in the 1930s, popular cultural forms inscribed global intrigues onto the spaces of Paris: 'One of the peculiar abilities of *Détective* is to open up the vistas of the glance between the spaces of the city and beyond, so that the distant seaport, the Spanish bordello or the colonial deserts can be mapped back on to it – a collapsing of cartographic distance and its problem of inaccessibility into the topography of city mnemonics' (Rifkin 1993, 162).

Images like those which adorned the covers of Hendrik de Leeuw's books convey something of this collapsed distance in at least two ways. First, they offer an image of a globalized district of sin, imaginable in each case as both familiar and exotically distant. Like other purveyors of what Pierre Mac Orlan had called the 'social fantastic' (Mac Orlan 1929/1989), De Leeuw traces, across his books, circuits of interconnected port cities, systems for the transnational traffic in women, and a generalizable urban topography of red light districts, opium dens and waterfront spaces of encounter. Second, the organization of these cover illustrations works to produce a sense of almost labyrinthian space, through the permeability of buildings (with their open windows and doors) and a sense of each image receding into twisted alleyways or foggy obscurity. These convolutions heighten the sense that each locale is interconnected with others, through the lines of metonymic flight built into each image's compositional style.

Criminal spaces may be magnified in other ways, through the ways in which they absorb within themselves the depth of historical time. This is another of the many insights of Dominique Kalifa's (1995) work on crime fictions of the French Belle Epoque. During this period, crime journalists, novelists and the creators of characters like Fantomas reached back to a pre-Hausmann Paris to recover a city of murky spaces and obscure trajectories. 'The medieval city', Kalifa suggests, 'imposed its structure on many feuilletons: its alleys, its mazes, its zigzag trenches formed the outlines of the story – and commanded the narrative web, plots, and successive connections and dead ends within each story' (Kalifa 2004, 192). Through their constant reintegration within fictional and journalistic narratives of crime, older topographies of the city were reinvigorated, used to strengthen the city's resistance to being fully understood. Mitchell Schwarzer has made a similar claim about US television crime programmes of the late 1950s, like *Naked City*. Shot on location, these programmes kept their gaze on the old New York of 'brick and stone' even as the city was being transformed by an architecture of steel and glass (Schwarzer 2004, 292).

Space is magnified here by the ways in which crime narratives integrate within themselves the sedimented residues of earlier architectures and forms of urban order. In his analysis of the case of London serial killer John Christie, who lived and buried his victims at 10 Rillington Place in North Kensington, Frank Mort shows how journalistic coverage of the murders and their geographical context operated as the excavation of lost historical phenomena: 'Much of the material environment for the murders was quintessentially nineteenth century, with corpses buried in the decrepit Victorian house and the narrow streets and terraces providing the dominant setting for the action. Many of the social actors in the affair were also presented through the language of nineteenth-century urban typologies' (Mort 2008, 328). More than most narratives, those of criminal investigation are able to capture the uneven development of urban species and environments. The frequently uncanny quality of such narratives has much to do with their capacity to resuscitate social types and situations long believed to have disappeared.

De Leeuw's books of the 1930s, *Cities of Sin* and *Sinful Cities of the Western World*, claimed topical relevance based on their echoing of League of Nations reports and other contemporary documents about international sexual commerce. Today, what binds them to their period is a sense of mystery and sensation closer to that of the pulp magazines, movie serials and other fantastic cultural forms of the time. In the 1930s, pulpy adventure narratives drew on a deep inventory of styles and motifs from the previous half-century. The raw materials of Victorian adventure, World War I intrigue, colonial revolt, political conspiracies, revolutionary upheaval and organized crime all piled up, in the popular culture of that decade, to make each story a dense wound set of interconnections. The visual renderings of place which accompanied these narratives overlaid the compositional clichés of Victorian melodrama upon the luridly colourful visual rhetorics of mass-produced print culture.

The shifts in de Leeuw's writings after World War II are symptomatic of the waning of such rhetorics; these shifts provide a bridge to our second set of images, those of the postwar American true crime magazine. In 1952, Hendrik de Leeuw published *Underworld Story,* an expose of criminal rackets and vice in the United States. *Underworld Story* was clearly intended to resonate with Senator Estes Kefauver's investigations into organized crime and municipal corruption within the US (Straw 1997). The inventorying of sex and vice, always prominent in de Leeuw's work, now came to organize itself around American examples exclusively. At the same time, the sense of reverie which had marked his books of the 1930s – the hallucinatory evocation of sensual abandon – was gone, replaced by a matter-of-fact reporting of dates and events. This shift in de Leeuw's attention and rhetoric mirrored a broader turn in the popular print culture of crime and vice – from the exotic and fantastic towards the highly specified, from tales told in the language of exploration and adventure to reportage which borrowed from journalism and the judicial investigation. One of de Leeuw's final books, *Woman, the Dominant Sex; From Bloomers to Bikinis* (1957) fits easily into a series of misogynistic polemics, from Philip Wylie's *A Generation of Vipers* (1942) through Lee Mortimer's *Women Confidential* (1960), which focussed almost exclusively on contemporary Western cultures.

The True Crime Magazine of the 1950s

The so-called 'true crime magazine' (or 'fact detective' or 'true detective' magazine) was born in the 1920s, in part as an offshoot of the true confession magazine. From among the confession magazine's multiple genres of sensation and confession (which included tales of fallen virtue and white slavery), crime quickly emerged as the focus of a specialized wave of new magazines. From the early 1920s until the death of the genre at the end of the twentieth century, these magazines offered ostensibly true stories of crimes and police investigations. To bolster their claims to documentary truth, these magazines employed photographs as the most common form of illustration, often printing official police mug shots or crime scene photographs. Vast numbers of the images published in these magazines, however, were imaginary reconstructions of crimes and crime scenes, staged by commercial photographers in studios and using professional models. As a result, the history of true crime magazine photography is interwoven just as tightly with those of fashion magazine illustration or print advertising imagery as with a history of the official documentation of crime.

In patterns I have described elsewhere, authentic photographs of crime scenes and criminals (taken by journalists or supplied by police departments) were arranged, within the true crime magazine, alongside clearly 'fictional' photographs of crimes staged in studios and posed by models (Straw 2006). Across the 80-year history of the true crime magazine, fashions in true crime imagery rippled across multiple titles and publishers, producing distinctive period styles. In the 1930s, the period of the genre's first great expansion, the typical covers of true crime magazines contained one or two human figures, painted with rough lines in compositions that conveyed fear or murderous rage. By the late 1940s, covers had come to be dominated by painted or touched-up photographs of the faces of solitary women. Throughout the 1940s, the covers of true crime magazines sought to convey a glamour from which all but the most oblique references to crime were banished. Set against backgrounds of solid colour, the cover models of the late 1940s true crime magazine occupied no recognizable dramatic, geographical or architectural space. Indeed, the development of true crime magazine cover illustration since the 1930s had seen the markers of distinctive spaces and of punctual moments of action gradually eliminated. The weakening of spatio-temporal visual markers was echoed in the tendency of 1940s true crime stories to unfold in relatively abstract, thinly textured historical moments and geographical places. Gone were the links to high-profile public crimes (like the capture of Bonnie and Clyde or killing of John Dillinger) which had characterized the true crime magazine of the 1930s. The stories recounted in 1940s magazines took place, for the most part, in generic locations familiar from crime fiction, like penthouse apartments or isolated country houses.

In the 1950s, the American true crime magazine elaborated a new relationship to place. This development was immediately evident on the covers of these magazines, which changed dramatically during the decade. Around 1953, magazines began to offer covers containing images of compressed dramatic action. Cover models, who had once gazed solicitously at a magazine's readers, now receded into scenes of action which involved multiple characters acting within detailed settings. On the covers of the most stylistically innovative of 1950s true crime magazines, like *Inside Detective* (published by Dell) and *Homicide Squad* (Sterling), the bright colours of the 1940s gave way to unglamorous, documentary-like black and white images. In this change, the status of spatio-temporal referents within these magazines was transformed as well. While celebrity criminals had served to ground the 1930s crime magazine in actuality, the magazines of the 1950s based their claims to realism on the ordinariness of tawdry settings and empty landscapes.

It is on the covers of these magazines that a new space of action first appears and quickly becomes a visual cliché of the genre. This space is that of the grassy field or gravely rural road, on which a body lays sprawled while other figures (friends, killers, policemen) crouch over it. The three images which accompany this chapter offer different versions of this visual composition, chosen from a larger corpus of similar images. In the 1950s, this dramatic space would become as common as others, more widely-known, which have long defined the visuality of crime, like the neon-lit city street or the fog-saturated urban harbour.

Race and Place

The human figures which inhabited this space were, virtually without exception, white. While this fixation on white characters and populations was typical of the true crime magazine as a genre, it was underscored in images characterized by high levels of illumination. In one of many columns written to help would-be writers of true crime stories, *Writer's Digest*, in 1938, had claimed the following: 'Picking your crime is restricted by a few rigid editorial taboos. Most fact detective editors will not consider the story of any crime perpetrated by a coloured person' (Monroe 1938). The consistency with which this prohibition was enforced is clear from the examination of a large corpus of these magazines, even if the precise reasons behind it are not. In the 1930s, the pulp fiction periodicals published alongside true crime magazines had been full of racially marked villainous characters and perilous situations. Likewise, the men's adventure and celebrity gossip magazines which flourished in the 1950s regularly offered up words and images which hinted at interracial romance or mobilized fantasies about interracial violence. The brazen directness with which other magazines marshalled anxieties about race makes the absolute whiteness of true crime magazines all the more noticeable. With very few exceptions, these magazines never engaged with the hot button issues of race, urban decline and poverty which ran through popular arguments about crime. Racially-based acts of violence, perhaps the most horrific American crimes of the twentieth century, were nowhere to be found in the true crime magazine. Well into the 1960s, these magazines remained fixated on bandit couples, rural fugitives and, increasingly, serial killers. These are all, statistically and stereotypically, the stock figures of white criminality.

The isolated spaces which recur on true crime magazine covers, from the mid-1950s onwards, are usually located on backroads or in rural areas. A fixation on such spaces is part of a broader flight, in these magazines, from urban glamour, towards hinterlands and small towns. This shift expresses a vision of crime as unfolding principally in a world of small-town white people. The well-lit covers of true crime magazines of this period offered a coherent stylization of rural white criminality, one distinct to the period and traceable to the influence of leading titles like *Inside Detective* (cf. Straw 2006). The lighting of cover photographs betrayed a prejudice which saw crime as a matter for poor whites, photographed to make their faces seem pale and under-nourished. Treatment of the Cutter family murders, covered by *Front Page Detective* in 1960 and the focus of Truman Capote's *In Cold Blood* (1965/1993), stands as a canonical example of this prejudice at work.

The isolated spaces of the mid-1950s true crime magazine cover invite a reading which sees them as visual devices designed to circumscribe the social meanings of crime. Images of rural backroads and white bodies detach crime from the thematics of race, poverty and urban life which increasingly, in the post-World War II era, shaped public debate over justice and the law in the United States. If this act of containment was not conscious or deliberate, it nevertheless expressed the urge to offer images of crime untroubled by lines of connection leading to larger social contexts – contexts that were less easily deciphered than the isolated patch of road or grass. We may contrast this impulse towards closure with the multiple gestures towards a variety of *elsewheres* on the covers to Hendrik de Leeuw's books.

Film Noir, Blanc, Gris, Soleil

The decline of shadowy stylization in true crime magazine illustration of the 1950s is part of a broader unravelling of the aesthetic associated, in the cinema, with film noir. Between a film like *Night and the City* (Jules Dassin 1950) and stylish crime films made 10 years later, like *Odds Against Tomorrow* (Robert Wise, 1959) or *A bout de souffle* (Jean-Luc Godard 1959), we see a general whitening of tone, a turn from shadowy spaces to scenes shot under bright sunlight in open, exposed settings. A major cause of this whitening was the increased use of on-location filming, itself influenced by developments in camera technology, film stock and lighting. One index of the dissolution of film noir is the tendency of films of the late 1950s to convey a new sense of menace through dramatic spaces marked by a bleached-out emptiness.

Film studies scholars have proposed a series of new terms to describe the mutations of a film noir style over time or across adjacent genres. One of the most widely adopted of these has been film gris (or 'grey film'.) For Thom Andersen (1985), who proposed the term, film gris circumscribes a cycle of American thrillers of the 1950s in which crime narratives were deployed in the service of leftish social commentary (see also Maland 2002). '*Gris*', here, functions in a double sense. As a secondary, hybrid colour, 'grey' serves here to designate the space of generic overlap between otherwise distinct genres (the noir thriller and the social commentary film), as if grey were produced in the overlaying of a post-war 'dark' cinema upon the illuminative conventions of the documentary. At the same time, film gris suggests the bleakness of grey spaces, of the impoverished small towns and highway stops (like diners and gas stations) which are, increasingly, the locations for crime (see also Dimendberg 2004). This dreariness, it is suggested, is often absent from the oneiric, shadowy worlds of film noir. In a further development of the term, Dana Polan uses it to characterize the grey, depressing landscapes through which characters of limited means and opportunity move in such films as *Thief's Highway* (Jules Dassin 1949) (e.g. Polan 2000, 136).

The term film blanc ('white film') has a more convoluted history. Unsurprisingly, it is most often used to describe an aesthetic that inverts the defining features of film noir. The most common use of film blanc has been to group together films whose transcendent optimism counters the stereotypical despair of film noir. (A website devoted to film blanc calls it the 'Cinema of Feel Good Fantasies': http://filmblanc.info/.) Ongoing exchanges, on the internet and elsewhere, have sought to specify this sense of the film blanc, using the term more narrowly for films concerned with the afterlife (e.g. Genelli and Genelli 1984). *It's a Wonderful Life* (Frank Capra 1946) is the most frequently invoked example of such films, interesting in part because it contains, embedded within it, scenes which are more conventionally *noir*. In this usage, '*blanc*' suggests both the optimism of religious faith and the cloudy visual blankness which, in so many films of the 1940s, stood for heaven or its antechamber, as in *Heaven Can Wait* (Ernst Lubitsch 1943) and *Here Comes Mr Jordan* (Alexander Hall 1941).

Other uses of the term 'film blanc' have employed it in a more literal sense, in reference to films whose dominant aesthetic is one of brightness. In her review of Paula Gladstone's 1978 documentary on Coney Island, *The Dancing Soul of the Walking People*, Judith Bloch writes that '[h]er film could be called a film blanc, its use of sunlight matching the *noir*'s use of shadow, its mood so much the daytime, seaside equivalent of the noir's urban light' (Bloch 1981, 62). In her reference to sunlight, Bloch anticipates ideas about 'sunshine *noir*' or 'film soleil' developed in recent years by Hoberman (2007), Holm (2005), and others. Central to all these moves is a recognition that the characteristic spaces of danger and crime in films are no longer those of the *noirish* city, organized around the play of light and shadow. On the contrary, a sense of menace is now conveyed most forcefully, in crime films, through the intense heat and light of spaces exposed to the sun.

Hoberman offered his term 'sunshine noir' to designate an aesthetic which has become prevalent in representations of criminality based in Los Angeles. Hoberman sees prefigurations of this style in the 'moody high-noon surrealism' of 1940s experimental filmmaker Maya Deren, or in the daytime ghostliness of early, LA-based experimental films by Kenneth Anger. In the Californian crime stories which follow, from *Sunset Boulevard* (Billy Wilder 1950) through *Chinatown* (Roman Polanski 1974) and beyond, sunshine is part of the structure of hypocrisy, hiding a deep cultural rot beneath bright, warm surfaces. Holm, in his pocket-sized guide to 'film soleil', develops similar ideas about the semi-independent crime thrillers of the 1980s and 1990s set in small southern towns. For Holm, the cycle of 'film soleil' begins with the Coen brothers' *Blood Simple* in 1984 and continues through other 'sunlit crime films' like *Kill Me Again* (1989) and *After Dark, My Sweet* (1990) (Holm 2005, 14).

This detour through arguments made in the contexts of film scholarship and criticism is necessary to our discussion of true crime magazine photography. It allows us to point to a broader reversal, between the late 1940s and the late 1950s, in the cultural meanings of darkness and light, brightness and obscurity. Across the thousands of photographs (staged and genuine) published in true crime magazines during this period, the consistency of this reversal becomes clear. The use of shadowy obscurity to evoke unease would become less and less common by the end of the 1950s. It gave way first to the grey, low-contrast look of black and white photography, as in the examples reproduced here. Then, by the 1960s, noirish chiaroscuro had further ceded its place, if only on covers, to garish colour photography and a flirtation with psychedelic effects which used excessive sunlight as the basis for distortions of light and composition.

The menacing strangeness of the brightly lit space stands as a puzzle in late twentieth century popular aesthetics. Thomas Elsaesser's astute analysis of R.W. Fassbinder's 1982 film *Veronika Voss*, a work he labels a film blanc, opens up several routes we might follow in taking up this puzzle (Elsaesser 1981, 114). If clean, white environments now seem oppressive, it is perhaps because their primary association is with a history of twentieth century institutions, a recurrent object of political critique in late modernity. Some of these institutions, like large bureaucracies, are simply dehumanizing; others, like hospitals and laboratories, may figure as places of cruelty and death, particularly through their association with totalitarian regimes. Amidst such new prejudices, arguments for the progressive character of shadowy obscurity find their conditions of possibility. Elsaesser notes how blacks and greys, in the antiseptic environments of modern institutions, may offer places of refuge, of 'security and respite.' He quotes French critic Yann Lardeau on the function of shadows: 'Without shadow a person cannot live, without it a person has no soul because no secrets' (Elsaesser 1981, 114).

We may trace this rehabilitation of the shadow across a number of recent interventions in cultural analysis. In all of them, shadowy obscurity becomes a zone in which human depth and freedom may thrive, rather than the space from which something emerges to threaten them. Rifkin notes how urban avant gardes in interwar France were drawn to obscurity as that which 'provides a glimpse of alternative trajectories and senses.' The darkness at the edge of visualized space may offer a sense of the endless possibilities latent within the everyday, rather than standing as a boundary which limits such possibilities (Rifkin 1993, 157). In his book *L'envers du visible: Essais sur l'ombre*, Max Milner (2005) explores the wide variety of ways in which shadows have functioned, like Simmel's secrets, to magnify the world. Shadows, he suggests, have made phenomena as diverse as human psychology, painterly space, and material objecthood all seem to possess a depth and volume that would be lacking in contexts of pure illumination. Shadows are no longer the signs of an ignorance or distortion which clarity and truth will banish, but, rather, guarantees of the incompleteness and multiple displacements characteristic of any image and of the identities it puts into play. A 'world without shadows', Ina Blom suggests, following visual artist Olafur Eliasson, is a 'world without difference' (Blom 2008, 119).

In a history of crime imagery, the benign character of the shadow in the present day is nourished by the sentimental familiarity which has attached itself to gothic, expressionistic or *noirish* films and photographs. This is in part because, as these styles have aged, the contrast between light and shadow within them matters less than the ways in which both together produce a visual texture comforting in its coherence and noble in its cultural lineage. The shadowless images of sun-drenched open spaces, on the other hand, or the bright emptiness of deserted noon-time highways, have snuck into our popular culture with no comparable pedigree. They partake of the menacing strangeness which, in late modern life, has attached itself to excessive levels of illumination and clarity.

References

Andersen, T. 1985. 'Red Hollywood', in Ferguson, S. and Groseclose, B. (eds) *Literature and the Visual Arts in Contemporary Society*. Columbus, OH: Ohio State University Press, reprinted with new material in Krutnik, F. et al. (eds) 2008. *Unamerican Hollywood: Politics and Film in the Blacklist Era*. New Brunswick, NJ: Rutgers University Press.

Bloch, J. 1981. 'Review' (untitled), *Film Quarterly*, 34:3, 61–3.

Blom, I. 2008. *On the Style Site: Art, Sociality and Media Culture*. Berlin and New York: Sternberg Press.

Capote, T. 1965/1993. *In Cold Blood*. New York: Vintage Books.

Dimendberg, E. 2004. *Film noir and the Spaces of Modernity*. Cambridge: Harvard University Press.

Elsaesser, T. 1981. *Fassbinder's Germany: History Identity Subject*. Amsterdam: Amsterdam University Press.

Genelli, T. and Genelli, L.D. 1984. 'Between Two Worlds: Some Thoughts Beyond the *"Film Blanc"'*, *Journal of Popular Film and Television*, 12:3, 100–11.

Hoberman, J. 2007. '"A bright, Guilty World": Daylight Ghosts and Sunshine Noir', *Artforum*, 45:6, 260–67.

Holm, D. 2005. *Film Soleil*. Harpenden: Pocket Essentials.

Kalifa, D. 1995. *L'encre et le sang. Récits de crime et société à la Belle Epoque*. Paris: Fayard.

Kalifa, D. 2004. 'Crime Scenes: Criminal Topography and Social Imaginary in Nineteenth-Century Paris', *French Historical Studies*, 27:1, 175–94.

Leeuw, H. de. 1933/1943. *Cities of Sin*. New York: Gold Label Books.

Leeuw, H. de. 1934. *Sinful Cities of the Western World*. New York: Citadel.

Leeuw, H. de. 1939. *Flower of Joy*. New York: L. Furman.

Leeuw, H. de. 1955. *Underworld Story*. London: Neville Spearman Limited and Arco Publishers.

Leeuw, H. de. 1957. *Woman, The Dominant Sex*. New York: T. Yoseloff.

Mac Orlan, P. 1929/1989. 'Elements of a social fantastic', trans. Robert Erich Wolf, reprinted in Phillips, C. (ed.), *Photography in the Modern Era: European Documents and Critical Writings, 1913–1940*. New York: The Metropolitan Museum of Art/Aperture.

Maland, C.J. 2002. 'Film Gris: Crime, Critique and Cold War Culture in 1951', *Film Criticism*, 26:3, 1–30.

Miller, M. 1995. *Shanghai on the Métro: Spies, Intrigue, and the French between the Wars*. Berkeley and Los Angeles: University of California Press.

Milner, M. 2005. *L'Envers du visible: Essai sur l'ombre*. Paris: Seuil.

Monroe, M. 1938. 'Selling the Fact Crime Article', *Writer's Digest*, February.

Mort, F. 2008. 'Morality, Majesty, and Murder in 1950s London: Metropolitan Culture and English Modernity', in Prakash, G. and Kruse, K. M. (eds) *The Spaces of the Modern City: Imaginaries, Politics and Everyday Life*. Princeton, NJ: Princeton University Press.

Mortimer, L. 1960. *Women Confidential*. New York: Copp Clark.

Polan, D. 2000. 'California through the Lens of Hollywood', in Barron, S. et al (eds) *Reading California: Art, image and identity, 1900–2000*. Los Angeles: University of California Press.

Porter, D. 1983. 'Backward Construction and the Art of Suspense', in Most, G.W. and Stowe, W. (eds) *The Poetics of Murder: Detective Fiction and Literary Theory*. San Diego, CA: Harcourt Brace Jovanovich.

Rifkin, A. 1993. *Street Noises: Parisian Pleasure 1900–40*. Manchester: Manchester University Press.

Schwarzer, M. 2004. *Zoomscape: Architecture in Motion and Media*. New York: Princeton Architectural Press.

Simmel, G. 1906. 'The Sociology of Secrecy and of Secret Societies', *The American Journal of Sociology*, 11:4, 441–98.

Straw, W. 1997. 'Urban Confidential: The Lurid City of the 1950s', in Clarke, D. (ed.) *The Cinematic City.* London: Routledge.

Straw, W. 2006. *Cyanide and Sin: Visualizing Crime in 50s America*. New York: PPP Publications, Andrew Roth Gallery.

Taussig, M. 1999. *Defacement: Public Secrecy and the Labour of the Negative*. Stanford, CA: Stanford University Press.

Wylie, P. 1942. *Generation of Vipers.* New York: Farrar and Rinehart.

Zhang, Y. 1999. 'Prostitution and the Urban Imagination: Negotiating the Public and the Private in Chinese Films of the 1930s', in Zhang, Y. (ed.) *Cinema and Urban Culture in Shanghai, 1922–1943*. Stanford, CA: Stanford University Press.

Films Cited

A bout de souffle (dir. Jean-Luc Godard 1959).
After Dark, My Sweet (dir. James Foley 1990).
Blood Simple (dir. Joel Coen 1984).
Chinatown (dir. Roman Polanski 1974).
The Dancing Soul of the Walking People (dir. Paula Gladstone 1978).
Heaven Can Wait (dir. Ernst Lubitsch 1943).
Here Comes Mr Jordan (dir. Alexander Hall 1941).
It's a Wonderful Life (dir. Frank Capra 1946).
Kill Me Again (dir. John Dahl 1989).

Night and the City (dir. Jules Dassin 1950).
Sunset Boulevard (dir. Billy Wilder 1950).
Thief's Highway (dir. Jules Dassin 1949).
Veronika Voss (dir. R.W. Fassbinder 1982).

Chapter 9

Walks in Spectral Space: East London Crime Scene Tourism

Chris Wilbert and Rikke Hansen

Objects can be presented in the courtroom, but spaces have always to be imagined, and represented; and representation has, from the early nineteenth century at least, been an art, controlled by psychological projection and careful artifice, more than a science.

Vidler 2001, 'X Marks the Spot'

After certain murders become cast as notorious or infamous in the media and public eye, we may be faced with all kinds of projections and 'careful artifice' – cinema, television, guided tourist walks, art and museum exhibitions, popular crime novels, simulacra of crime scenes in tourist attractions, and more, that mesh together with often odd affects. Some of the trails of this careful artifice we outline in this chapter in relation to tourist walks. As we focus on what comes 'late' to experience and long after the event, it therefore seems appropriate to start our account of crime scene tourism close to the end.

And so we find ourselves on a guided walk in the east end of London together with 50 perhaps 60 other people. It is the most popular tourist walk in the city, one that focuses on the notorious 'Whitechapel Murders', committed by an unknown infamous figure that became known as Jack the Ripper. As this tour is moving to a close, our guide makes one of his last stops to address us. On this spot, he explains, the body of Annie Chapman, the murderer's second victim, was discovered on 8 September 1888. He tells us that at that time residents in the surrounding buildings charged a penny from anyone wanting to look at the deceased's body through their windows. Even after the corpse was moved, the guide informs us, people were still charged money to gaze at the scene of the crime. It is difficult not to see the irony here, as we have also paid a – relatively larger – fee to visit, not only this, but all five of the murder sites.

Monetary inflation aside, this is, however, a complex visitation, in which a series of non-spectacles are kept alive by the thousands of people who visit them every year. In crime fiction it is the detective who untangles the web of traces. On these tours, the guide often plays a similar role; however, it is a use of space that also differs from the forensic logic of detective novels. Much has of course changed in the London areas of Whitechapel and Spitalfields since 1888 and, as we walk through the streets, the tour guide encourages us to imagine that which we can no longer see. This leads to a *performative articulation* of the site in which distinctions between presentation and representation blur or collapse. Importantly, as the example above shows us, this is an articulation which itself takes place in a *post-performance* space, in which traces and memories are not the only elements called upon or performed by the guide, but where stories of voyeurism and spectatorship have themselves become sufficiently fetishized to be quoted within the narrative of the tour. As such, we maintain a distinction between performance and performativity, in which the former indicates a clear separation of stage and off-stage and the latter a notion of performance-in-the-expanded field. The term 'performative' thus implies a breakdown between that which is clearly marked as 'performance' and that which is not; it also indicates the simultaneous existence of 'saying' and 'doing', of 'representing' and 'acting out' within one and the same articulation or utterance (see Austin 1962).

This chapter considers an aspect of leisure/tourism that has its focus on the events of murder, crime scene tourism, and uses of media in east London. Whilst our main focus is on Jack the Ripper walks, an attraction which has existed for more than four decades, we also briefly focus on Kray Twins guided walks, which have similarly contributed to the mythologization of the east end of London as a site of crime and deviance. What we point to is how such guided walks are suffused with media representations that when also put to use by participants on walks may give rise to dissonance, boundary crossings, and develop sometimes bizarre linkages and affects across an arc of cultural-economic realms.

To understand the workings of the kinds of site making alluded to above, we approach the notion of a 'performative articulation of site' from two sides. Firstly, we propose that the production of murder scene attractions cannot be separated from wider contemporary fascinations with crime scenes. This fascination may range from fictionalized and true crime novels, to spontaneous memorializations of crime scenes such as sites of mass murder or violent deaths of public figures, as well as more personalized loved ones, or a shared belonging to national or ethnic groups that often work performatively as a call for some form of responsibility (Santino 2006; Margry and Sanchez-Carretero 2007). Such contemporary fascinations with fictionalized or actual crime scenes also include the attraction of tourists to other sites of disaster and destruction, a phenomenon sometimes referred to as 'thanatourism'. Nor can the allure of the crime scene be separated from contemporary mediascapes. Recently, television series such as *Crime Scene Investigation* (CSI) have contributed to the way we might read urban spaces, in this case the cities of Las Vegas, Miami and New York, as sites which can be mastered through a forensic gaze. As for the Jack the Ripper murders, to speak his 'name' is already to evoke media connotations since the unknown murderer himself received his title from the printed press at the time of the killings, and newspaper representations from that time have also informed many later films and television adaptations. In the case of the Krays, these were gangsters already notorious for staging themselves within the media during the 1950s and 1960s.

This leads us, secondly, to consider how such tourist-media spaces cannot be separated from a certain incongruity of history. So, while an external contextualization may suggest that these tours are representative of a cohesive *zeitgeist*, their internal multiple narratives, with their borrowings from past and present, fact and fiction, add complexity to this logic. Here we follow Niru Ratnam (2005, 72) who observes that: 'history is neither a juggernaut bearing down upon us nor a conveyor belt spread out in front of us; instead history returns in dribs and drabs, and sometimes not at all'. Furthermore, such temporary 'quotations' may in this context also be used to 'render strange' those spaces which most of us might normally consider everyday: alleyways and social housing estates, spaces which are picked out during the Whitechapel murders walks either because they coincide with the actual murder sites or because they contribute to a historical east end atmosphere by carrying other, more symbolic, connotations with them, such as 'deviance' and 'poverty'.

London's Crime Scene Tourism

Crime scene tourism may at first seem by some to be in bad taste – practices of distinction through embodied dispositions of taste of course being omnipresent in tourism (as they are in wider consumption practices). Moreover, how and why there is such a thing as crime scene tourism when very little is to be seen at the crime scene – the bodies and their traces have after all long gone, the scene itself has often been utterly transformed through the passage of time – remains very much an open question. Tourism is normally thought of as an activity where the tourist's gaze is attracted by the extraordinary or distinct (Urry 1990). Yet, we may find ourselves standing in a dilapidated building used as a temporary car park being told this is the scene of a gruesome crime. Of course this may still be a distinct place – not because of how it looks – but because of the historical associations that the guide convinces us to imagine.

London, like other cities, has many spaces that are connected through acts of violence of the past made present in the now. Indeed, it can be argued that London has its own spectro (or spectral) geographies, being also populated by ghosts, the murdered, their perpetrators, and more. Such spectrality points to how the past may be still very much alive and at work, but the allegorical notion of the spectral also potentially confounds any settled orderings of past and present, not only working to displace 'place and self through the freight of ghostly memories; it works to displace the present from itself' (Wylie 2007, 172). Likewise, for Steve Pile (2002) cities are where 'ghosts' gather and are gathered in more concentrated ways than other spaces and often force us 'to recognize the lives of those who have gone (before)' in a wide variety of ways.

Pile argues that the city is an intense site for the spectral gathering of affects in which we cannot begin to understand the social senses of space unless there is room for feelings, emotions, and how apparitions are to be found to still occupy, or haunt, certain urban spaces in the now. Of course the pertinent aspect of this latter point for us is how consumers of these east end walks are invited to then consume urban heritage crime spaces by companies that offer tourist walks as part of a wider heritage industry that has increasingly embraced 'affect' as a distinctive selling point of experience economies (Pine and Gilmore 1999). As such, tourism walks focused on crime scenes might also be seen as a summoning of 'ghosts of place' in which certain macabre events have been commodified through heritage practices for tourists and other leisure visitors (Mayerfeld Bell 1997, 828).

Yet, for David Lowenthal one of the benefits that traits of the past found in heritage have for people is a sense of termination, that what happened in the past has ended in a way that helps people locate themselves in a linear narrative that connects, past, present and future (Ashworth and Graham 2005, 8–9). Spectrality, from the above reading is a sense of time-space where such easy orderings may be undermined or complicated. So we might ask if Jack the Ripper walks challenge such linear narratives and displace the present, or if they instead provide a sense of termination, a reassuring sense that the past is in the past? To this the answer is they can do both and more – and as such have certain ambivalences – partially dependent on the narratives of the guides and how these are interpreted and made meaningful by tourists. These walks are of course not separable from wider media forms, such as the burgeoning literature of London psycho-geographical wanderings of Iain Sinclair (1987) – amongst others – that repeatedly invoke the Whitechapel murders and the strange connections found by such persistent callings upon these ghosts of the past in present spaces.[1] Yet, it is the media of Hollywood and British cinematic representations from which narratives, plots, and characters may be deployed by tourists or guides that is important to these walks.

As part of the myriad ways that the past becomes brought up into the present it is increasingly the case that we see publics *convening* around past sites of violence or representations of violence, murder, and more general instances of death in themed historical museums, guided walks, memorials, or other sites, as tourists and leisure visitors (though it must be emphasized that such practices are not new as such). A range of terms have been developed to

---

1   See Barker (2006) for a critique of Sinclair's often disparaging depictions of east-end people and culture, and Baker (2005) who claims Sinclair was instrumental in how the Jack the Ripper murders have become central to the literary psychogeography of Whitechapel, building on earlier unsavoury fascinations with these murders and crime in the east end of London by Guy Debord and the Situationist International, that seems to echo that of some 1960's 'counterculture' celebrations of the Kray Twins as supposed anti-authority figures.

describe this convening, or the places of trauma where convening occurs. We have heard of dark tourism (Lennon and Foley 2000), tragic tourism (Lippard 1999), thanatourism (Seaton 1998), fatal attractions (Rojek 1993) and traumascapes (Tumarkin 2005), amongst others. Some of these have drawn attention to the differences between 'authentic' sites where events took place and more distant sites where events are represented – for example, museums that are in other locations from where events took place such as the Holocaust museum in Washington DC (Miles 2002; Williams 2007). Yet, hard and fast distinctions between authentic sites and distanced representational sites can often be rendered problematic as fact, fiction, and marketing blend, fold, re-produce and inflate such crime scene spaces in often odd and sometimes troubling ways (see Charlesworth 2003).

Contemporary Jack the Ripper guided tourist walks in the east end of London, which visit the crime scenes of five prostitutes murdered between 31 August and 9 November 1888, are just one popular focus of convening around crime scenes.[2] These walks are the most popular guided walks in London with current estimates of up to 100,000 annual participants.[3] However, tourism to the east end of London, or even the scene of the Ripper murders is not new. Since at least the early 1800s wealthy and educated elites would go 'slumming' in the spaces of poverty of the east end either as a form of evening's entertainment or to visit and view the work of philanthropic institutions, or through undertaking philanthropic work (Koven 2004; Walkowitz 1992, 52ff).

Motivations of tourists visiting the east end of London in the twenty-first century have undoubtedly changed since the late nineteenth century. Robert Mighall (2003) has noted how a certain mythologized heritage notion of criminality (fed by various media representations) has come to inform many people's cultural maps of London in recent years. Moreover, the popularity and uses of stories of crime in certain areas of the city have increased just as gentrification of these areas has gathered apace over the past 20 years, and indeed these can be seen as processes that are loosely tied together. In some housing developments attempts have been made by developers to link gentrified and luxury housing with a heritage-inscribed luridness of these places – in ways that link in to a renewed fascination of a (now sanitized and heritagized) past underworld and underbelly of London (Mighall 2003, 274).

---

2   That there were five murders connected with 'Jack the Ripper' is a fairly recent consensus. Yet, in 1889 it was thought that perhaps as many as 11 murders might be attributed to this figure.

3   These are only estimates made by some walking tour companies. No data has been collected to verify these estimates as several different tour companies undertake tours.

Fatal Media Attractions

This use of narratives of past crimes to give en 'edgy' feel to gentrified and 'improving' areas of the city also reflects a contemporary media fascination with crime – though again this is not a new phenomenon. The naming of the murderer as 'Jack the Ripper' was most likely invented by a journalist as part of the massive media furore surrounding the murders in 1888 (Walkowitz 1992). This intimate connection of the Whitechapel murders to mass media forms continues today and examples of 'Ripper' media include, amongst others, dedicated websites and numerous dramatic renditions of the story. Within this particular public realm hoaxes and speculations regarding the murderer's identity constantly circulate, and have done in the printed press since 1888 and this has continued since then within different media, meaning that this realm had, and still has, a sense of indeterminacy and doubt at its core (Coville and Lucanio 1999).

Andreas Huyssen (2003, 19) has noted that in recent times it has become clear that to talk about memory is to speak about audiovisual representations of events through television, cinema, radio, and the internet, which 'compel us to think of traumatic memory and entertainment memory together as occupying the same public space.' Following on from this we argue that Jack the Ripper walks might be thought of as forming a part of a wider theatricalization of experience in which certain cultural memories are circulated, *acted out* and *performed* in the streets of the city and beyond in media flows.

Public fascination with the Jack the Ripper murders in literature (and in the then emergent cinema) began soon after the murders of 1888.[4] Marie Belloc Lowndes' hugely popular book *The Lodger* appeared in 1912. Silent movies also used fictional representations of the 'Ripper', though not all alluded to the murderer by this name. One of the first films focusing directly on the Whitechapel murders was Alfred Hitchcock's 1926 adaptation of *The Lodger*. More recently we have seen further merging of factual and fictional figures in cinematic forms such as the detective Sherlock Holmes seeking Jack the Ripper in *A Study in Terror* (1965). Also noteworthy is the film *Dr Jekyll and Sister Hyde* (1971) that merges the Jekyll and Hyde myth (a play which was running in a west end theatre at the time of the murders) with Arthur Conan-Doyle's theory that the Ripper murderer was a woman. Most recently we have had the Hughes Brothers adaptation of Alan Moore's graphic novel *From Hell* (2001) starring Johnny Depp.

The Whitechapel Murders have also moved beyond cinema to include regular television debates, documentaries, fictionalized dramatizations and books about the identity of the murderer that help reproduce interest in, or at least background knowledge of, the murders and murder sites in the east

---

4   Some fictional literature on these murders did emerge in the months when the murders were occurring.

end of London. One of the latest is crime fiction author Patricia Cornwell's (2002) speculation that the painter Walter Sickert was the Whitechapel killer. Cornwell received much criticism from so-called *Ripperologists* and art critics alike, with the latter up in arms about her vandalizing a painting in the search for fingerprints and DNA to back up her theory. Contrary to the author's supposed intentions, her 'discovery' has, however, led to increased popularity of Sickert's work with major exhibitions of his paintings being held in recent years.[5] Lisa Tickner notes how Cornwell's destruction of Sickert's work rests on the transformations of 'monuments' into 'documents' (2007, 55ff). Here Tickner cites Erwin Panofsky's famous observation that 'everyone's "monuments" are everyone else's "documents" and vice versa', implying that monumentality and documentary value are not inherent qualities within objects but relational concepts born from the context within which the objects are placed, and thus subject to historical transformation. Broadening the scope here, it is noteworthy that both the media sites and the walking tours dedicated to Jack the Ripper work through this 'twin pull' of celebration and exposure. Where Cornwell's approach to uncovering the real murderer differs from most amateur detectives however, is her declared determination to obtain justice for the victims, despite her very literal interpretation of Sickert's interest in scenes of homicide and his pursuit of depicting 'life as he saw it'.

What occurs in and through the strangely tortuous mingling of such practices of representation of these murders is a kind of inflation in the scope of the landscape of murder. Firstly, we have a re-imagining and performance of this landscape in different but interpenetrating media, reproducing it as a product in which the anonymous murderer, and his victims, emerge from, and move back into, a distributed locale (Seltzer 1995). Within these mediated landscapes the story of the Ripper is circulated into ever more diverse films, novels, and also other hugely popular London based tourist attractions such as the London Dungeon and Madame Tussaud's waxwork museum where simulacra of the 1888 crime scenes are displayed for dramatic effect. Walking tours inevitably become indivisible from the compass of these diverse media landscapes, whilst fact and fictions multiply, are re-sorted by tour guides, tour companies and self-styled 'Ripperologists', for example on web sites (such as http://www.casebook.org), from where they may then feed back into other mass media formations. Such a notion of extended landscape is therefore an on-going process that has less to do with landscape as a 'way of seeing' and more a focus on how landscapes are performed through corporeal, imaginative, and technological mobilities (Haldrup and Larsen 2006).

---

5   These exhibitions include: 'Walter Richard Sickert: The Human Canvas' at Abbot Hall Art Gallery, Kendal, 2004; 'Walter Sickert: The Camden Town Nudes' at The Courtauld Gallery, London, 2007–2008; and 'Modern Painters: The Camden Town Group', of which Sickert was a prominent member, at Tate Britain, London, 2008.

A much smaller tourist guided walking economy (and here we mean economy as an ordering and performance of heterogenous elements) than the Whitechapel murders focuses on the Kray Twins – east end gangsters jailed for life in 1969 for two murders (one of which was executed in Whitechapel). These notorious criminals partially modelled themselves on Chicago gangsters of 1930s movies and literature, and have also been the focus of film, television and many books that have fed back on each other.[6] *The Krays* (directed by Peter Medak 1990) is perhaps the only film biography. Yet, the Kray Twins also played the media whilst actively engaged in their violent criminal activities by intentionally seeking a constant visibility in the press, deliberately being photographed with celebrities, actors and actresses in night clubs and other situations. As Dick Hebdige (1974, 25) has argued:

> The Krays were ... the darlings of the media of the sixties. Feted and filmed whenever they emerged from the womb of the Underworld, they exercised their privileges as celebrities with an adroitness and a sophisticated awareness of the importance of public relations matched only in the image-conscious field of American politics.

Some of the images from this time regularly surface in the contemporary British press – in particular the portraits taken by the fashion photographer David Bailey. Yet such images of the Krays have also seemingly become clichés of a certain 'criminal type' and it is as both clichés and parody that Kray-style criminal brothers or twins have studded many other recent British gangster films, TV soap operas – such as the BBC's *EastEnders* – and in comedy sketches such as Monty Python's Piranha Brothers (Jenks and Lorentzen 1997, 88).

One obvious difference between Jack the Ripper murders and the Kray Twins is that the identities of the latter were not only well-known at the time of their lives, but put on display through the twins own public 'performances', an act of self-staging through which they made a claim to a social role, or a set of roles, already occupied by actors who came before them, real-life and fictional ones, as well as sowing a seed for further interpretations of this position. Jenks and Lorentzen (1997, 90ff) argue that part of the Krays fascination factor was their 'twinness', which unsettled and transgressed our definition of identity-through-difference. With Jack the Ripper, the attraction may rest on a similar, yet different sense of identity transgression. Whilst the serial killer's true identity is still a mystery, the public imagination, from the time of the murders until today, has created an image of a gentleman of the upper classes transgressing and literally feeding on a working- or lower-class neighbourhood, a smartly dressed, charismatic man in a cape and top hat,

---

6 In recent years gangster murders of the 1920s and 1930s have become a focus of much more entertainment based crime scene tourism in Chicago where 'guides' dress up and speak as gangsters – or rather as Hollywood versions of these (Becker 2005).

Figure 9.1   A scene of a crime? A Kray Twins guided walk in east London

Source: authors own.

often depicted as being of higher intelligence. In terms of the Kray Twins, their charisma has often been connected to their embodiment of a working class pride, sense of honour and agency. Furthermore, as Hebdige (1974) points out, the 1960s, the time of the Kray's reign, was also an age in search of a suitable media image of anti-authority via the celebration of criminality. This is a fascination which nevertheless happened and continues to happen within the confines of capitalism's celebration of the self-made business 'man'. However, we are not suggesting that such fascinations are themselves without agency, as shall become clearer below when we turn our focus towards the internal narratives of the tourist walks.

Performing the Crime Scene

Mainstream tourism, perhaps especially guided tourist walks, are also often viewed by some *other* tourists as forms of passive consumption and experience of places and events. However, one walking tour argues for an opposing view – claiming that tourists who went beyond the tourist bubble of central London out to the east end Jack the Ripper walks are more like 'adventure tourists'. However this notion of the adventurous tourist perhaps works more as a form of tourist marketing flattery than anything else (see Jansson 2007: 15ff). Even so, we argue that more may be going on in around tourist guided walks than such 'adventure tourist' flattery makes claims for, and in focusing on this we change our register to focus on what tourists experience, perceive and receive, and also what they do.

The city-spaces of Whitechapel and Spitalfields can be approached as open stages for performances of differing historical and spatial narratives by a variety of walking tours, guides and media. In these performances of place, the tours work through the city using recognisable existing or imagined 'authentic' material traces of late nineteenth century Whitechapel and Spitalfields to dramatize their story. Shifting register here – to that of participants – let us take examples of some of the places and features visited on these walks.

After meeting in early evening at an underground station we are guided along and across busy streets, past a church known as a place where prostitutes circulated in the 1880s. We walk along Whitechapel High Street, then move up Brick Lane, a place known for its Bangladeshi community, the latest in a long line of immigrants who settled in this area, and their 'Indian' and 'Balti' restaurants, which themselves are increasingly threatened by gentrification and spiralling property rents. We are directed to long gone pubs where victims of the Ripper supposedly drank. Links are made to other migrant communities that settled in the east end through the use of materials – showing us, for example, now gentrified Huguenot housing – or moving down 'authentic' Victorian narrow alleyways and being shown a vague Star of David on a wall, in order to discuss the thousands of East European Jews who during the 1880s had fled the pogroms of eastern Europe and settled in the east end. We wander around the outside of churches, are shown iron grills in recesses in walls, put in, we are told, to stop thieves hiding in the dark, this part of the city in 1888 being unlit off the main streets at night-time. Also included, of course, are places that are thought to be the real sites of the murders – otherwise anonymous alleyways and streets that would not usually draw attention except as places to best avoid of an evening. On at least one instance we are on a walk when another group run by a rival company arrives in the same street, but our guide explains that the other group were looking at the wrong place regarding where a bloody apron had been found after the double murder of 30 September 1888. On another walk the somewhat eccentric guide uses nineteenth century pennies (four of which were needed by the women to get a room in a doss house for the night) and old photographs from the time to illustrate his narrative. At the crime scene of Catherine Eddowes – the fourth victim – he tells us that this was the first time photography had been used at a murder by a police force in England, whilst showing us copies of these pictures as well photos of the murder scene of the final victim – Mary Kelly. In between the stops where the guide tells us stories, we, as tourists, can also see, hear and smell other aspects of places and things that draw our attention that may not be part of the usual tourist repertoire of attractions.

Yet of course much has changed in these areas since 1888 and few obvious buildings and streets from that time have survived. Proximity of the areas of the 'Whitechapel murders' to the City of London has recently brought huge changes through processes of gentrification and property speculation. Prior to this, the areas suffered destruction by bombing during World War II and saw subsequent building of cheap social housing as part of several phases of slum clearances, as well as progressive de-industrialization that widens our scope in to globalization of the textile trades that once dominated this area, amongst other emergent political-economic practices. Lucy Lippard (1999, 126) argues that in places where traumatic events have occurred, where traces have disappeared 'leaving only the voids to speak' we fill the blanks with our own experiences, associations, and imagery. Of course, on these walks the guides are there to *aid* us in filling in these blanks and uncovering the traces of the city, the sites of murder and the many layers of myth, memory, and history that obscure them. Jack the Ripper walks work, we argue, as a kind of *street theatre* – with the stage stretching far beyond the actual street itself. Such examples of 'fatal attractions' can be engaged in performative terms – as something being 'acted out', rather than merely being 'represented' within tourism and different media forms. The important point here is the way such *acts* rarely take place on one single stage, but happen in the interplay *between several* different stages on which cultural memories are played out, thereby shaping the meaning-making processes at play within these walks.

Tours, Narratives, Landscapes

Tourist-performances cannot simply be described in terms of 'on location' encounters, but are created through the complex intertwining of cultures, bodies, images and texts – developing what Soja (1996) terms a thirdspace. One aspect of this is seen in the ways narrative structures of cinema have also contributed to the 'narrativization' of the walking tours – even though the guides tend to claim the opposite – and while most walks do not follow a linear logic from murder one to murder five, most tours end up at the site of the final killing – that of Mary Kelly. This follows Carol J. Clover's (1987) notion of the 'Final Girl', a phenomenon that she argues characterizes the slasher genre, the final girl being the last victim, the gravitational point towards which the story steers. In slasher films, this woman is usually only an *intended* victim, since she is the one who 'undoes' the line of murderous acts. In Ripper cinema, this structure has been appropriated either by 'adding' a sixth victim who got away, as in *Jack the Ripper* (1959), or by suggesting that Mary Kelly was confused with someone else and therefore escaped, which is the case in *From Hell*. Since most of the tour guides spell out that they only stick to 'facts', continually pointing out the differences between cinematic myth and real life, no such 'Final Girl' speculations are incorporated directly into the narratives, but Kelly is nonetheless described as an individual, and thus shown to be different from the other prostitutes in the story.

That tourists bring media-based knowledges to these tourist landscapes is evident from the questions posed by some of them to the guides, which tend to centre on information that tourists have ordered, gathered, and introduce from outside sources. Listening to tourists' questions and internal discussions on walks it is evident that some tourists *reconstitute* certain references from films and other media sources that they may have consumed at home into the spaces where they are engaged in tourist activities in order to make sense of these spaces. So, on these guided tours what some of the tourists appear to do is to reconstitute cinematic knowledges into the performances of the walks by asking questions concerning the significance of certain *plots* or *props* featured in films or other media. In these performances the image-spaces of the media intersect with the practices of tourist-audiences on the ground and change the ways experiences are produced in this city space. These processes are complicated by the manner in which everyday distinctions between guide-as-performer and tourists-as-audience are rendered as both normal and estranged. For example, tourists perform for each other – from the basic ways that they show competences to listen and act out their roles as tourists, to more idiosyncratic ways where on several occasions we witnessed female tourists posing on the ground as victims and being photographed by friends. Tourists also perform and become actors – to be looked at and critically observed by other local users of the space – as they move through the city streets. On some rare occasions young people try to make fun of the group or shout at the people on the walks. On a Kray Twins walk local British Asian children on one street shouted 'go home immigrants', a bizarre but amusing reaction to what was predominantly a white middle-class group of Londoners and tourists. There are therefore somewhat differing notions of estrangement taking place in and around the walks, as guides render other peoples everyday places strange through invoking 'ghosts' of the past for tourists, a making of place that is not always wholly welcome by non-tourists. Indeed other local users of the spaces often seem to find the staging of these places for crime scene tourism a strange one, in which tourists also become performers and, at times, figures of wry amusement or irritation.

Figure 9.2   A Jack the Ripper walking tour in east London

Source: authors own.

On most of the walks we undertook that were led by more serious guides (that is, ones not led by 'actors'),[7] the guides continually use occasions when tourists ask about cinematic representations to distance themselves and the

---

7   Actors are employed as guides by at least one walking tour company. Such guides tend to perform in a different way from more 'serious' guides – some adopting 'cockney' or 'posh' accents to perform the historical narratives of the walks.

walks from 'myths' pushed forward by film and other media. For example, this occurs when guides point out that the film *From Hell* was shot away from the 'authentic' location on a film set near Prague. This distancing of the walks from cinematic representations can also be seen in the ways some guides explain how the London fog – a ubiquitous background for nearly all Jack the Ripper films – was based solely on cinematic myth-making. Indeed, on occasions where people appeared to ask too many film-related questions some of these more serious guides became visibly annoyed. Such annoyance may well emerge from the frequency of the same kinds of questions from tourists – 'What was Johnny Depp like?' being a strangely familiar one (Johnny Depp went on one of the most popular Jack the Ripper walking tours prior to his role in *From Hell*). Such annoyance with tourist's questions about films and television plots and story lines can also be seen to be invasions of the 'authentic' historical narratives that serious guides are seeking to perform in ways where television and cinema are viewed as having relatively low cultural value.

By situating the walks in *opposition* to other supposed 'inauthentic' accounts of the Jack the Ripper story many of the guides inadvertently show how meaning is created through *difference*. On a Kray Twins tour, a guide started the walk by stating that he never worked with tourists, only visitors, in what could be interpreted as an attempt to bring the group closer together through an acknowledgement of the participants' ability to distinguish different levels of 'taste'. We are here reminded how meaning-making takes place *between* concepts, not within them. This does not just mean that Ripper *fact* requires the existence of Ripper *fiction* to become 'authentic', but that the 'true' story narrated by the tour guides also has to draw on fiction in order to evoke mental images in this interplay between media, history and site. Such incongruity is illustrated in a reply to a question from a tourist regarding the significance of the top hat often associated with the Ripper's silhouette, where a guide replied that the murderer was more likely to have worn a deerstalker hat 'like the one Sherlock Holmes wore'.

Returning to the End

Crime scene tourism, such as that of the Jack the Ripper murder walks, or the less well-known Kray Twins walks, are defined by a more complex economy than may initially be thought – where strangeness shifts and flows dependent on whether we focus on tourists, guides, or other users of the local places of the walks. This making strange of landscapes is therefore relative and relational. Groups of tourists become performers themselves with other users of the landscapes often stopping to watch them, sometimes wondering what they are doing, what they are looking at. This is especially true of the Kray Twins walks that take place in spaces even further outside the normal tourism repertoire than the increasingly gentrified spaces of the Whitechapel murder walks. But these landscapes are also media saturated and tourists and guides engage in differing strategies in engaging with media in the narratives and performances of the walks that interweave categories of home and away, real and imagined, past and present in a variety of ways.

Crime scene tourism that focuses on the spaces of the Whitechapel murders and Kray Twins walks tend to give a relatively wealthy tourist and leisure audience a heritagized view (in a sense of heritage as a practical *use* of the past in the present for particular purposes) of the social conditions, poverty, and gruesome murders of a lived time-space beyond these travellers experiences or imaginations. The walks also allow tourists to experience areas made to appear as if off the beaten track, everyday spaces for many people, rendered strange to a tourist audience. This can be seen in the ways many tourists peer in to houses, shops, look at run-down buildings as if they were exotic signifiers of a remote exotic or archaic culture. In this sense such tourist walks – focused as they are in generally working class, poor areas of the city (though with some jarring pockets of gentrification) – have an underlying estrangement in terms of social class voyeurism that is some distance away from nineteenth century 'slumming' but which still has undertones of those journeys.

There is an unmistakably gendered moral dimension to these walks. Indeed, after undertaking many walks, the Whitechapel murder walks seem peculiar to us as researchers in that many tourists (and some more actorly guides) do not appear to have much of a sense that they are returning to the scene of crimes of serially murdered women – seemingly because the 'iconic' memorialization of the murderer seems to have taken precedence over other possible uses of this past in the present.

Moreover, in our investigations of the internal narratives of the tours we have found these narratives difficult to place in terms of those strangely (in)separable notions of times, epochs, or experiences that so generally structure ideas of past and present. On the one hand, the narrative practices of the walks seem to adhere to the logic of the *modern*, and Jack the Ripper has indeed been called a thoroughly modern criminal, with the serial nature of his killings echoing that of a machine. However, to the extent that we may choose to see the tours dedicated to the murderer as an unsavoury 'monument' to his 'legacy', we should also note, paraphrasing Lewis Mumford, that no monument can be modern (see Young 2000, 94). Yet the modern, with its connections to notions of renewal and change, is also dependent on a continuous unveiling of its own past, whilst simultaneously fixing the past *as* past. On the other hand, the allegorical impulse, a characteristically *postmodern* drive according to Craig Owens (1984), permeates the story-telling on these tours, with the stories themselves turning the city into a palimpsest of sorts, with one set of stories overlaying another. Yet, there may also be something thoroughly *not* postmodern about the narratives of the Ripper walks, for example in the ways they ignore issues of gender politics by fetishizing and trivializing male violence towards women. And then again might such fatal attractions lean toward the *anti-modern* precisely because of their attempts to keep the past open? This is not simply a question of choosing between historical positions but an acknowledgment of the way the borders between them remain porous. Modern, postmodern, and anti-modern are approached here more as doings or modes and less as *isms*, periods, or concepts, turning the question from a 'what' to a 'how' and substituting a chronological understanding of historical attitudes for an investigation of the spatial performative playing out of the relations between pasts and presents.

References

Ashworth, G. and Graham, B. (eds) 2005. *Senses of Place: Senses of Time.* Aldershot: Ashgate.

Austin, J.L. 1962. *How to Do Things With Words.* Oxford: Oxford University Press.

Baker, B. 2005. 'Secret City: Psychogeography and the End of London', in Kerr, J. and Gibson, A. (eds) *London: From Punk to Blair.* London: Reaktion Press.

Barker, J. 2006. 'Reader Flattery – Iain Sinclair and the Colonisation of East London', *Meta Mute.* http://www.metamute.org/?q=en/reader-flattery.

Becker, K. 2005. 'Most (Un)wanted: Gangster Tourism in Chicago', in Picard, D. and Robinson M. (eds) *Tourism and Performance: Scripts, Stages, and Stories.* Sheffield Hallam University: Centre for Tourism and Cultural Change (CD Rom).

Charlesworth, A. 2003. 'Landscapes of the Holocaust: Schindler, Authentic History and the Lie of the Landscape', in Robertson, I. and Richards, P. (eds) *Studying Cultural Landscapes.* London: Arnold.

Clover, C. 1987. 'Her Body, Himself: Gender in the Slasher Film', *Representations*, 20, 205–28.

Cornwell, P. 2002. *Portrait of a Killer: Jack the Ripper, Case Closed.* New York: Putnam.

Coville, G. and Lucanio, P. 1999. *Jack the Ripper: His Life and Crimes in Popular Entertainment.* London: MacFarland and Company.

Haldrup, M. and Larsen, J. 2006. 'Material Cultures of Tourism', *Leisure Studies*, 25:3, 275–89.

Hebdige, D. 1974. *The Kray Twins: A Study of a System of Closure.* Birmingham: Centre for Contemporary Cultural Studies Occasional Paper 21.

Huyssen, A. 2003. *Present Pasts: Urban Palimpsests and the Politics of Memory.* Stanford, CA: Stanford University Press.

Jansson, A. 2007. 'A Sense of Tourism: New Media and the Dialectic of Encapsulation/Decapsulation', *Tourist Studies*, 7:1, 5–24.

Jenks, C. and Lorentzen, J. 1997. 'The Kray Fascination', *Theory, Culture and Society*, 14:3, 87–107.

Koven, S. 2004. *Slumming: Sexual and Social Politics in Victorian London.* Princeton, NJ: Princeton University Press.

Lennon, J. and Foley, M. 2000. *Dark Tourism.* London: Continuum.

Lippard, L. 1999. *On the Beaten Track: Tourism, Art and Place.* New York: New Press.

Margry, P. and Sanchez-Carretero, C. 2007. 'Memorializing Traumatic Death', *Anthropology Today*, 23:3, 2–3.

Mayerfeld Bell, M. 1997. 'The Ghosts of Place', *Theory and Society* 26, 813–36.

Mighall, R. 2003. 'Crime and Memory in the Capital', in Kerr, J. and Gibson, A. (eds) *London: From Punk to Blair.* London: Reaktion Press.

Miles, W. 2002. 'Auschwitz: Museum Interpretation and Darker Tourism', *Annals of Tourism Research*, 29:4, 1175–8.

Owens, C. 1984. 'The Allegorical Impulse: Towards a Theory of Postmodernism', in Wallis, B. (ed.) *Art After Modernism.* New York: Godine.

Pile, S. 2002. 'Spectral Cities: Where the Repressed Returns and Other Stories', in Hillier, J. and Rooksby, E. (eds) *Habitus: A Sense of Place.* Aldershot: Ashgate.

Pine, J. and Gilmore, J. 1999. *The Experience Economy.* Boston, MA: Harvard Business School Press.

Ratnam, N. 2005. 'This Is I', in Butt, G. (ed.) *After Criticism: New Responses to Art and Performance.* Oxford: Blackwell.

Rojek, C. 1993. *Ways of Escape.* Basingstoke: Macmillan.

Santino, J. (ed.) 2006. *Spontaneous Shrines and the Public Memorialization of Death.* New York: Palgrave Macmillan.

Seaton, V. 1998. 'War and Thanatourism: Waterloo 1815–1914', *Annals of Tourism Research*, 26:1, 130–58.

Seltzer, M. 1995. 'Serial Killers (II): The Pathological Public Sphere', *Critical Inquiry*, 22, 122–49.

Soja, E. 1996. *Thirdspace: Journeys to Los Angeles and Other Real and Imagined Places.* Oxford: Blackwell.

Tickner, L. 2007. 'Walter Sickert and the Camden Town Murder', in Wright, B. (ed.) *Walter Sickert: The Camden Town Nudes.* London: Courtauld Gallery.

Tumarkin, M. 2005. *Traumascapes: The Power and Fate of Places Transformed by Tragedy.* Melbourne: Melbourne University Press.

Urry, J. 1990. *The Tourist Gaze.* London: Sage.

Vidler, A. 2001. *Warped Space: Art, Architecture, and Anxiety in Modern Culture.* Cambridge, MA: MIT Press.

Walkowitz, J. 1992. *City of Dreadful Delight: Narratives of Sexual Danger in Victorian London.* London: Virago Press.

Williams, P. 2007. *Memorial Museums: The Global Rush to Commemorate Atrocities.* Oxford: Berg.

Young, J. 2000. *At Memory's Edge: After Images of the Holocaust in Contemporary Art and Architecture.* New Haven, CT: Yale University Press.

Chapter 10

The Soul of the City:
Heritage Architecture, Vandalism and the
New Bath Spa

Cynthia Imogen Hammond

Introduction

This chapter explores the strangeness of a boldly modernist building arriving in the heart of a city renowned for its historic architectural image. In October 2004 and again in January 2005, vandals attacked the Thermae Bath Spa, a state-of-the-art facility then under construction in the UNESCO World Heritage Site of Bath, England. Smashing the image of the city as reflected in the spa's glass curtain wall, the attacks made visible the strange incursion of the spa into the purportedly public space of the city. Using this violence as a point of rhetorical departure concerning the fraught urban context of Bath, this chapter examines the ways in which architecture can bring hegemonic struggles to the surface of the urban realm, creating a strange space of convergences between past and present, civic ambition and local resistance, architecture as icon and architecture as communication.

Since the 1960s the iconic and the communicative have been fundamental expressions of architectural discourse and practice. They have developed largely under the mantle of postmodernism, and frequently in opposition to one another. In its contemporary form, the iconic is often closely linked to authorship and to civic ambition (Sklair 2006), while the communicative has a strong basis in the critique of International Style modernism and the return to the surface and ornament of architecture, and to the deliberate embrace of buildings and their surfaces as signs and signifiers, often with didactic purpose (Venturi and Scott Brown 2004). This chapter charts a course between the way that the new Bath Spa operates in promotional and journalistic discourse as an icon, and the way that its design is, in fact strangely discreet. Mobilizing a stealth modernism, the Spa appears to refuse the appellation, 'icon', operating smoothly as communicative mirror of its social and cultural context. Smoothly but for critical moment(s) of rupture – the violence enacted upon and the discord surrounding the spa – in which the estranged relations between newness, heritage, power and cityscape in this particular location come sharply into focus.

At the close of 1997, at the peak of the 'heritage crusades' (Lowenthal 1998), the city of Bath, located in the southwest of England, embarked on a plan to revive its thermal spring waters, whose reputation for spiritual and healing powers has shaped the city since antiquity. The use of these waters had been suspended since 1978 when a young girl's death from meningitis was thought to have been the result of swimming in the King's Bath, a pool whose use in the city dates back to the Roman occupation of England. In the intervening two decades, bereft of its 2000-year tradition of bathing in the hot springs, Bath was for many a 'city without a soul'. With the approach of the millennium, bolstered by research that proved the city could again draw from the hot springs without danger (Kellaway 1991), the city applied for lottery funding to develop a new spa. In November 1997 the Millennium Commission awarded the city a grant of £7.78 million, 60 per cent of the total original budget, to create a spa complex to open in 2002. Architects Nicholas Grimshaw and Associates in collaboration with engineers, Ove Arup and restoration architects, Donald Insall and Associates, sought to create a modern, state-of-the-art health and recreational spa, incorporating five heritage buildings in one of the city's most celebrated Georgian (neoclassical) sites.

Almost a decade later, the opening of the new spa was five years behind schedule and £30 million over budget. Grimshaw's design brought dramatic spatial and visual changes to this corner of Bath. At the same time, the for-profit handling of the spa by a Dutch company, Thermae, destabilized the popular, local understanding of the hot springs as the sacred property and collective inheritance of the citizens of Bath. The excessive delays and expenditures of the project frayed local tempers and heated existing tensions between residents and the annual, itinerant, tourist population, whose disposable income and appetite for novelty were an important part of the new spa's conceptualization, justification and planning. Since the spa's opening on 7 August 2006 the complex has been heralded, alternatively, as a roaring success and as Bath's biggest flop.[1] The deeply divided stakeholders, the controversial design and the polarized reception to the building have brought the project repeatedly to the forefront of the local and national press. In contrast to this antagonism and discord, the Thermae Bath Spa complex itself is, architecturally, strangely unobtrusive.

Glass, Glance, and Gaze

As a modernist glass envelope, the primary visual impact of the spa lies in its effect as a mirror-like, reflective surface (Figure 10.1). Tucked into the interstitial space between historic buildings in an already compressed section

---

1   News headlines have included 'Regeneration accolade for spa project' (*Bath Chronicle* 2007b) and 'Worst financial fiasco in Bath's civic history' (Mapstone 2008).

Figure 10.1   New Royal Bath, Thermae Bath Spa, Bath, England, 2006. Photograph by Michael Maggs, 2007

of the city, the new (as opposed to retrofitted) sections of the Thermae Bath Spa almost disappear; they are not what one sees. In contrast to the established relationship of building to site, or figure to ground, in much 'iconic' western architecture, there is no axis, no approach to the spa at ground level that allows one an uninterrupted or complete view of the complex. Unlike other notable, glazed buildings, such as the Crystal Palace (J. Paxton 1851) or I.M. Pei's glass pyramid addition to the Louvre (1989), the Thermae Bath Spa has little in the way of negative, or framing space, which allows the visitor or the user to contemplate the architectural form as a kind of sculpture, set apart from other buildings. The new spa's design even sidesteps the conventions of more ordinary architecture, such as the privileging of the façade or entrance in order to signify a building's purpose and uniqueness. Strategically, the spa slips its operations and its architecture into and behind existing buildings, cityscapes and views; much of the new architecture is obscured by the surrounding, eighteenth-century urban fabric. Indeed, the only fully visible new-build façades – the south and east elevation of the New Royal Bath – are overwhelmed, suffused with the reflection of the historic architecture that enfolds and constitutes the spa. In other words, when gazing upon the spa, what you see is Georgian Bath. And equally, thanks to the expansive, glazed volume of the interior ground storey, when gazing out from within the spa, Georgian Bath is everywhere you look.

This circularity of effect has a strong architectural precedent in Bath; indeed, the city is famous for it. John Wood the Elder (1704–1754), Bath's best known architect, both alone and with his son, John Wood the Younger (1728–1782) produced some of England's most famous Palladian neoclassical architecture and town planning. The Queen's Square (begun 1728), the King's Circus (begun 1754) and the Royal Crescent (begun 1767) are all monumental housing schemes unified through palatial, integrated façades and a strikingly introverted geometry. Each scheme can be understood, spatially and visually, through its name: the Crescent curves; the Square has four, inward-facing elevations arranged around a central square; the Circus is a tightly controlled ring of attached Georgian houses, whose intimacy borders upon claustrophobia. These marvellous, rationalized settings – some of the best preserved architecture in Bath – proposed to their Enlightenment-era subjects not just a site to gaze upon, but a place in which the gaze would always refer back to the place of gazing. As primary Bath tourist destinations in and of themselves, these buildings suggest a preferred reading for the city as a whole. Akin to the experience of being inside the new Bath Spa, the effect of Wood's designs is most clear in the King's Circus; no matter where one stands, what one sees is Georgian domestic architecture – and people looking at that architecture.

The design of the Thermae Bath Spa communicates a certain longing for the visual and architectural interiority of Bath's beloved squares and circles. In this longing, the building reveals its actual function as a parergon, a frame that supplements and shapes the adored or venerated object. In this way, the surface of the Thermae Bath Spa performs much like an eighteenth-century ocular device called the 'Claude glass'.[2] Named for the landscape painter, Claude Lorraine (1600–1682), the convex piece of polished glass, often slightly tinted with sepia, was small enough to fit in a satchel or even a hand. The Claude glass was popular among Enlightenment-era tourists and amateur artists, who brought it with them on their travels. Standing with their backs to the sites they had come to visit, the traveller was supposed hold the Claude glass up to eye level, and see themselves reflected within the small, curved mirror. In so doing, tourists could visually place themselves in the dramatic architectural or pastoral setting of another era, or of another culture. The Claude glass served a secondary function as well: to eliminate anything from this idealized vision that might detract from the image of self in setting. Thanks to the sleek, aqua-tinted skin of the Thermae Bath Spa, visitors to Bath today have a parallel opportunity to visually immerse themselves in the image of Georgian Bath, the era of the town's celebrated heyday, the period between 1688–1830. Simply walking past the spa offers the opportunity to gaze upon oneself, framed, it would seem, by same architecture that had once been the setting for the likes

---

2   A Claude glass has some overlap with a 'Claude mirror', which was another name for the instrument of the occult known as the 'Black Mirror' (Maillet 2004).

of Jane Austen, Tobias Smollett, Thomas Gainsborough and Sarah Siddons. In this encounter the building all but disappears, and becomes secondary to the subjective, embodied experience of self and city.

This mirroring, this opportunity for self-gazing is bound up with what Rob Shields describes as the 'virtual' aspects of the personal or interior image of the urban. As Shields writes, the virtual city is formed through

> [...] idealizations but not abstractions; real but not accurate visions of an actually-existing object [...] In part, the power of the virtual derives from the mythic and memorial – urban pasts which were once actual but now reside in memory as recollections which are neither mere fantasies (abstractions), nor projections of the imagination, but virtualities which, like a daydream or déjà vu, may be experienced as more real than the concrete world of everyday life. This is not simple recollection but re- creation, sometimes via rituals of re-actualization [...] producing not only, for example, a 'tourist gaze' but prosthetic memories in which memory is welded to flesh [...] These go far beyond the empirical experience of an urban place in and of itself. (Shields 2004, 25–6)

For Shields, the 'strange quality of visualcity' lies in this nexus of memory, image, myth and the simultaneity of past and present in the city, all of which intersect with the material and political conditions of that place. In Bath, such relationships are underscored by the local history of deliberate and enthusiastic urban image-making, a practice whose impact is now international, thanks in large part to the designation of Bath as a World Heritage Site (WHS) in 1987. For UNESCO, beyond the city's substantial contribution to the history of English town planning, 'the architectural quality of Bath lies above all in the excellence of the façades and the urban and landscape spaces that they enclose [...] Historically, Bath is important, not for events of momentous importance, but as a setting for social history [and ...] the residents and visitors to this national health resort' (UNESCO 2006). As a carefully cultivated but no less passionately debated 'setting', diminutive Bath has provided ample fodder for 'visual citations' of its urban realm, whose excellent façades form a privileged surface on which to play at finding (oneself in) history. The visual evidence of Bath's Georgian surface is so saturated as to be almost cinematic; the neoclassical, pale golden stones of Bath offer the viewer what Shields calls the 'arrested mobility of the eternal moment',[3] a kind of eternal return afforded by the city in which one can return to a privileged moment in the past, again and again – as long as the will and the funds remain.

This visual effect of the new spa is underscored by the incorporation of five historic buildings into the complex: neoclassical architecture becomes not only the image reflected on the surface of the spa; it is also the frame that

---

3   Shields is here discussing London, but the parallels are substantial (25).

encapsulates and interpolates the spa. In this way, the curtain wall of several hundred, aqua-tinted Italian glass plates reiterates the city's representation of itself first and foremost as a Georgian city, ready to welcome visitors to its easily consumable slice of history. The result is not so much an 'iconic building' – a hope articulated in local, official discourse.[4] Rather, through a skilled architectural gesture of self-effacement the spa presents an iconic image of tourists, residents and clients, all set against the backdrop of Bath's famous neoclassical architecture.

In October 2004, as the furore over protracted building delays and dismally exceeded budgets began to mount, anonymous individuals smashed several of the expensive glass panels of the curtain wall. Again, less than three months later, the BBC reported the following headline: 'Vandals in £120,000 attack on spa' (BBC 2004 and 2005). *Management Today* described a scene on the building site in spring 2006: 'a group of builders in hard hats examine a shattered pane of the sleek tinted glass on the exterior of the building' (Garrett 2006). A year and a half later, a *Daily Mail* article reported that 'vandalism and subsidence' were the reasons given for the replacement of all 274 windows, for a total cost of £700,000 (Fryer 2006; Weaver 2005). In addition to leaky pipes, cracked foundations, peeling paint in the spa pools, ongoing legal struggles and mistakes in drilling the crucial boreholes (Sharp 2006; Garrett 2006; Weaver 2005), the faulty and often vandalized curtain wall constituted a major blow to the vision of the city council that approved the initial plans. It is significant to note that in shattering the panels, vandals of course destroyed – temporarily but repeatedly – the pristine mirror image of the city as reflected in the tinted glass. What follows seeks to explore how this destruction brought into relief the strange incursion of the purportedly public spa into the very core of the heritage zone, and thus, the defamiliarization of the Georgian city of Bath.

The Spa

The ghost of John Wood the Elder haunts the Thermae Bath Spa project beyond the return to the visual reflexivity of his designs. The enfolding of architecture, discourses of healing and notions of the sacred that is present in Grimshaw's new spa also has a crucial antecedent in the time and work of Wood. The elder Wood's hospital projects, St John's Hospital (1728) and the Royal Mineral Water Hospital (begun 1738) as well as an unbuilt health centre

---

4   Bryony Jones (2006) quotes city councilor, Nicole O'Flaherty, Executive Member for Tourism, Leisure and Culture as saying, 'In the middle of a World Heritage site … we now have an unashamedly modern building, an iconic building'. See also the Bath and Northeast BANES press release, 2006.

were where the architect's dreams of a sacred city of healing and culture came closest, perhaps, to fruition. As early as 1727 Wood proposed,

> [...] larger and renovated baths, still open to the air but shielded from both wind and prying eyes [...] the whole complex, together with toilet facilities, the areas for the gentle exercise so often urged by Bath physicians [...] would comprise the Imperial Gymnasium, the *heart of Bath* [...] (Mowl and Earnshaw 1988, 141, emphasis mine)

Despite decades of effort, Wood was never able to combine this idea with his preferred site: 'one near to but slightly below the outpouring of the mineral springs' (ibid., 141). But Wood's fundamental concepts, as well as his ideal site, are to be found in the design, programme and situation of the Thermae Bath Spa today. The Thermae Bath Spa is located less than five minutes' walk from the centrally located Abbey courtyard, flanked by the High Gothic Bath Abbey (1500) and the neoclassical Great Pump Room (Thomas Baldwin 1795), which is also the location of the remains of the Roman baths that give the city its name. Hot springs rise up through three sources in close proximity, the King's Spring, which sources the Roman baths (the Roman hydro-engineering still works today), the Cross Spring and the Hetling Spring. The latter two springs are the site of the Thermae Bath Spa, to which specially-drilled boreholes bring an uncontaminated portion of the 1.3 million litres that course underground every day. The cool blue glass of the new spa ascends through the surrounding stone cityscape in an architectural gesture that evokes the passage of the spring water on its 10,000-year journey; what fell once as rain on the Mendip Hills, 25 km from Bath, now rises as geothermally heated water through the limestone faults of the Bath region (BANES 2002).

This geology appears to have inspired the new spa complex's most obvious intervention in the urban landscape. The New Royal Bath is a cubic glass volume that rises up behind existing Georgian architecture, providing views from its rooftop pool to the surrounding city. This bath is where the bulk of the programme of the new spa takes place, and where the mirror-like qualities of the project are most evident. The architects describe their design thus,

> An elaborate architectural lighting scheme ensures that this interplay of light, form and reflection continues into the evening and through the nighttime hours [...] The building's outer envelope [the glass curtain wall] follows the line of Bath Street. Within this envelope, the spa facilities (pools, gymnasiums and massage suites) are contained in a freestanding stone cube on slender columns. The dimensions of this cube relate directly to the plan of the adjacent Hot Bath building ... (Grimshaw Architects 2006)[5]

---

5   The New Royal Bath, located at the juncture of Bilbury Lane and Beau Street, replaces an earlier swimming pool, the Beau Street Baths. Built in 1923, well outside

The Hot Bath (J. Wood the Younger 1777), once known as the Old Royal Baths, is a Grade II listed heritage building and is the second-largest part of the complex. It is situated immediately adjacent to the New Royal Bath to the west, connected internally to the New Royal Bath via a private, underground access. This simple, neoclassical building once had an 'ingenious arrangement of dressing-rooms and small private plunge-baths, symmetrically disposed around the octagonal central bath' (Ison 1980, 37). The values of symmetry, austerity and order find an echo of sorts in the new Thermae complex, with the reiteration of the Hot Bath's dimensions in the central, stone core of the New Royal Bath. This, perhaps, is where the use of the past as a design inspiration ends in the new project. Today, the Hot Bath is programmed within the larger complex for private spa beauty and rejuvenation treatments and packages. These treatments, and therefore access to the refurbished Hot Bath, are purchased for additional fees. Thus, the New Royal Bath, which offers scented steam rooms, one rooftop and one street-level pool, is the least expensive and arguably most public part of the new spa. Residents are entitled to a discounted entry fee to the New Royal Bath (approximately 20 per cent) and to the Cross Bath (approximately 50 per cent) (Thermae Bath Spa, n.d.).

Three former houses, 7, 7a, and 8 Bath Street, are now connected to the spa through purpose-built, semi-glazed bridges or walkways. These former houses have been folded into the new complex while retaining for the most part their Georgian façades. The architects removed two Georgian ground-floor shop fronts, and glazed the entire first storey to create a modern entrance way for the new spa. The frameless glass panels on this elevation are the same aqua blue as those used on the New Royal Bath. A gift shop, a café, offices and meeting rooms now occupy the interior spaces of the former houses. It is where these houses literally touch the new-build aspects of the project that the design challenges are most apparent, and where the palimpsest-like qualities of the new spa are most evocative (Figure 10.2). The architects write that the listed building parameters led to 'a direct and interplay between new and old.' In order to resolve height differences between existing buildings, the architects used 'split level planning and the clear articulation of the connecting spaces as transparent bridge links which open up vistas along, across and through the complex' (Grimshaw Architects 2006).

Not only do these bridges offer encounters with the juxtaposition of historic and modern architecture, they also emphasize the importance of the exterior views, again, to the surrounding heritage cityscape, as do the many portholes, and the substantial glazing.

---

the Georgian era, these baths seem generally to be considered to have only 'minor' architectural importance (Ison 2). The lack of fanfare accompanying their destruction – as compared to the considerable public resentment towards the removal of two Georgian shopfronts – points to the relative value assigned to neoclassical buildings in Bath.

Figure 10.2   Detail of bridge walkways between historic fabric and New Royal Bath, Thermae Bath Spa. Photograph by the author, 2007

Two additional buildings complete the new spa complex: the Hetling Pump Room (once a place to 'take' or drink the spring waters) is now an interpretative centre for the spa, located to the west over a narrow cobblestone lane. Apart from some careful interventions via lighting design, the Hetling Pump Room hardly betrays its refurbishment. Finally, the Cross Bath (J. Palmers 1793) is the jewel in the crown of the Thermae Bath Spa (Figure 10.3).

Figure 10.3   Cross Bath, Thermae Bath Spa, east front. Photograph by Thomas Strickland, 2007

It is a Grade I listed building, whose redesign interprets the historic façade and footprint with a new, partially open roof plate and bathing pool. Located several meters from the main entrance of the Thermae Bath Spa, the small and delicate Cross Bath sits alone in a small courtyard, dwarfed by surrounding buildings. The restored serpentine façade, blind recesses and Corinthian pilasters, which can be viewed from all elevations, communicate the considerable appeal of the Georgian era. For the Georgian Group, a conservation organization that awarded the Cross Bath top prize for restoration in their 2006 competition, the merits of the restoration do not erase the memory of 'the grim post-war years during which Bath was sacked by philistines'; they are, however, 'an encouraging indication that sanity has returned [to Bath]' (The Georgian Group 2006).

The Soul of the City

In his work on tourist writing, Derek Gregory (1999) uses the term, 'scripting' to describe the complex ways in which cities and sites come into discursive existence as imaginary geographies. At a given site, the interrelation of local self regard, views of that place from abroad (and by outsiders) and the texts produced, result in mediated, material consequences for that place, be they organizational, architectural, spatial or representational. For Gregory, scripting is a metaphor that 'accentuates the production of space [...] [foregrounding] the performative [...] and [bringing] practices into view' (ibid., 116), Scripting is a useful notion for considering the ways that written accounts of Bath, and of the new Spa have helped to bring the project not only into representation, but also into the often conflicting concepts of local identity, national conservation mandates, and international heritage. The discrepancies as well as the consistencies of the 'script' – as found in journalistic accounts, academic writing, informal online discussion groups and promotional materials – are crucial to understanding how it is the spa came to be, given the fierce local antipathy to architectural modernism.

From its earliest conception, the project was wrapped in the language of healing, revival, the true nature of Bath (bathing) and, most significantly, the restoration of the soul of the city via the spa. News articles published during the six months prior to the official opening of 7 August 2006 echoed this idea. 'For almost 30 years, Bath has been a city without a soul,' proclaimed one article, concluding in the words of a local councilwoman, '[the new spa] is simply perfect for Bath – it reconnects the city with its soul' (Jones 2006). The 'rejuvenating benefits' of spring bathing for visitors was explained in a news release on the promotional *Visit Britain* website: 'Continuing the Roman belief that the natural springs offer medical assistance to the ill, Thermae Bath Spa will provide visitors with physical, emotional and spiritual healing' (2006). The journal, *Management Today* provided a cautionary note. 'Bath may have restored its soul, but at considerable cost to its pride and its coffers' (Garrett 2006). In layman's terms, Vincent Crump (2006) concluded, 'Bath just wasn't Bath without a proper bath'. Indeed, this sort of observation is central to the history of bathing in Bath, whereby the waters become intimately related to the city's physical and cultural development, its sense of the past and its hopes for the future.

Substantial archaeological work at the site now known as Bath has revealed that cultural development attached to the sulphurous springs dates back to Celtic times, if not earlier. The springs, naturally heated to 45 degrees Celsius, as well as the Avon River seem to have been the reasons that this location was originally inhabited. What may be traces of sacred buildings dedicated to a Celtic water goddess, Sulis, have been found beneath the archaeological remains of the Roman bathing complex. The Romans developed their extensive baths according to custom on the same site, incorporating local rites and religious observances in the first century CE (Cunliffe 2000). Thus Sulis became 'Minerva' and the town 'Aquae Sulis' (the waters of Sulis) was created.

Falling into decline after the departure of the Romans in the fifth century CE, Bath gradually settled into a new identity as a small, wool-producing town during the medieval era. Much of the Roman bathing complex was either built over or dismantled in order to put the land and building materials to new use. Despite the almost complete disappearance of the built traces of the former Roman occupation during this time, the waters that fed the Roman cauldarium, tepidarium and frigidarium continued to flow beneath the surface of the city, sourcing the public baths constructed in the late medieval and Elizabethan eras. By the early eighteenth century, the Roman past of Bath was generally acknowledged, but was considered a curiosity or romantic flourish of the town's history (Cunliffe 2000). This attitude would begin to change dramatically throughout the eighteenth and into the nineteenth century. The years 1688–1830[6] marked the period in Western Europe that seized upon antiquity in a completely novel way. The 'long eighteenth century' (O'Gorman 1998) sees the emergence of the modern disciplines of archaeology, art history, and 'scientific' architectural practices, which based their aesthetics upon the careful, if not always accurate, study of historical precedents. It is this period that sees the revival of Bath as a town of great repute and the importance of its waters enthusiastically reaffirmed (Neale 1981).

It is during this period that the architecture of Bath begins to be mobilized as the culturally significant frame for the naturally significant hot springs. It is also the moment when the fiscal possibilities of both become harnessed, mutually, to a simultaneous expression of the past and the future. Through key figures and popular accounts, Roman history, Celtic mythology and Arthurian legend collided with both the age of Enlightenment and the speculative building market, which by its very nature has an eye to the future. The Georgian era is also the time when the notion of Bath as a place of healing or spiritual sustenance becomes firmly entrenched, not least through the effusive writings of John Wood the Elder, who saw in the city the possibility for a new Rome (Wood 1765/1969). For Wood, Bath's healing waters were at the symbolic as well as the physical centre of the city. Significantly, in Wood's buildings and writings, architecture is a hierophany: a means through which the sacred reveals itself.

As interest in Bath's antiquity grew, its popularity as a spa faded. Bath's appeal to the elite as a temporary stop on their annual migration from London waned as more upper-middle- and middle-class residents stayed year

---

6   The dates also reflect the first significant visit of a member of the British royal family to Bath, Queen (then Princess) Anne in 1688, an event that prompted a fashion for spring-water cures. 1830 is the date agreed upon to mark the end of the Georgian period in architecture in England (Ison 1980).

round (as opposed to living in Bath only during the social 'season') (Davis and Bonsall 1996). By 1850, sea bathing had replaced hot springs as the health treatment of choice for the elite, making the hot spring spas seem crowded and unsanitary by comparison (Corbin 1994). Bath's predominantly female population during the Victorian era contributed to the perception that this was a dull city of 'old maids', where women who had passed their bloom came to live out their lives in gentile poverty. Ironically, it was also the Victorian era that saw increased pressure on those aspects of the city perceived to be morally unstable, such as prostitution and itinerancy (Davis 1990). As for the architecture, the stone with which Bath was built in the eighteenth century could not withstand the ravages of an increasingly industrial age; the easy-to-cut porous limestone stained quickly, looking bruised and dirty within a few generations. According to photographic documentation, by the end of World War II, the houses of Bath were almost completely black (De'Ath 1998). The combination of coal dust and acid rain wrecked havoc on much of the neoclassical detailing of the city's architecture. By the 1960s, in contrast to the wave of modern building and 'New Town' planning that spread over England, the architecture of Bath seemed, to some, fusty, ruinous and impossible to maintain. Seeking to modernize, sections of the city were razed, particularly the working-class districts to the south and west of the historic Georgian core and baths. A modern bus station, mock-Georgian pedestrian shopping mall and technical college sprang up in their wake.

Despite these attempts at architectural modernization and adaptation, Bath struggled to maintain pace with modern clinical methods. The National Health Service, which had subsidized hydrotherapeutic treatments in the Cross and Hot Baths, suspended their funding in 1976, believing the springs to have no more effect than tap water. By 1978, after the death from meningitis noted above, Bath's reputation as hopelessly old-fashioned seemed incontrovertible. The hot spring spas – still well used and beloved at the time – were closed within the year, and remained locked to visitors and locals for over two decades. In 1991, the Cross Bath became the headquarters of the Springs Foundation, a group dedicated to preserving the Celtic and sacred aspects of the hot springs of Bath, while the Beau Street Baths (now the New Royal Bath) and the Hot Bath sat boarded up nearby. The Roman Baths could still be visited, but not used for bathing. With these changes and closures – brought on by modernism and modernization – Bath's image as a place of healing and spiritual sustenance faltered, and folded in on itself.

Collective Memory, Collective Amnesia

[M]useums and heritage centres … seem to have as their aim the cultivation of nostalgia, the production of sanitized collective memories, the nurturing

of uncritical aesthetic experiences, and the absorption of future possibilities
into a non-conflictual arena that is eternally present. (Harvey 2000, 168)

The ten-year project of the Thermae Bath Spa, its incidences of vandalism,
its reception in local and national press, and its unique, stealth approach
to modernism offer powerful points of contemplation for the strangeness
(perhaps even the violence) of how new meets old in a place that is obsessed
with a certain idea of the past. While the conjoining of historic buildings
with modern or contemporary architectural solutions was the primary design
strategy of much postmodern architecture and therefore not unique, what
makes this subject of pressing concern is the larger context of the Thermae
Bath Spa and the 'World Heritage City' of Bath. The 'heritagization' of
urban centres and the 'commodification of pastness' (McIntosh and Prentice
1999, 590) in cities are growing phenomena that bear critical examination,
particularly under the present circumstances of dramatic urban growth and
economic disparity between citizens, both at a time when 'urban regions
have become, more than ever before, landscapes of power' (Swingedouw and
Kaïka 2003, 12). In Bath, the landscape of power is undeniably a landscape of
heritage, which has been inextricably linked to the scripting of Bath as a city
of pastness, a city of culture, and a city of healing.

The closing of the baths in 1978 took place in the context of a new phase
in Bath's history: destruction. As described above, since the end of World
War II, the city had been attempting to modernize, in part by destroying what
was considered 'minor' Georgian streetscapes and buildings. Residents began
to galvanize around these shared losses. As in other cities worldwide, the
elimination of historic architecture was the impetus behind the drive to protect
and preserve architectural heritage (Urry 2005). The Bath Preservation Trust,
established in 1934, stepped up its activities in the late 1960s and 1970s, acting
as a watchdog and advocate for the Georgian heritage of Bath. A reflection of
and instigator to further preservation efforts was Adam Fergusson's *The Sack
of Bath: A Record and an Indictment* (1973), which detailed the devastation
of 'minor' Georgian architecture as a result of post-WWII modern town
planning and development. Significantly, modernism – in so far as it affected
the Georgian fabric – was widely considered to be a form of 'vandalism'
against the historic city (Ison 1980, xi; Fergusson 1973). The result of these
and other efforts to prioritize the older architecture of Bath was that by
the beginning of the 1970s and increasingly until today, 'the eighteenth-
century fabric achieved an architectural ascendancy unparalleled since the
period of its creation' (Borsay 2000, 184). In large part through the efforts
of its residents, Bath's twentieth-century identity began to form, as a site of
grassroots, activist preservationism, bolstered by the re-imagining of the city
as a beautiful treasure to be protected from the ravages of modernism.

The slow and complex struggle to recognize the historic importance of Bath
had a major victory in 1987 when it was named a World Heritage Site. This

designation added authority and weight to an increasingly powerful movement, in Bath and beyond, to support historic preservation in architecturally significant cities. Capitalizing on the WHS designation, Bath experienced a major urban renaissance in the 1990s. With some major exceptions such as the Roman baths and Bath Abbey, programmes of restoration in the late 1980s and throughout the 1990s addressed primarily the façades of the cityscape (Borsay 2000, 194).

These programmes, combined with a growing culture of architectural tourism, have had uneven effects in the city. After expensive cleaning and controversial sandblasting, the architecture once again gleams pale gold in the sunlight. But increased traffic and tourist demand for resources and space put heavy pressure on the small city. Although the heritage programme of Bath is becoming more inclusive, the architecture of other eras, such as the late Victorian period, has not received the same attention as that of the long eighteenth century (Jackson 1991). The precious inheritance that is Bath's architecture must peddle itself in order to keep the town afloat. A 2006 BANES press release paints a clear picture: 'Tourism generates £400 million per year in the local economy [...] [this] industry accounts for 8,000 jobs in Bath and North East Somerset and it is the responsibility of the [Bath City] Council to protect employment rates.' As a World Heritage Site, Bath today has international renown. An effect of this fame, however, is the power the conservation agenda holds in Bath, resulting for some scholars in the 'control and manipulation of effective public opinion' (Borsay 2000, 195). What this means, bluntly, is that the activist spirit that once had saved Bath from the wrecking ball has been assimilated into the landscape of power, supporting the city's 'nurturing of uncritical aesthetic experiences' (Harvey 2000, 168).

Bath is, nevertheless, a place of ambivalence. Despite the removal of telephone wires, shutters, Victorian window-boxes and anything else that is considered out of 'character' for a properly Georgian place, the city breathes discontent. As soon as one steps away from the pedestrian walkways and sandblasted façades, issues of homelessness and urban poverty become all too apparent. In addition to these very real contemporary issues, to what extent do the forgotten and subsumed histories of this place remain in its shadows? Local historian Trevor Fawcett has suggested that profits of the slave trade funded at least a portion of Bath's prized neoclassical architecture (Fawcett 1993). One of Bath's most venerable residents, Lady Selina Hastings, the Countess of Huntingdon (1704–1791) owned slaves in Georgia. Their labour contributed to her financial empire, which she used to build Methodist chapels through England in the mid-eighteenth century, including one in Bath (Hammond 2006).[7] The imbrication of architecture in such historical realities plays a minor role, if any role at all, in the presentation of Bath's heritage

---

7   Hastings' gothic revival chapel (1765) now houses the Building of Bath Museum, dedicated to the history of neoclassical architecture in Bath.

in most tourist sites and experiences in the city. But the architecture itself remains heavy despite its brightness; it carries these histories although they remain just out of sight, caught only occasionally in the corner of one's eye, in the accident of a glance, or discovered only in the determined searches of an historian.

The Thermae Bath Spa is a major building project that is unequivocally centred in the city, spatially, economically, and discursively. What can it reveal about existing tensions and historical continuities in the city? To what extent might it be said to mediate the struggle in Bath to be both modern and eternally, perfectly historic? How might this mediation be strange, or what strangeness was it that brought 'vandals' to act, not once but four times upon its glass skin? Because of the tradition of preservationist activism in Bath, local resistance to any architecture that signifies modernity and futurity operates in Bath as a kind of counterculture of sorts, a renegade yet conservative stance against the forces of change. Thus residents who opposed a 'great glass lump' (*Bath Chronicle* 2007a) being built in the heart of the Georgian city see themselves as defenders of the 'Bathness of Bath' (Marks 2007). Vocal, involved and effective, this constituency successfully mobilizes the architectural culture of the city not only to resist change, but arguably, also to maintain an image of the city that fulfils a strong local sense of civic identity. 'There may be a case for exciting new buildings in other, run down outskirts of the city', writes one resident, 'but Bath is not Bilbao' (Hyett 2007). What is important in such observations is not only the obvious wish to keep Bath as it is but also the way in which what was once the parergon for Bath's waters – architecture – has now become the prized ergon, that which must remain pure, as art, as history, as the trace of genius.[8]

Of course, not all residents and critics wish to keep Bath in the past, architecturally or otherwise. For design critic, Stephen Bayley (2007), the preservationists in Bath are a group of 'nimbys' and 'petulant lowbrows', whose crusade it is to protect 'a brainless, airless, lifeless pastiche of bogus classicism'. Relatedly, debates within the community about what the city should look like do often centre upon oppositions of past and present, and the staged versus the 'real'. The compressed urban conditions of Bath are such that major projects like the Thermae Bath Spa bring these debates to the surface. The recently rejected proposal by architect Eric Parry to extend the Holburne Museum is another prime example of a design with strongly modernist tendencies becoming the fulcrum upon which Bath's innate identity is rehearsed, debated and solidified. Bath City Council rejected Parry's design in summer 2007, with waves of public support for the decision and outrage

---

8 The conflation of the city in toto with John Wood the Elder is common. Although many architects and builders helped to create Bath, the most recent publication available on the UNESCO website mentions only John Wood the Elder, and credits the success of the city to his 'genius' (2006).

against yet another inappropriate design proposal following the verdict. The loss of the Holburne extension could mean a permanent closure for the institution, which has been housed in a late-eighteenth century hotel since 1916 (Bayley 2007). As such, the rejected design has underscored the fact that in Bath, the majority would rather lose a beloved museum than see a modern extension keep it alive. But the anticipated closure of the Holburne has allowed counter positions to surface, with the new Spa project lurking between the lines of online discussion groups. Christopher Frayling, Chairman of the Arts Council in Bath wrote in 2007, 'I love Bath, but that doesn't mean you can't have faith in your contemporaries to create buildings of quality that have a future.' Citing the city's history of architectural and planning innovation, another writer asks, 'why now, at the beginning of the 21st century, are we contemplating turning Bath into a museum?' (Anon 2007), while the following anecdote expresses an important facet of the local consequences of dealing daily in the past:

> People in Bath are quite protective about the buildings ... It just makes things difficult for everyone ... I recently heard an American tourist ask a passer-by whether or not Bath closed in the winter. They assumed it was some Georgian theme park, which isn't the impression that Bath people would want to give of this city. (Unal 2007)

This nuancing of the hegemonic scripting of Bath helps to explain how Bath has come to be a contested example of the 'city-as-museum' while demonstrating how residents themselves contribute to and debate that phenomenon, through their affective relationship with the architectural heritage in question. This context brings us back to significance of the temporarily devastated surfaces of the new Bath Spa. Although the actions remain anonymous and branded as 'vandalism', they must also be understood as the thin end of the wedge of discontent and resentment in a city which promises the past but delivers instead the constant mediation of an idea of the past. The spa, and with it, the vandalism brought crucial, local struggles into representation in a temporary but profound defamiliarization, or making-strange, of the image and the space that is Bath.

Conclusion

A theory of architecture emerged in the eighteenth century, known as *architecture parlante*. 'Speaking architecture' was thought to have 'reformative qualities capable of reshaping society. The outer surfaces of the building now conveyed its functions more literally, making the totality of its surfaces and profiles a comprehensive site of representation' (Leatherbarrow and Mostafavi 2002, 10). The buildings of Bath are, many of them, speaking buildings, holding their desired reading out in front of them in the form of their faces, their façades. These rational, symmetrical, perhaps even noble buildings, climbing up and spilling down Bath's seven hills were intended to be containers for the best society possible. That ideal society never occurred, but the buildings have persevered. It is no surprise that many citizens of Bath long for architectural purity. But the past is not one thing only; the visual calm of Bath belies the bitter struggles over the city's image, its spaces, and its identity. Peter Borsay (2002, 9) writes, '[f]rom its very beginning the Georgian city [of Bath] generated images built around a matrix of issues; health, environment, society, consumption, morality and order'. As the twentieth century has progressed, however, these images have come into contact with contemporary contradictions and ironies, perhaps the greatest being that the culture the preservationists would wish to protect is the same culture that the tourist industry would consume and, inevitably, transform.

Against this matrix of issues, images and contradictions, the designers of the Thermae Bath Spa were in an impossible situation. They had to create a building that would be unapologetically modern, that would create space for future possibilities, but that would not disrupt the local conviction (if not consensus) that Bath should not change. The solution was remarkable. The new spa manages to be a modernist building challenging the historic cityscape with industrial aesthetics and building materials at the very same time that it acts a literal and figurative mirror, held up to and appearing to sublimate itself to the placid face of the city. An enormous Claude glass – who was it for? What undesirable elements did it exclude? A glass building in a town of stone is unbearably strange when the former has come to signify vandalism of the most criminal variety. Framed in and reflecting Georgian architecture, the new spa did its job too well. It mediated the relationship between tourists, their destination, and the residents who also did their job too well, in fighting for a picture-perfect, eighteenth-century city. The mirrored walls of the spa revealed the superficiality of the image of Bath, and the subtle but relentless excision of the complexity of the city. The perfect surfaces of the spa, the perfect images of the city that they produced, were not what they citizens wanted after all. They were bound to be broken.

Two hundred years ago, Bath was one of the most important hydropathic centres in Europe. Today, it is one of Europe's most thriving heritage destinations, and has its sights set on being the newest and best spa destination in the European Union. The reopened spa waters needed a building of great architectural significance, whose appearance would appeal to the cosmopolitan traveller. But at first glance, the Thermae Bath Spa is not seen at all. The glance, which reveals the strangeness of the building, becomes the gaze, refocused not on the sleek lines and surfaces of the new spa, but on the sun baked neoclassicism that surrounds and is reflected in the mirror-like planes and curves of the spa's exterior. Set into this majestic, shimmering image of history, one also finds oneself, and one's camera. The picture, then taken, contains all that is unarguably lovely, all that is obscured, all that is strange about the incursion of a modernist building in the heart of heritage Bath. It encompasses larger issues of space, control, and the inevitable ruptures between the image of a given place and its manifold realities.

References

Anon ('Angry'). 2007. Response of 14 September to 'Was Bath right to reject the Holburne Museum scheme?', *Building Design* (published online 3 August 2007), <http://www.bdonline.co.uk/comments.asp?storycode=3092724>, accessed 20 August 2007.

BANES (Bath and Northeast Somerset Council). 2002. *The Hot Springs of Bath: Geology, Geochemistry, Geophysics*. Bath: Bath and Northeast Somerset Council.

BANES (Bath and Northeast Somerset Council). 2006. 'Bath and North East Somerset Council welcomes Spa opening' (press release), Bath and Northeast Somerset Council (website) (published online 3 August 2006), <http://www. bathnes.gov.uk/bathnes/media/press+releases/2006/Council/Bath+and+ north+east+somerset+wecomes+spa+opening.htm>, accessed 20 March 2008.

Bath Chronicle, The. 2007a. 'Curse of Bath strikes again?' (published online 11 June 2007), accessed via Highbeam Research, <http://www.highbeam. com>, 20 March 2008.

Bath Chronicle, The. 2007b. 'Regeneration accolade for spa project' (published online 9 June 2007), accessed via Highbeam Research, <http://www.high beam.com>, 20 March 2008.

Bayley, S. 2007. 'Is Bath Britain's most backward city?', *The Guardian* (published online 16 September 2007), <http://arts.guardian.co.uk/art/architecture/ story/0,,2170069,00.html>, accessed 20 March 2008.

BBC News. 2005. 'Vandals in £120,000 attack on spa' (published online 7 January 2005), <http://news.bbc.co.uk/2/hi/uk_news/england/somerset/4155221.stm> accessed 29 January 2007.

BBC News. 2004. 'Vandals hit troubled spa project' (published online 12 October 2004), <http://news.bbc.co.uk/2/hi/uk_news/england/somerset/3737110.stm>, accessed 29 January 2007.

Borsay, P. 2000. *The Image of Georgian Bath, 1700–2000: Towns, Heritage, and History*. Oxford: Oxford University Press.

Corbin, A. 1994. *The Lure of the Sea: The Discovery of the Seaside in the Western World, 1750–1840*. Berkeley and Los Angeles: University of California Press.

Crump, V. 2006. 'At last, Bath gets its spa', *Sunday Times* (published online 30 July 2006), <http://travel.timesonline.co.uk/tol/life_and_style/travel/desti nations/england/article693963.ece>, accessed 20 March 2008.

Cunliffe, B.W. 2000. *Roman Bath Discovered*, 3rd edn. Stroud: Tempus.

Davis, G. 1990. 'Beyond the Georgian Façade: The Avon Street District of Bath', in Gaskell, M. (ed.) *Slums*. Leicester: Leicester University Press.

Davis, G. and Bonsall, P. 1996. *Bath: A New History*. Keele: Keele University Press.

De'Ath, P. 1998. *The Archive Photographs Series: Bath: The Second Selection*. Stroud: Chalford Publishing.

Fawcett, T. 1993. 'Black People in Georgian Bath', *Avon Past*, 16, 3–9.

Fergusson, A. 1973. *The Sack of Bath: A Record and an Indictment*. Salisbury: Compton Russell.

Frayling, C. 2007. 'Response, nd, to Bayley, S. 'Is Bath Britain's most backward city?'' *The Guardian* (see above), accessed 20 August 2007.

Fryer, J. 2006. 'The Steamy Truth about the Roman Bath', *Daily Mail* (published online 4 August 2006), <http://www.dailymail.co.uk/news/article-399017/The-steamy-truth-Roman-Bath.html> accessed 20 March, 2008.

Garrett, A. 2006. 'Spa Wars', *Management Today* (published online 1 April 2006), <http://www.managementtoday.co.uk/search/article/550579/spa-wars/> accessed 29 January 2007.

Georgian Group, The. 2006. 'Architectural Awards: Restoration of a Georgian building in an urban setting' (website – page updated 2006), <http://www.georgiangroup.org.uk/docs/awards/winners.php?id=4:52:0:1>, accessed 20 March 2008.

Gregory, D. 1999. 'Scripting Egypt: Orientalism and the Cultures of Travel', in Gregory, D. and Duncan, J. (eds) *Writes of Passage: Reading Travel Writing*. London: Routledge.

Grimshaw Architects. 2006. 'Project Data, Bath Spa', <http://www.grimshaw-architects.com> (home page), accessed 27 March 2008.

Hammond, C.I. 2006. '"Dearest City, I am Thine": Selina Hastings' Architectural Vision', *Women's Studies: An Interdisciplinary Journal. Special Issue: Women and Architecture*, 35:2, 145–69.

Harvey, D. 2000. *Spaces of Hope*. Edinburgh: Edinburgh University Press.

Hyett, J. 2007. 'Response of 4 August to "Was Bath right to reject the Holburne Museum scheme?"' *Building Design* (published online 3 August 2007), <http://www.bdonline.co.uk/comments.asp?storycode=3092724>, accessed 20 August 2007.

Ison, W. 1980. *The Georgian Buildings of Bath, from 1700–1830*, 3rd edn. Bath: Kingsmead Press.

Jackson, N. 1991. *Nineteenth Century Bath: Architects and Architecture.* Bath: Ashgrove Press.

Jones, B. 2006. 'The City that got its soul back', *BBC News Channel* (published online 31 July 2006), <http://news.bbc.co.uk/1/hi/england/somerset/5225872.stm>, accessed 20 March 2008.

Kellaway, G. 1991. *The Hot Springs of Bath: Investigations of the Thermal Waters of the Avon Valley*. Bath: Bath City Council.

Leatherbarrow, D. and Mostafavi, M. 2002. *Surface Architecture*. Cambridge: Cambridge University Press.

Lowenthal, D. 1998. *The Heritage Crusade and the Spoils of History*, 2nd edn. Cambridge: Cambridge University Press.

Maillet, A. 2004. *The Claude Glass: Use and Meaning of the Black Mirror in Western Art*. New York: Zone Books.

Mapstone, M. 2008. 'Worst financial fiasco in Bath's civic history', *Western Daily Press* (published online 2 January 2008), accessed via *Highbeam Research*, <http://www.highbeam.com>, 20 March 2008.

Marks, S. 2007. 'Response of 3 August to 'Was Bath right to reject the Holburne Museum scheme?'' *Building Design* (published online 3 August 2007), <http://www.bdonline.co.uk/story.asp?storycode=3092724>, accessed 20 August 2007.

McIntosh, A.J. and Prentice, R.C. 1999. 'Affirming Authenticity: Consuming Cultural Heritage', *Annals of Tourism Research*, 26:3, 589–612.

Mowl, T. and Earnshaw, B. 1988. *John Wood, Architect of Obsession*. Bath: Millstream Books.

Neale, R.S. 1981. *Bath 1650–1850, A Social History; or a Valley of Pleasure Yet a Sink of Iniquity*. London: Routledge and Kegan Paul).

O'Gorman, F. 1998. *The Long Eighteenth Century: British Political and Social History, 1688–1832*. Oxford: Oxford University Press.

Pearman, H. 2003. 'Nicholas Grimshaw in Bath: A New Spa for the 21st Century', *Gabion: Retained Writing on Architecture*, <http://www.hugh pearman.com/articles4/bathspa.html>, accessed 20 March 2008.

Sharp, R. 2006. 'Five years late, £30m overspent, mired in legal rows. Finally, Bath Spa opens', *The Observer* (published online 30 July 2006), <http:// travel.guardian.co.uk/>, accessed 27 March 2008.

Shields, R. 2004. 'Visualicity: On Urban Visibility and Invisibility', *Visual Culture in Britain*, 5:1, 23–36.

Sklair, L. 2006. 'Iconic Architecture and Capitalist Globalization', *City*, 10:1 (April), 21–47.

Swingedouw E. and Kaïka, M. 2003. 'The Making of "Glocal" Urban Modernities: Exploring the Cracks in the Mirror', *City*, 7:1, 5–21.

Thermae Bath Spa. 2007 (probable year). *Thermae Bath Spa: Britain's Original and Only Natural Thermal Spa* (brochure). Bath: Thermae Bath Spa.

Unal, N. 2007. 'Response, n.d., to Bayley, S. "Is Bath Britain's Most Backward City?"', *The Guardian* (see above), accessed 20 August 2007.

UNESCO. 2006. 'State of Conservation of World Heritage Properties in Europe: City of Bath, United Kingdom', *Periodic reporting; Section II Summary* (published online), <http://whc.unesco.org/en/list/428/documents/>, accessed 10 August 2007.

Urry, J. 2005. 'Gazing on History', in Kleniewski, N. (ed.) *Cities and Society*. Oxford: Blackwell.

Venturi, R. and Scott Brown, D. 2004. *Architecture as Signs and Systems.* Boston, MA: Harvard University Press.

Visit Britain. 2006. 'Thermae Bath Spa opens in Britain' (press release), <http://www.visitbritain.us/press/news/news-releases/2006/nr060728thermae-spa.aspx>, accessed 20 March 2008.

Weaver, M. 2005. 'Ill-fated spa hit by new delays', *The Guardian* (published online 13 January 2005), <http://www.guardian.co.uk/society/2005/jan/13/urbandesign.arts1>, accessed 20 March 2008.

Wood, J. (the Elder). 1765/1969. *An Essay Towards a Description of Bath*, 3rd edn. Bath: Kingsmead Reprints.

PART 3
Secrets and Wonders of Media Spaces

Introduction to Part 3

André Jansson and Amanda Lagerkvist

There is something inherently magic about mediation. While most media forms, as Vincent Mosco (2004) points out, reach their greatest social impact once we come to regard them as mundane and take them for granted, their transcendental power is usually the source of initial spectacle and fascination. In her study of how new electronic communication devices were introduced to the public during the nineteenth century Carolyn Marvin (1988) stresses the affinity between technological optimism and a metaphysical, even Biblical discourse. In a number of popular technical journals, as well as in other forms of public media, the performative spaces of for instance electrical light installations and sky-projections were described with words such as 'fairyland' and 'enchanted place'. The following quote from the documentation of the Chicago Columbian Exposition of 1893 is most telling: 'And now, like great white suns in this firmament of yellow stars, the search lights pierced the gloom with polished lances, and made silver paths as bright and straight as Jacob's ladder. [...] The white stream flowed toward heaven until it seemed the holy light from the rapt gaze of a saint, or Faith's white, pointing finger!' (quoted in Marvin 1988, 172–3).

The achievement of every medium is to *overcome*. A medium helps us overcome the divides between ourselves and our spatial and temporal surroundings. We can see and hear things from afar spaces. We can watch and listen to things from the past. What technological media *cannot do* is to bring us into divine spaces or into the future – although new media are often treated precisely as futurological, heavenly gadgets in themselves (see also Gitelman and Pingree 2003). However, this magical affordance is inherent to the most traditional meaning of the term 'medium': a person with para-psychological abilities. Individuals who can bring us in touch with people in the lands of the dead, or tell us about events in the future, eradicates the utmost obstacle to overcome, our own earthly existence, and thus also signifies the utmost form of magic.

The *magic of media*, whether we speak of objective techniques or metaphysical phenomena, thus stems from two communicating sources of energy: the direct power of the *spectacular* or *wondrous* on the one hand, and the more subliminal power of the *secret* or *unknowable* on the other. The art of magic, as Marcel Mauss (1972) once pointed out, is to make people believe in the unbelievable through putting up an overwhelming performance, replacing reality by images, while hiding the underlying techniques: 'A magician does nothing, but makes everyone believe he is doing everything, and all the more so since he puts to work collective forces and ideas to help the individual imagination in its belief' (ibid., 141–2). Magic, understood through the lens of *secrets and wonders*, is thus a social phenomenon – and so is the abolishment of magic. As Marvin's (1988) explorations show, while early electronic communication devices (both their performances and their technological assets) were often spectacularized in their infant stage, once the technological 'trick' was immersed by popular knowledge, there was really no space left for magic to grow. Nevertheless, in media society there will always remain spaces and processes that are hidden or secret – spaces whose insides are just rarely exposed to the public gaze. As Nick Couldry (2003) argues, the much hidden character of media production nourishes the symbolic power of (mass) mediation in the modern era.

This part of *Strange Spaces* will deal precisely with the above-mentioned dualism. It also deals with the interplay between material and representational media spaces. First, we encounter concrete spaces that become magical primarily through their status as sacred and/or hidden, but whose operations we simultaneously tend to understand as spectacular. The first two chapters deal with two such concrete emplacements of media power. Staffan Ericson discusses the modern enchantment woven in-between the monumental architecture and the scientifically designed studio interiors of the 1932 BBC Broadcasting House in London. Applying examples from architectural discourse as well as literary fiction, notably the detective story *Death at Broadcasting House* from 1934, Ericson unveils the mysterious insides of one of media history's most groundbreaking purpose-built civic centres. As shown here, mystification and strangeness play a key role in the cultural conflation of mediation and centralization, but the relationship between myth and reality remains an open question. A similar argument is developed in André Jansson's analysis of two other historical manifestations of mediated symbolic power, the Operations Control Centre and the International Broadcasting Centre at the Montreal *Expo 67*. Through their visual exposure of normally hidden technological processes these pavilions evoked ambiguous viewing positions that also reproduced dominant communication ideologies in society. As Jansson suggests, strangeness is here closely tied to a particular understanding of the future as an era of intensified ephemeral flows, which also invoke a sharpening competition for fixing these very flows.

Secondly, Johanne Sloan's chapter about lunar and aeroplane motifs in early twentieth century postcards, examines mediated spaces where a magic aura of new communication technology is discursively woven through spectacular transcendence and associations with divine mythologies. Sloan points to how the new postcard society, just like new cinematic genres, provided a link between the historical dream of lunar travel and the domestication of new technological wonders. Images of airborne women in lunar environments were circulated and appropriated as part of everyday visual culture, thereby normalizing at the same time a certain form of vernacular visuality and the management of social contacts at a distance. Sloan's analysis furthermore points to the mechanisms of disenchantment, as the historically specific genre of moon-postcards gradually lost its aura, and effectively disappeared during the militarization of the air during World War I.

Thirdly, we will explore two cinematic spaces marked by strangeness through secrecy and in-betweenness: the museum and the underworld. In Steven Jacob's chapter we are taken on a grand tour of cinematic museums and art galleries, encountering iconoclasts, thieves, spies, secret lovers, mummies, and other figures who are haunted, hiding or in transit, seeking shelter in the museum. These spaces, Jacobs argues, have occupied an enduring position among film-makers as places of conflict, estrangement and death – caught in-between the public and the clandestine. An even more haunted and mythologized space is the underworld. David L. Pike's chapter, which also ends this book, provides a vertical view of the world, in which the underworld is cinematically reproduced as a model of the 'out of place' in modern society. Pike makes us see how the cinema throughout its modern history has nurtured and exposed the fears attached to anything beneath us, from sewers to cellars, but also adapted the underground as a spatial refuge for ventilating the many contradictory and often secret dreams that saturate modern life. In the underworld we reach the point at which strangeness through the historical work of myth and mediation becomes the normal state.

References

Couldry, N. 2003. *Media Rituals: A Critical Approach*. London: Routledge.

Gitelman, L. and Pingree, G.B. 2003. *New Media 1740–1915*. Cambridge, MA: MIT Press.

Marvin, C. 1988. *When Old Technologies Were New: Thinking About Electric Communication in the Late Nineteenth Century*. New York: Oxford University Press.

Mauss, M. 1972. *A General Theory of Magic*. London: Routledge and Kegan Paul.

Mosco, V. 2004. *The Digital Sublime: Myth, Power, and Cyberspace*. Cambridge, MA: MIT Press.

Chapter 11
Death at Broadcasting House

Staffan Ericson

Death at Broadcasting House (DaB) is the title of a detective novel, first published in 1934. It is written by a pair of BBC insiders, one of them Val Gielgud, Head of Production for Drama at the time. The genre is the 'whodunit', or classical detective story (Cawelti 1976), often associated with Agatha Christie. In this one, however, there are some interesting departures from the rules. While the crime of a classical detective story is situated within the private sphere, disrupting order by placing dead bodies in the midst of our family circle, this one involves a murder at the heart of a mediated centre: the studios of Broadcasting House, i.e. the first purpose-built headquarters of the BBC, inaugurated in London in 1932. During the live broadcast of a radio play, one of the actors, isolated in one of the talk studios, is strangled to death. While the task of a classical detective usually involves tracking past events via material clues and eyewitness accounts from the scene of the crime, this detective faces an intriguing dilemma: while millions have listened in to the live performance of a murder, no one has seen anything, not a single clue was left in the studio. To explain what happened, detective Spears must reconstruct the locality of a crime that has registered only in the ether.

An exterior description of the scene of the crime (Figure 11.1) opens this story.

Broadcasting House has been called a good many names, and described as a good many things. Names and descriptions have varied from the complimentary to the scurrilous, and almost from the sublime to the ridiculous. The building has been compared with a ship, with a fortress, with a towering cliff. It has been called 'Majestic'; 'A Worthy Edifice, well-fitted to house the marvels it contains'; 'A Damned Awful Erection' (by various architects who would have liked to have had hands in the building of it); 'Sing-Sing' (by certain frivolous members of the BBC staff who had visited Berlin and heard this term applied to the new building off the Deutsches Rundfunk in the Masurenallee); 'One of the Seven New Wonders of the World' (by a patriotic daily newspaper). Broadcasting House, in short, has been extravagantly lauded and ludicrously damned. But one thing about it remains: if you walk northwards from Oxford Circus for more than fifty yards you cannot miss it. The Round Church ceases to be the dominant architectural figure of the landscape. You stop. Your eyes travel slowly

Figure 11.1　Postcard Broadcasting House, 1940s, private collection

upwards from the bronze entrance doors; pause for a moment questioningly at Prospero and Ariel; continue by way of the flower-bordered balcony of the Director-General's room, past one row of windows after another, to the trellised metal towers upon the roof, and the flagstaff with the Corporation's flag flattering against the sky. You think. Announcers ... News Bulletins ... Dance Band Music ... the Prime Minister speaking from the Guildhall ... the Derby ... Wimbledon ... Gillie Potter ... Christopher Stone ... Walford Davies ... Symphony Concerts ... Talks ... Plays ... Microphones ... Machinery ... Actors ... Engineers ... 'It's the hell of a big place anyway', you murmur to your companion, with a certain lack of conviction. (DaB, 7f)

'Response to Tradition', a British modernist manifesto published in 1932 in *The Architectural Review*, opens with an imagined scene from the same site. The writer is architect Wells Coates, the actual designer of the studios in which the fictitious murder takes place.

A foreigner, let us say, on a tour of London walks to the top of Regent Street, and finding there four or five architectural critics standing about, points to a building on the corner and says: 'What is that?'
　　The variety of possible responses to his question might include that it was a building which 'expressed its purpose', or its 'construction'; or that it displayed very bad manners indeed; or that it 'expressed' an important aspect of the national life; or that it was an example of unsymmetrical design; or that it 'looked like a ship' and was built of Portland stone; or he might be

told what he probably wanted to know, that, indeed, it was 'Broadcasting House'. Such a collection of verbal responses does not suggest an unfair picture of the state of architectural criticism today. (Coates 1932a, 165)

In the above quotes, the imagined effect of the physical appearance of this building is similar: an abundance of words, but a failure of meaning. And media houses,[1] in general, are strange sights: often imposing, at times monumental. Their external authority suggests the presence of centralized power, though not really of any traditional kind (like the church, or the state). They are presented to us as civic centres; being surrounded by parks or squares, inviting outsiders to come closer. But unlike other features in modern cityscapes (like the shopping centre, stadium, cinema, museum), they are not really sites for collective experiences. Most of us will never see their interiors; as the boundaries between 'inside' and 'outside' are upheld by electronic gates and security guards.

Approaching Broadcasting House some 80 years after Coates and Gielgud, the multiplicity of meaning is still there. We are reaching the northern end of Regent Street, one of the busiest streets in London (according to some, the first and longest shopping street in the world). Its termination is accentuated by a shift from the horizontal to the vertical. The line of sight is tilted upwards, by three objects pointing into the sky: dominating the central perspective is the spire of John Nash's nineteenth century All Souls Church. Slightly to the left, a reproduction of BBCs radio mast from the 1930s. Slightly to the right, a work of art, added in the ongoing extension: the glass ramp of a beam of light, to be shot into the sky in commemoration of lost journalists, at regular news hours.

What we are looking (up) at, then, is a triad of attempts in supra-terrestrial communication. Or, with Mircea Eliade (1957/1987), a triad of attempts in the founding of 'sacred space'. According to Eliade, 'profane space' is formless, fluid, homogenous, chaotic. Sacred space is real, absolute, orienting, organizing. Since the days of primitive habitation, man has attempted to found his world in the latter, by constructing sites (houses, sanctuaries, temples), that are a) situated at the *Centre of the World,* b) *opening a link of communication* between different cosmic planes (earth and heaven, life and death). Through history, such links have taken the form of poles, posts, pillars, spires. This particular triad has some suggestive interferences: The radio mast next to the light ray memorial: a reminder of how the word 'medium', in the early days of sound recording and radio, was entangled with nineteenth century spiritualism, i.e. 'the art of communicating with the dead' (Durham Peters 1999). The radio mast next to the church spire: a reminder of how the 'natural',

---

1 This chapter is a part of the research project 'Media Houses: On Media, Architecture and the (Re-)centralisation of Power', financed by the Baltic Sea Foundation.

interior qualities of sound – listening, hearing, speaking – have been linked to divinity and salvation, for thousands of years (Sterne 2003).

The symbols of art and religion guide the visitor entering the building. In the entrance hall, the final destination *sans* accreditation, we find the following latin inscription:

> This temple of the arts and muses is dedicated to the ALMIGHTY GOD by the first Governors of Broadcasting in the year 1931, Sir John Reith being Director General. It is their prayer that the good seed sown may bring forth a good harvest, that all things hostile to peace or purity may be banished from this house, and that the people, inclining their ear to whatsoever things are beautiful and honest and of good report, may tread the path of wisdom and uprightness.

So what sort of 'temple' is this? Who were these 'Governors of Broadcasting'? What sort of powers did they serve? What sort of knowledge did they disseminate? To direct our attention towards questions like these, Harold Innis introduced the distinction between 'space-biased' and 'time-biased' media:

> A medium of communication has an important influence on the dissemination of knowledge over space and over time and it becomes necessary to study its characteristics in order to appraise its influence in its cultural settings. According to its characteristics it may be better suited to the dissemination of knowledge over time than over space, particularly if the medium is heavy and durable and not suited to transportation, or to the dissemination of knowledge over space than over time, particularly if the medium is light and easily transported. (Innis 1951/2006, 33)

The upholding of religious tradition, and the emphasis on sound, listening, orality, were for Innis typical traits of *time-binding* media. And this stone temple certainly appears heavy, durable, 'not suited for transportation'. On the other hand, it is built for the weightless dissemination of radio, that 'ethereal medium par excellence' (Milutis 2006, x). Defiance of gravity was also signified in the exterior decorations of the building: the corporation flag, flattering from the rooftop, portraying 'an azure field representing the ether [...] broadcasting being represented by a golden ring encircling the globe' (*Broadcasting House* 1932, 13). For the entrance, the BBC asked Eric Gill for a series of sculptures of Ariel, that 'spirit of the air' from The Tempest. For the entrance hall, a sculpture of the biblical sowerman (the parable used by Durham Peters (1999) in *Speaking into the Air*, for linking Christianity and early mass communication), literally 'broadcasting' his seeds.

That such speech also could be *space-binding*, in terms of territorial-political control, was obviously not lost to those Governors of Broadcasting. For the first Christmas at Broadcasting House, Sir Reith asked the British king,

George V, for a speech. It was scripted by Rudyard Kipling, and broadcast live:

> Through one of the marvels of modern science, I am enabled, this Christmas day, to speak to all my people, throughout the Empire. I take it as a good omen that wireless should have reached its present perfection at a time when the Empire has been linked in closer union. For it offers us immense possibilities to make that union closer still.[2]

In temporal terms, radio, much like television, has been understood as a promoter of actuality and ephemerality. But one may note that King George, with this very speech, initiated the still running series of Royal Christmas messages, an annual 'media event' (Dayan and Katz 1992), indicating the active nature of broadcasting media in the organizing of cultural *tradition*. And that in hearing the voice of King George today, we are routinely registering a form of *permanence* that was once regarded as one of the more fascinating novelties of sound recording. In a chapter titled 'The Voices of the Dead', Detective Spears gets a sudden break: he incidentally learns that the play is to be retransmitted by the BBC, across the British Empire.

> 'Re-transmitted?' repeated Spears. 'Do you mean to say that – By Jove!' [...] Spears smacked his fist down in the desk in front of him. 'You mean you've got the play recorded?' he said, and even in his voice there was a thrill of excitement. 'You mean you can hear the actual scene over again?' 'We can hear that scene', said Caird, 'not only over again, but over and over again. As often as you like. I wonder if the murderer thought of that?' (DaB, 38f)

What the murderer (and the detective) also had to consider, was that the performed scene was not confided within our regular sense of situational geography:

> You see Spears, this not an ordinary case. You know what broadcasting is. It gets the public in their homes. There are nearly six million people who feel as if anything that happens inside Broadcasting House has happened by their own firesides. (DaB, 121)

To *The Architectural Review*, devoting a full issue to the opening of Broadcasting House, this building was not fixed by its geographical location:

> When it is the centre of a great public service, it has a double significance. People will come to London to see it; instead of swooning at Savoy Hill,

---

2   CD-disc (1997): *Radio Collection BBC: 75 Years of the BBC*, BBC Worldwide Ltd.

they will see what to them will be the new Tower of London. On the covers of magazines, on films, in catalogues, in guide books, in all the many means of publicity the new BBC building, its studios, its gadgets, its engineering devices, will appear. Since it is to be the new Tower of London, a focal point and a trade-mark for Broadcasting, it becomes something more than a mere block of offices, enclosing a sound factory. Like the Tower of London itself, it becomes a national monument. For this reason *The Architectural Review* devotes a whole issue to Broadcasting House, as it did to the Shakespeare Memorial Theatre. But where the latter must draw its audience to Stratford the former can entertain the world. (*Architectural Review* 1932a, 43)

'The new building parts the roads like a battleship floating towards the observer', noted *The Architectural Review* (ibid., 46) in its editorial article. The metaphor has stuck with the building, though the external referents are quite few: the rounded external shape, the antennas/masts, the porthole-like windows at the top. But the ship, according to Michel Foucault (1967/1998), is the 'heterotopia par excellence' – i.e. the sort of place that, on the one hand, *does have* real existence (unlike the utopia), but that, on the other hand, remains 'outside of all places', being strangely linked to, but also contradicting or neutralizing, *all* other sites.

And maybe one way to attach the time/space-polarity to Broadcasting House, would be the dialectics of 'inverted determinism', suggested in Blondheim's (2003) reading of Innis: once a culture is dominated by space-biased forms of communication (like broadcasting), it is also threatened by the discontinuity of time. To compensate, it develops strongly time-biased concerns (like the building of broadcasting temples). When this happens, the process of dissemination is reversed: signals of the media, spreading 'into the wind' (the title of Sir Reith's autobiography), are reintroduced as signals of architecture (mass, solidity, locality). And the ubiquitous is provided with an interior.

Interiorizing the Modern

In terms of architectural styles, Broadcasting House represented both old and new to the editors of *The Architectural Review*.

The finished building [...] represents the outcome of a struggle between moribund traditionalism and inventive modernism. Struggles of this kind are not as a rule conducive to good architecture, and still less to good decoration. But in this case, fortunately, the struggle ended in a victory which largely favoured the modernists; for when it came to organizing the interior of a building of such a necessarily complex plan, the combined brains and help of a corps of architects and engineers were required. (Byron 1932, 47)

In other words, *tradition* was here represented by the *exterior*: the solid stone walls, the fenestration with small glass panes, suggested the appearance of a medieval, concentric fortress. While the *modern* was represented by the *interiors*: the site of the new technology, occupying the 22 studios, piled in an inner tower, with the huge control room sitting on top. All designed by a team of young, radical architects/designers, one of them Wells Coates, another Serge Chermayeff. According to V.H. Goldsmith (1932), chairing the BBC's 'Studio Decorating Committee', the assignment of these men followed the outcome of a 'battle', in which principles of modernism – fitness for purpose, avoiding all decoration not following function – defeated more 'anachronistic' ideas. Decorating the studios as 'period pieces' (Venetian style, Jacobean style, etc, to represent different phases of history), or as locations of the Empire (an elephant head for India, a log cabin for Canada, etc) was really not an option for institution as 'enlightened' as the BBC, a medium as 'fresh' as broadcasting.

At the time, Coates and Chermayeff were at the centre of a movement (cf Cohen 2006) committed to breaking the 'retrospective stupor'[3] of British design, and introducing an architecture embracing the new era of 'Steel and Communication' (Coates 1932a). Between 1931 and 1933 these two men visited the German Bauhaus, published articles, participated in exhibitions, formed organizations, designed interiors, and completed houses, still standing as modernist classics. The interior of Broadcasting House was one of their first assignments. It certainly succeeded in providing the shock of the new to a Lord Gerald Wellesley, Fellow of the Royal Institute of British Architects:

> The interior of Broadcasting House is the most important example of untraditional decoration yet completed in this country. The accumulated rubbish or wisdom of the ages has been washed away, and something which is definitely and entirely new has taken its place. Such a phenomenon has never occurred before in the world's history. [4]

The 'struggle' between old/exterior and new/interior had been a topic of interest in Siegfried Giedeon's (1928/1955) *Bauen in Frankreich*, and the work it soon came to inspire: Walter Benjamin's (1999) *Passagen Werk*. Looking back at industrial constructions of the nineteenth century – arcades, railway stations, factories – Giedeon and Benjamin saw the traditional and the modern caught in transition:

---

3 Chermayeff's choice of words, from a 1931 speech on 'A New Spirit and Idealism', delivered at Heal's store for modern furnishing in London.
4 http://www.miketodd.net/other/bhhistory/bh_1932b.htm.

Outwardly, construction still boast the old pathos, underneath, concealed
behind the old facades, the basis of our present existence is taking shape.
(Giedeon 1928/1995, 87)

By the 1930s, however, such historicizing masks should be long gone. To
Giedeon and Benjamin, and to Coates, in his manifesto, stone was the material
of old architecture. With steel, glass, concrete, the piled-up wall was no longer
an essential element of the structure. From there on, Giedeon claimed, modern
housing should strive for 'the greatest possible overcoming of gravity' and for
'maximum openness':

Corbusier's houses are neither spatial nor plastic: air flows through them!
[...] There is only a single, indivisible space. The shell falls away between
interior and exterior. (Giedeon 1928/1995, 169)

This conception of modern architecture is revisited in the most ambitious
current study of the relations between media and architecture: Beatriz
Colomina's (1994/2000) *Privacy and Publicity: Modern Architecture as Mass
Media*. This book declares that the relation between inside and outside, private
and public, has been drastically changed by the presence of the media:

It is actually the emerging systems of communication that came to define
20th century culture – the mass media – that are the true site within which
modern architecture is produced and with which it directly engages. In
fact, one could argue (this is the main argument of the book) that modern
architecture only becomes modern with its engagement with the media.
(Colomina 1994/2000, 14)

Colomina measures the media's impact on modern architecture in terms
of the strategies developed by two canonical figures: Adolph Loos and Le
Corbusier. She demonstrates how the architecture of Le Corbusier is not only
produced *for* the media (that is, for symbolic, immaterial reproduction, rather
than construction on site) but also *by* the media (that is, is heavily influenced
by the surrounding media culture; ads, newspaper clippings, films). But there
is an even more fundamental connection at stake:

The building should be understood in the same terms as drawings,
photographs, writing, films, and advertisements; not only because these are
the media in which more often we encounter it, but because the building is a
mechanism of representation in its own right. (Colomina 1994/2000, 13)

In other words, the book's deepest concern is properly announced in the subtitle:
the possibility of thinking modern architecture *as* mass media. The private
homes of Le Corbusier exemplify how the use of windows may transform

the home into a camera, directed towards outer forms of life, transgressing demarcations of inner and outer through light and transparence.

With Broadcasting House, though, something else must surely be going on. For the stony shell of tradition – heavy, closed, monumental – is still there. And while the 'falling away' of the shell of *domestic* space may be linked to the breaking up of distinctions between private and public (a main argument in Colomina's book), Broadcasting House is neither private nor domestic. It is planned as the space of a *public service*. And while Colomina and others tend to link media and architecture in *visual* terms (as when comparing Corbusier's interior plans to the mechanisms of a camera), Broadcasting House is planned as a 'sound factory'.

To Lt Col.Val Myer, chief architect of Broadcasting House, the task of the programme was clear:

> In the case of Broadcasting House, we had first to consider its functions. These are twofold; the actual broadcasting, and the administration of broadcasting. Obviously, the studios, Control Room, and the accommodation of technical equipment come first, with the actual studios as the most important factor of all. Accordingly, it was the planning of the studios which had to be the key to the whole scheme. (*BBC Handbook* 1932)

And the main concern in that scheme was sound insulation. Every single studio had to be acoustically sealed off from the outside world (and from all other studios). Val Myer's general solution was to erect an inner brick tower for the studios, and to wrap the administrative offices around that tower (Figure 11.2). With this plan, the interiorizing of the new seems less the result of the concealment of tradition, than by a series of functions involving *both* media and architecture.

Firstly, the plan realized a major *technological* step in the development of housing. Since the open air could not 'flow' through Broadcasting House, it had to produce its own. With the embalming of the inner tower, each studio had to be supplied with artificial (and quiet) ventilation, humidity, temperature, lighting. According to Kenneth Frampton (1983), historian of architecture, light and climate control should be regarded as technology's most decisive influence on the modern building: from that point on, connection to local context may be cut, truly universal standardization applied. Broadcasting House was the first building in London to realize this possibility, as a direct result of the demands of media production.

Secondly, this plan reflects an *ideal model* of the *media institution* at the time. Howard Robertson, Principal of the Architectural Association School, compared the organization of the building with a medieval castle:

> There is the central Donjon, the Keep, which is the inner core, and round it are more public apartments, the service ways, like the outer ring of the

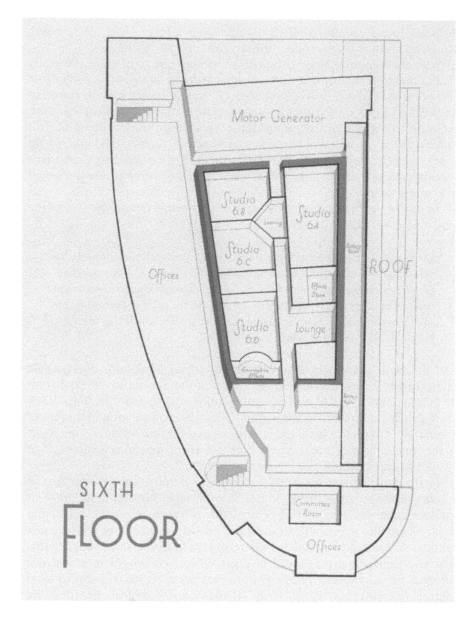

Figure 11.2   Plan, third floor. *Broadcasting House* **(1932), British Broadcasting
Corporation**

defence. In the new building the public might invade the corridors, and even the offices, but the staff of the BBC could take refuge in their inner fastness, lock themselves in, live, cook, eat, circulate, and even produce music, plays, and noises-to-taste, without in any way being disturbed. (Quoted from *The Architectural Review* 1932a, 43)

The need for interior design is intrinsically related to this model:

The new Broadcasting House is unique. It is in essentials a factory for the production and reproduction of sound, but it is not sufficient that it shall behave perfectly as a machine: it is considered vital that the artist shall derive inspiration from his surroundings, and in consequence some form of permanent interior finish or decoration is regarded as an integral part of the scheme. In the coordination of this and a chaos of complicated mechanical equipment lay the architects' task. (*The Architectural Review* 1932b, liv)

To Howard Robertson, this task exceeded the usual demands on design:

Architecture must become an aid to well-being. The designer, with no existing manual to help him, must improvise himself psychoanalyst. (*Broadcasting House* 1932, 23)

Thirdly, in terms of a 'machine', this plan suggests alternative ways of thinking '*architecture as mass medium*'. Returning to the plan (Figure 11.2), we may recognize a familiar shape: a human ear, with the studio tower in the position of the inner ear. The effect may be unintentional, nevertheless: the plan spatially reproduces what, according to Jonathan Sterne (2003), is the ultimate model of sound reproduction technology since the eighteenth century. Through various mechanisms (the eardrum, the bones of the inner ear, the auditory nerve), the human ear has the capacity to turn incoming sound into something else, and that something else back into sound. The imitation of this 'transducing' mechanism was the key to various experiments leading to the telephone (using electricity and phone lines), the gramophone (using tracks and styluses), the radio (using electromagnetic waves, transmitters/receivers). In this sense, Broadcasting House may actually qualify as a medium, 'in its own right' (Colomina 1994/2000): one large-scale, spatially organized, *tympanic machine*.

Making the building perform like one was a prime task for the team of interior designers, according to V.H. Goldsmith:

When it is remembered that the position of every piece of ventilating equipment, every lamp, every signal light, microphone lead, bell push, observation window, telephone, and every piece of furniture was fixed dependent on the precise needs of the programme in each individual studio

[...] that in addition thereto every material used, from quality of paint to nature of fabric, was subject to restriction as to its sound-absorbing or reflecting qualities, its position and area, it will be realized that never have interior designers had to solve a problem more severely conditioned. (Goldsmith 1932, 55)

While some of the studios reproduced pre-existing sites with public functions – the concert hall, vaudeville theatre, chapel, library – Coates was committed to spaces solely defined by the functions of broadcasting: the control rooms, the studios designed for drama, talk, news, sound effects, gramophones. Including their equipment: Coates designed the air-suspended microphone fitting (Figure 11.3), picking up sonic effects from the walls, floors and tanks of the effect studio, to produce 'every conceivable noise' (note the six surfaces of the table). And the way these studios were interconnected: Coates also designed the Dramatic Control Panel (Figure 11.4), mixing incoming sounds from up to eleven studios, into one broadcasted play.

Coates explained the function of this device himself, in *The Architectural Review*:

> At the dramatic control panel table sits the producer, who gives the actors in other studios their cues by switching on cue lights controlled by the keys on the dramatic control panel, or governs the volume of the sound going out to the ether by turning the control handles which, if necessary, can cut out any studio or actor. Thus, when dramatic effects like rain, wind, or the hoofs of a horse are required, a switch will give the cue to the dramatic effects studio on another floor, and the turning of a control handle will increase the sound or diminish it until the producer cuts it out by turning the handle back. (Coates 1932c)

In *Death at Broadcasting House*, detective Spears eventually realizes that to solve this crime, he must understand 'the inside of that box of tricks'. The producer of the play explains why the performers must be physically separated:

> 'Well', said Caird, 'the chief reasons why we use several studios and not one, are two. The first is that by the use of separate studios, the producer can get different acoustic effects for his scenes. That is to say, in a small studio like 7C, which is built as to exclude all echo, you get the effect of a closed room or a dungeon – as in the scene when Parsons was killed. Whereas in a fairly large studio like 6A, you can get the effect of greater spaciousness – as in the scene previous to the murder scene, a ballroom. Secondly, the modern radio play depends for its 'continuity' – if you understand the film analogy – upon the ability to 'fade' one scene at its conclusion into the next. In an elaborate play, therefore, the actors require as many studios as the varying acoustics

Figure 11.3 Sound Effects Studio. *Broadcasting House* **(1932), British Broadcasting Corporation**

of the different scenes require, while, in order to avoid their being confused by music or extraneous noises, sound effects have a studio of their own, gramophone effects one more, and the orchestra providing the orchestral music yet another separate one'. (DaB, 78)

A sound effects man explains his way of producing 'natural' sounds:

Figure 11. 4 Dramatic Control Panel. *The Architectural Review* (1932), August

> You see, in the old days when we started with sound effects, we did our best
> to make the real noise in front of the microphone. At Savoy Hill, I believe
> it's true that people fired blank cartridges along the corridors, and even
> assembled the greater part of an aeroplane and then dropped it from the
> ceiling of the studio to get the effect of an aeroplane crash. Now we know
> better. We wreck ships by crumpling match boxes and create avalanches with
> a drum and few potatoes. (DaB, 61)

The overall attention that the BBC paid to the studios when erecting their
first headquarters, and particularly the type of tasks executed by Coates,
lends support to a bearing argument in Jonathan Sterne's (2003) book on *The
Audible Past*:

> Without studios, and without other social placements of microphones in
> performative frames that were always real spaces, there was no independent
> reproducibility of sound. [...] This is contrary to the often-made claim that
> reproduction decontextualizes performance and deterritorializes sound [...]
> From the very beginning, recorded sound was a studio art. (Sterne 2003, 236)

In such studios, people performed for the machines, not for audiences. The machines were built to reproduce sounds, not eavesdrop on existing, 'original' ones. In contrast with discourses that stress the 'liveness' and 'fidelity' of sound reproduction, Sterne claims that distinctions between copy and original, artificial and real, were irrelevant in the early days of mass-distributed sonic events. In contrast with discourses which attributes 'no sense of place' (Meyrowitz 1985) to broadcasting, Sterne claims that such events were the product of a reproducibility to which 'location was everything'.

This argument runs parallel to the plot of *Death at Broadcasting House*, and to a theme in Walter Benjamin's (1936/1968) classic essay on 'The Work of Art in the Age of Mechanical Reproduction'. Here, Benjamin exemplifies the loss of 'aura' by a visit to a film studio: a site where an actor performs for the machines, and a visitor never loses sight of the technical equipment. The film spectator, though, sees none of it:

> The mechanical equipment has penetrated so deeply into reality that its pure aspect freed from the foreign substance of equipment is the result of a special procedure, namely the shooting by the specially adjusted camera and the mounting of the shot together with similar ones. The equipment-free aspect of reality here has become the height of artifice; the sight of immediate reality has become an orchid in the land of technology (Benjamin 1936/1968, 233)

According to Benjamin, this 'procedure' was unimaginable before film. And according to V.H. Goldsmith, the studios of Broadcasting House could not duplicate the practice of film.

> In the film studio, not only do production and rehearsal precede a 'shot', as they do for some of the work in broadcasting, but 'shots' can be repeated until the desired result is achieved. This repetition of 'shot' is not possible in broadcasting, hence every help must be given to the artist in his one and only actuality. (Goldsmith 1932, 53)

In other words, Benjamin's 'equipment-free reality' was here to be performed in real time. This was the rationale behind the organization of the studios of Broadcasting House. And part of the dilemma of detective Spears: this crime had only left traces in equipment-free reality. Listening to the recording of the broadcasted play, as edited by the Dramatic Control Panel, the localities of the reproduced sounds, as well as their status as artificial or real, were already indistinguishable. The final solution, the detective concludes, is 'inseparably connected with the methods and ingredients of broadcast play production', and thus with 'the geography' and 'inside working of Broadcasting House' (DaB, 96). Like the murderer before him (constructing his perfect alibi), the detective must reproduce the spatial conditions of a 'land of technology', organized by Coates and his colleagues.

Liquidating the Interior

The preferred scenario for the linking of media and architecture (cf. Rice 2007, ch. 5), is usually that of the 'home' (an actual, pre-existing site), being infiltrated, disturbed or challenged by 'the media' (immaterial, site-less). In the career and mind of Coates, designing the broadcasting studios of the BBC more or less coincided with a rethinking of domesticity.

In an article from 1932, 'Furniture Today and Furniture Tomorrow', Coates claims that the modern architect should not be concerned with various styles and fashions, but with 'the organization of a new service'.

> The natural starting-place for this new service must be the scene in which the daily drama of personal life takes place; the interior of the dwelling – the PLAN – and its living-equipment, the furniture. (Coates 1932b, 31)

Coates pursued this ideal as chief architect for ISOKON (Isometric Unit Construction), a company set up in 1931 to produce 'unit dwellings', with inbuilt furniture and accessories. The most renowned result was *Lawn Road Flats*, projected and completed between 1930 and 1934. This complex, located in Hampstead, London, consists of 30 'minimal flats' (from 18 to 30 m²), with adjacent 'communal' areas – club, bar, roof-top terrace, garden, garage – including 'very full domestic service'[5] (washing, cleaning, cooking, shoe-shining). When completed, Lawn Road Flats attracted some interesting tenants – among them, Agatha Christie, and the the international avant-garde of design: off and on during the thirties, Lawn Road Flats housed members of the German Bauhaus – Marcel Breuer, Walter Gropius, Lazlo Moholy Nagy – some providing assistance in ISIKON's designs. When Coates died in 1958, the memorial in *The Architectural Review* claimed that Lawn Road was 'nearer to the *machine à habiter* than anything Le Corbusier ever designed' (Richards 1958).

There are some tangible correspondences between Coates' organizing of studios (as in Broadcasting House) and his notion of a 'machine for living' (as in Lawn Road Flats). In terms of 'the plan', Coates organizes his dwellings as *minimal* units, externalizing their points of access (stairs) and social functions (communal spaces), serving them with all necessities (water, heating, air) from the outside, and securing that 'conducted sound had been reduced to an

---

5   Publicity material for Lawn Road Flats, quoted from Cohn (1999, 159).

absolute minimum'.[6] Visually, the open air-solutions of Le Corbusier could not be more distant (and complaints of claustrophobia were soon heard from the tenants of Lawn Road Flats and the artists of BBC). The idea of a 'total design' was equally applicable to the studio and the dwelling (during the late thirties, Coates designed several of the most popular British radio sets for home use). To Coates, supplying the proper 'living-equipment' meant obliterating the 'old-world dwelling-scene' of our parents:

> How barbaric their habit of overloading was! How seldom did an object stand in the place which correlation points to it! How obstrusive their pictures and ornamental bric-à-brac! (Coates 1932b, 32)

To Coates, tables, chairs and beds were no more 'personal belongings' than heating systems or bathtubs. He presented his most radical proposal of 'unit dwelling' in 1947, as a (never realized) plan for production of room units, ordered in parts off the shelf, complete to the last light switch, transportable to the countryside for the weekend. Already in 1932, Coates was arguing for a notion of modern dwelling not fixed by location:

> The love of travel and change, the mobility of the worker himself, grows with every opportunity to indulge in it. The 'home' is no longer a permanent place from one generation to another. The old phrase about a man's 'appointed place' meant a real territorial limit; now the limits of our experience are expanding with every invention of science. (Coates 1932b, 32)

In other words, Coates' idea of modern dwelling was committed to that 'structure of feeling' that Raymond Williams (1974/1997) has used to explain the historical development of broadcasting: *mobile privatization*. With the notion of a 'furniture of tomorrow', Coates presented the architect with the task of providing 'supplies, equipment for the living of a free life':

> There is an important distinction to be realized between what is 'possessed' as an adjunct of personal vanity or wealth (a 'museum-piece' you are told, with a smack of satisfaction) and what is merely included for use in the dwelling-scene for what its efficiency and formal significance is worth in the daily drama and routine of life. In the latter case the article is not valued as a 'personal possession' so much as a means, a medium, for the liberation of individual values and appetencies which alone are the truly 'personal possessions' of a man. (Coates 1932b, 33)

---

6   Presentation of Lawn Road Flats in *The Architectural Review*, August 1934, quoted from Cohn 1999, 167.

Such articles were to be 'machine-made', and affordable for 'the people'. They would reveal 'the colossal pretence that has stood for 'art''. In other words, while referring to furniture as a 'medium', Coates made similar predictions, and used similar distinctions, as Walter Benjamin's art work-essay, written a few years later. In his *Arcades Project*, Benjamin had already registered the loss of aura through the history of our dwellings. Benjamin starts with the bourgeois apartments of the nineteenth century:

> In the style characteristic of the second empire, the apartment becomes a sort of cockpit. The traces of its inhabitant are moulded into the interior. Here is the origin of the detective story, which inquires into these traces and follows these tracks. (Benjamin 1999, 20*)*

With the turn of the century, this sense of the interior was lost. To Benjamin, the 'artistic visions' of Jugend provided a 'mirror image of the world of commodities', not a space of refuge for the individual. The challenge of the modern was to embrace this experiental loss. In an article from 1933, 'Experience and Poverty', Benjamin (1933/1996–2003) describes the contemporary world as a barbaric, impoverished space, where a 'naked man ... lies screaming like a newborn babe in the dirty diapers of the present' (ibid., 733). Still, this world must provide housing for its citizens. Benjamin is fascinated by the 'stations for living' imagined by poet Paul Scheerbart: 'adjustable, movable, glass-covered dwellings of the kind since built by Loos and Corbusier' (ibid., 733): 'Objects made of glass have no 'aura'. Glass is, in general, the enemy of secrets. It is also the enemy of possession' (ibid., 734).

Such a 'liquidation of the interior' (see '1939 Exposé', in Benjamin 1999, 20) does not lament lost experience, but executes the necessary break with the (always illusory) 'dream image' of private life.

> If you enter a bourgeois room of the 1880s, for all the cosiness it radiates, the strongest impression you receive may well be, 'You've got no business here!' And in fact you have no business in that room, for there is no spot on which the owner has not left his mark – the ornaments on the mantelpiece, the antimacassars an the armchairs, the transparencies in the windows. A neat phrase by Brecht help us out here: 'Erase the traces!' is the refrain in the first poem of his *Lesebuch für Städtebewohner* [...] This has now been achieved by Scheerbart, with his glass, and Bauhaus, with its steel. They have created rooms in which it is hard to leave traces. (Benjamin 1933/1996–2003, 734)

This is the sort of room in which a dead corpse is placed, in *Death at Broadcasting House*. The producer and the writer discover it, after the broadcast:

> In the far corner, almost under the microphone standard, lay a man's figure unnaturally crumpled [...] Behind the three of them the door shut

automatically. 7C was a studio with special acoustic treatment removing all natural echo, and at that moment Rodney Fleming felt acutely the oppressive, almost sinister atmosphere of the room with its single shaded light, its thick carpet and queerly padded walls. The ventilation was perfect, but he felt he wanted to draw unusually deep breaths. (DaB, 17)

The Scotland Yard arrives, and the room is properly photographed and searched. But the detective is left without any traces to track.

'You can see for yourself, sir' he said. 'This room's as bare as a board. The carpet's too thick to take any impression, and whoever did this job knew too much to leave anything behind him. Here are the contents of the pockets, sir.' [...] There was something indescribably wretched and forlorn about the little pile of coppers: the paper packet of ten Players cigarettes, three quarters empty; the indubitable pawn-ticket; the soiled handkerchief; the three loose keys on a piece of knotted string: the chubbed stump of pencil; and the shabby pigskin pocket book. (DaB, 33)

These are the possessions of modern man, the traces of a crime committed in the interior of the modern. According to the detective novel, Broadcasting House is not only a marvel of technology, but a scary place to be. If the task of the designer was to 'improvise himself psychoanalyst', Coates and partners may have succeeded all too well. The building was soon to produce a stronger image of interior terror. In George Orwell's *1984* (1949), Room 101 is a chamber for psychological torture, the room where our imaginary fears will suddenly and inexplicably materialize. The original Room 101 is believed to have been located in Broadcasting House, where Orwell worked during the Second World War. In the novel, this room contains 'the worst things in the world'. That is, no specific objects at all, since the worst thing imaginable will vary individually: to be buried alive, to drown, or, the fear of protagonist Winston, to have your face eaten by rats. And perhaps this is the scariest aspect of Room 101: once you enter, your torturers will reveal that they have access to your most private space, your worst inner fears.

Interiorizing Social Space

In 2000, the BBC decided to put 'architecture once again at the heart of [its] strategy' (Jackson 2003, 14). The old Broadcasting House was to be transformed into one 'huge, highly efficient global broadcasting machine'.[7] To Greg Dyke, Director General at the time, the BBC had neglected what the

---

7   The words of John Smith, at the time serving as the BBC's Director of Finance. http://news.bbc.co.uk, 31 October 2000.

founding fathers had grasped: the symbolic importance of buildings. But while the old Broadcasting House had spoken of a 'self confident organization with a clear vision of its role in the world', it could not, according to Dyke (2003), 'reflect the values and ethos of the modern BBC':

> As a building it's patriarchal, even frightening. [...] today, the BBC needs buildings that connect with our audiences, not buildings that frighten them.

'For the first time', declares the publicity material, 'the BBC in central London will have a public face accessible to all – where broadcaster and audience can meet directly' (BBC 2006). After a limited competition, architect Sir Richard MacCormac was awarded this task. His strategy was not so much outright transparency, as the framing of so-called *interstitial* spaces (Jackson 2003, ch. 4). In natural geography, the interstitial is the shoreline, in architecture, it is the gap between walls, neither outside nor inside. With MacCormac, the interstitial is defined as social space, an interface for meetings. From this follows his re-interpretation of the function of the building: not in terms of what the factory produces, but in terms of the social and symbolic organization that it houses (MacCormac 2005).

How is this notion translated into architecture? When describing the interior organization of the new extension, MacCormac (2004) tends to refer to public, outside places: the market place, the high street, the forum, the thoroughfare. Places where people interlock, become physically and visually aware of each other (like they will, on the inside of the new Broadcasting House, on stairs, circulation routes, breakout areas). What is being interiorized here is not so much modern space, as (old) public space. To express the interstitial externally, McCormac enhances the opposites (inside/outside, light/heavy, convex/concave, opacity/transparency) through which the 'in-between' may be experienced. The new spire of the artwork, for instance, has the same geometric dimension as that of the church, but is turned upside down, and made transparent. The heavy convexity of the old building is countered, inverted, by the air-light concavity of an open-air square or theatre (Figure 11.5), surrounded by a special type of glass, producing a sensation of volume: opaque during the day, transparent during the night, when the inner light transforms the whole building to a theatre. What the new building encircles and embalms, is not so much the production areas of the artists, as this open-air square: 'the heart of the BBC.'

For the BBC to return to Broadcasting House, the shell had to be opened up. The irony being that the exterior that is now preserved as the BBC's proud heritage from the 1930s – the Portland stone, the sculptures by Eric Gill – was already at that point signifying tradition. While what was considered modern and forward-looking at the time – the design of the interiors, Coates' studios – had more or less vanished within a decade. Stronger forces of technology

Figure 11.5   Computer image of site on completion, MJP Architects (MacCormac Jamieson Prichard)

called for reorganization: the arrival of television (with new spatial needs for production), the outbreak of the war (with bombs and rockets threatening central functions, from above).

What remains is photography: in 1932, the BBC published a book with over 100 pictures of the brand new interiors of Broadcasting House (including Figures 11.3 and 11.4). To architect MacCormac, these pictures appear 'spooky – like a German expressionist film-set' (ibid.). An abandoned film-set, one might add. For what is lacking in them is people: not one single trace of a living soul – not even a corpse! – in any one of these studios. Benjamin (1931/1980) once noted the same absence in the Parisian scenes of nineteenth century photographer Eugene Atget: his street exteriors were 'empty', 'voiceless; the city in these pictures is swept clean like a house which has not yet found its new tenant' (210), the city resembles 'the scene of a crime' (215). A sight of 'healthy alienation', according to Benjamin, neither artistic nor realistic, but a forerunner to the 'constructed' nature of surrealist photography, and hence, to Benjamin's notion of 'equipment-free reality'. With the pictures from the interiors of Broadcasting House we are looking at, not equipment-free reality, but the material and technological conditions of its production. Spaces and machines for the reproduction of the 'voices of the dead' that are, precisely, 'voiceless', uninhabited, suggestive of unspoken crimes and terrors.

Before finally demolishing the interior structure of the Broadcasting House, in preparation for the ongoing redevelopment, the BBC asked British artist Rachel Whiteread for a cast of its most infamous space: Room 101. It is an odd memorial, a countermove to Benjamin's 'liquidation of the interior', to the Brechtian motto: 'Erase the traces!' In Whiteread's Room 101, the features of interior space are inverted; the metaphor for the worst thing in the world is transformed into an object – 'blank' and 'ghostly', but material. When asked what would be in her own Room 101 (that is, her worst fear), Whiteread (2003) replied: 'Outer space'.

> Because there are no walls, no parameters, nothing to relate me to the earth, nothing to stop you going off. I would find that frightening.

About as frightening as Mircea Eliades's 'profane space' – borderless, fluid, chaotic. Maybe Whiteread's casting of interiors is contemporary man's version of founding sacred space, of establishing the centre of the world. To reconnect with ourselves, rather than the gods, what is needed is not the raising of spires, but the solidification of social space.

References

Architectural Review, The. 1932a. 'The New Tower of London' (editorial article), August.

Architectural Review, The. 1932b. 'Trade and Craft: Notes on Materials at Broadcasting House' (anonymous author), August.

BBC. 2006. Press release, April, http://www.bbc.co.uk/pressoffice/keyfacts/stories/bh_development.shtml.

BBC Handbook. 1932. Quoted from http://www.miketodd.net/other/bhhistory/bh_1932b.htm.

Benjamin, W. 1931/1980. 'A Short History of Photography', in Trachtenberg A. (ed.) *Classic Essays on Photography*. New Haven, CT: Leete's Island Books.

Benjamin, W. 1933/1996–2003. 'Experience and Poverty' ('Erfahrung und Armut'), in *Selected Writings*. Cambridge, MA: The Belknap Press of Harvard University Press.

Benjamin, W. 1936/1968. 'The Work of Art in the Age of Mechanical Reproduction', in *Illuminations*. New York: Schocken Books.

Benjamin, W. 1999. *The Arcades Project (Das Passagen-Werk)*. Cambridge, MA: The Belknap Press of Harvard University Press.

Broadcasting House. 1932. London: British Broadcasting Corporation.

Blondheim, M. 2003. 'Harold Adams Innis and his Bias of Communication', in Katz, E. et al. (eds) *Canonic Texts in Media Research*. Cambridge: Polity Press.

Byron, R. 1932. 'Broadcasting House', *The Architectural Review*, August.

Cawelti, J. 1976. *Adventure, Mystery, Romance. Formula Stories as Art and Popular Culture*. Chicago: University of Chicago Press.

Coates, W. 1932a. 'Response to Tradition', *The Architectural Review*, November.

Coates, W. 1932b. 'Furniture Today and Furniture Tomorrow', *The Architectural Review*, July.

Coates, W. 1932c. 'The Dramatic Control Panel No 1', Plate IV, *The Architectural Review*, August.

Cohen, D. 2006. *Household Gods. The British and their Possessions.* New Haven, CT: Yale University Press.

Cohn, L. 1999. *The Door to a Secret Room. A Portrait of Wells Coates.* Aldershot: Scolar Press.

Colomina, B. 1994/2000. *Privacy and Publicity. Modern Architecture as Mass Media.* Cambridge, MA: MIT Press.

Dayan, D. and Katz, E. 1992. *Media Events. The Live Broadcasting of History.* Cambridge, MA: Harvard University Press.

Durham Peters, J. 1999. *Speaking Into the Air.* Chicago: University of Chicago Press.

Dyke, G. 2003. Speech given at the British Property Federation Conference 2003, http://www.bbc.co.uk/print/pressoffice/speeches/stories/dyke.

Eliade, M. 1957/1987. *The Sacred and the Profane. The Nature of Religion.* Orlando, FL: Harcourt Inc.

Foucault, M. 1967/1998. 'Of Other Spaces', in Mirzoeff, N. (ed.) *The Visual Culture Reader.* London: Routledge.

Frampton, K. 1983. 'Towards a Critical Regionalism: Six Points for an Architecture of Resistance', in Foster, H. (ed.) *The Anti-Aesthetic.* Seattle: Bay Press.

Giedion, S.1928/1995. *Building in France, Building in Iron, Building in Ferro-concrete (Bauen in Frankreich).* Santa Monica, CA: Getty Center for the History of Art and the Humanities.

Gielgud, V. and Marvell, H. 1934. *Death at Broadcasting House.* London: Withy Grove Press.

Innis, H.A. 1951/2006. *The Bias of Communication.* Toronto: University of Toronto Press.

Goldsmith, V.H. 1932. 'The Studio Interiors', *The Architectural Review*, August.

Jackson, N. 2003. *Building the BBC: A Return to Form.* London: BBC.

MacCormac, R. 2004. Notes from a talk on Broadcasting House, February, http://www.cityofsound.com/blog/2005/12/notes.

MacCormac, R. 2005. 'Private View II: When Art Meets Architecture', *TATEetc*, Issue 5, Autumn 2005.

Meyrowitz, J. 1985. *No Sense of Place.* New York: Oxford University Press.

Milutis, J. 2006. *Ether. The Nothing that Connects Everything.* Minneapolis: University of Minnesota Press.

Orwell, G. 1949. *Nineteen Eighty-Four.* London: Secker and Warburg.

Rice, C. 2007. *The Emergence of the Interior. Architecture, Modernity, Domesticity*. London: Routledge.

Richards, J. M. 1958. 'Wells Coates 1895–1958', *The Architectural Review*, December 1958.

Sterne, J. 2003. *The Audible Past: Cultural Origins of Sound Reproduction*. Durham, NC: Duke University Press.

Whiteread, R. 2003. *Independent on Sunday*, 16 November.

Williams, R. 1974/1997. *Television: Technology and Cultural Form*. London: Routledge.

Chapter 12

Communication Clinics: Expo 67 and the Symbolic Power of Fixing Flows

André Jansson

Putting something behind glass walls almost automatically draws our attention to it. Transparency speaks to our curiosity and fascination, even to our voyeuristic impulses. But it also conceals, separates and reduces. It invokes a dichotomy between the object and the spectator. Can we believe in what we see? And what determines how to look, and what to believe?

During the 1967 World Expo in Montreal people could visit two peculiar exhibits of media power: *The International Broadcasting Centre* (IBC) and *The Operations Control Centre* (OCC). The peculiarity of these pavilions lay foremost in their lack of performances. Their aim was to show how new media – surveillance and broadcasting – operated 'behind the scenes'. Through huge panorama windows visitors could gaze into studios and control rooms, where humans and machines were in constant operation – observing, recording, measuring and producing flows of people and information. Not only were the sights strange. Also the viewing positions were highly ambiguous. Visitors at the IBC and the OCC were both drawn into and excluded from mediation. The strategy of showing 'the truth behind the scenes' reproduced the very magic it was said to reveal.

But what was this magic all about? The aim of this chapter is to unpack the ways in which the strangeness of these pavilions was fuelled by the dominant ideologies of modern communication. I will particularly point to the fact that the heterotopian character of places such as the IBC and the OCC contributes to the sacralization of the media as sites of flow monitoring in general, and their capacity to *fix* flows in particular. This sacred position, in turn, represents the flip-side of the modern mythology of flow, mobility and circulation (of people, goods and information) as signs of 'development'. The accentuation of more or less ephemeral flows, paired with the accentuated *public belief* in the social and technological pervasiveness of these flows, accentuate the exclusivity of those agents in society who hold the capacity to invoke fixity, whether in reality or in a representational manner.

My analysis entails three steps. Firstly, I will present a theoretical framework based on three critical views of symbolic power: Michel Foucault's theory of the clinic and the *observing gaze*, Nick Couldry's thesis regarding *the myth of mediated centre*, and Tim Cresswell's identification of two competing moral geographies, the *metaphysics of fixity and flow*. Secondly, I will apply this framework for reconstructing the extraordinary site of Expo 67 in terms of a *flow-space*, and pinpoint the IBC and the OCC as spectacular exhibits of *flow monitoring*. Thirdly, I will deconstruct the viewing positions invoked by the OCC and the IBC, showing how spatial design, notably the prominence of glass architecture, in conjunction with media discourses and modern ideologies fostered a sense of *serialized displacement*.

Fixing the Flow: The Spatial Ideology of Modern Communication

In his history of the modern clinic, Michel Foucault disclosed the scientific mythology through which medical statements were elevated as objective *truths*. From the late eighteenth century and onwards the expressions of medical expertise came to represent an incontestable view of physical conditions, based on systematic observations of symptoms, which concealed the discourse through which clinical utterances were spoken. According to Foucault, the *clinical gaze* was the new visual regime that transformed and 'gave speech to that which everyone sees without seeing' (1963/1994, 115). It was the magic medium between appearances and diagnosis, operationalized as 'measured language' invoking a secluded realm of *fidelity* and *fixity* under conditions otherwise plagued by uncertainty and anxiety among ordinary people (ibid., 114). The clinic itself was, and is, if we follow Foucault's later writings, a *heterotopia* – an enclosed, exceptional space adhering to a regime different from that of the everyday space outside it, but still working as a machinery for societal maintenance and control (Foucault 1967/1998). The common understanding of hospitals as strange harbours of secrecy and clarity, fears and hopes, morbidity and vividness, which runs deep into the experience of modernity, is therefore intimately tied to the exclusive mastery of the clinical gaze.

While Foucault's theory of the clinical gaze has influenced certain strands of visual culture (see Sturken and Cartwright 2001) it has rarely been adapted within the broader field of media studies. Yet, it entails an understanding of modern society and its obsession with the systematic gathering of earthly truths that shows interesting parallels to for example Nick Couldry's (2003) more recent analyses of the 'myth of the mediated centre'. Couldry argues that the power of modern media above all lies in their potential to define and position the *centres of social life*. Departing from a neo-Durkheimian view of social community as produced through *ritual action*, and a Bourdieuan view of social dominance as produced through *symbolic violence*, Couldry finds that the power of the media stems from a shared, naturalized understanding of what is mediated as that which is important. In a mediatized society, in which people lead their lives in media saturated environments, media representations are no longer regarded merely as separate events, but interweave with all kinds of ritual behaviour. Nevertheless, people continually assign mediated processes a certain relevance. What is mediated is customarily given the status as something important, and therefore reproduces the social understanding of where the centres of society are located – geographically, ideologically, culturally.

In this way, Couldry contends, symbolic violence, as a means of social dominance, is '*inherent* to the media's operations but it can only be unpacked through a theory of the specific rituals and ritualisations that sustain it on a day-to-day basis' (ibid., 41, italics in original). This is to say that the power of the media is reproduced not merely through the content of mediations, but also, and more fundamentally, through the rituals that sustain the institution of mediation as such. These observations lead Couldry to identify what he calls 'the myth of the mediated centre' – a double myth, based on the beliefs a) that society has a centre and b) 'that 'the media' has a privileged relationship to that 'centre', as a highly centralized system of symbolic production whose 'natural' role is to represent or frame that 'centre'" (ibid., 45).

The media's socially privileged position, notably through the institution of journalism, is acquainted to the clinical gaze inasmuch as it manifests itself precisely as a machinery of fidelity and fixity, whose views and selections of complex events in the 'real world' follows and re-invokes a certain symbolic order. As with medical expertise, however, the logic of this order, as well as the rules and techniques underpinning professional procedures (e.g., news gathering and evaluation), are alien to the everyday audience – as are the concrete sites from which the circulation of information is governed. There are few opportunities for ordinary people to grasp what goes on behind the scenes, beyond the camera lens, or before and after the broadcast. Nevertheless, they are continuously led to believe in what is being told about the world, and also about themselves as a controlled, measured audience (cf. Baudrillard 1983; Bourdieu 1996/1999). As Jonathan Crary (1999) asserts, the media holds a pervasive power to fix people's attention, and therefore 'individuate, immobilize, and separate subjects, even within a world in which mobility and circulation are ubiquitous' (ibid., 74). In a series of newspaper articles during the Gulf War, Jean Baudrillard (1991/1995) pulled this argument to its extreme, questioning whether the war was actually taking place, and, if so, whether it was a war and not merely a calculated television spectacle, aiming to draw people's attention away from other issues.

While very different in kind, media industries and medical clinics thus share a certain form of symbolic power, based on their authority to 1) observe and re-articulate visible conditions in terms of fixed 'diagnoses' and 'reports', and 2) govern people's attention. This authority is further enhanced through the concealment, even mystification, of their inner logics. While firmly incorporated societal entities with ritualized ties to everyday life, they are at the same time regarded as centralized heterotopian spaces separate from everyday life. However, while clinical heterotopias open and close their gates according to a regular logic, bringing patients and clients successively deeper into their bewildering labyrinthine spaces, the sites of media production are only occasionally accessible to the public. Inside experiences of media industries are exceptional and estranging almost by definition (see Couldry 2000).

But is the mediated centre just a myth? And why is there in modern society such an obsession with centrality? In his ambition to pinpoint the myth of the mediated centre, Couldry pays little attention to how mythological structures are actually producing and produced by economic-material processes. If we want to grasp the epitomization of electronic mass media in the modern era, rituals and mythologies are only one aspect of a very complex development. For instance, while the competitive logic framing the transition from what Ronald Robertson (1992) has called the *Struggle for Hegemony Phase* (1920s–1960s) to the *Uncertainty Phase* (1960s–1990s) of globalization, integrates a myth of the media as symbols of centrality and legitimizers of power, it is through the increased global flows of people, capital and information, reinforced by the media themselves and associated with an increasing mediated sense of liquidity and uncertainty (cf Bauman 2000), that the mythological epitomization of the media can occur. The symbolic power of the media rests not only upon their ritually sanctioned authority to narrate what is important in society (and this authority is being transformed in an increasingly networked and digital media landscape), but also upon their infrastructural position as generators of deterritorializing information circuits, and their insistence on this development as essential for societal *development* at large.

In order to understand how the symbolic power of the media is reproduced, I argue, we must therefore closely consider their role as *fixtures* in a *flow-space*. This is to acknowledge their re-territorializing role. The term flow-space refers to the spatial integration of various kinds of flows – of people, goods, information, etc. – and I apply it here in order to distinguish my view from Castells' (1992) well-known understanding of the 'space of flows', which deals more exclusively with abstract information flows. In modern times the composition of flow-space is generally understood as increasingly liquid – a condition that the media are themselves materially generating, while at the same time advocating as a more or less natural force of modern development. But one must not underestimate the material underpinnings of this development. A fixture, then, is to be understood as a strong point, or a node, in a network of flows, operating as *a socially and territorially recognized determinant of pathways of circulation*.

What must be stressed here is that in a media society the myth of the centre is inseparable from any 'real' centre, in the same way as media representations are inseparable from the 'realities' they (seemingly) represent. Furthermore, the mythologized modern ambition to centralize the power of fixing flows, is dependent upon a complementary mythology elevating mobility and fluidity as the taken for granted key to 'development'. The media reproduce not only flows as such, but also, through their representation of society, the mythology of flow, and the mythology of fixtures as sites of power. The more flows there are, and the more ephemeral and omnipresent they are thought to be, the more significant (economically, politically, culturally) becomes the power to control flows and construct places of media power, that is, social and infrastructural fixtures.

We can gain a broader historical understanding of this contradictory condition if we consider it as the outcome of the negotiation of two competing moral geographies – what Tim Cresswell (2006, ch. 2) has termed the *metaphysics of fixity and flow*. These moral geographies, Cresswell (ibid., 34) argues, runs deeper than ideological differences, and 'act as the bedrock for remarkably consistent sets of assumptions that work across political and theoretical divides.' The first orientation, also termed *sedentarist* metaphysics, corresponds to ideas that raise the home and stable roots as fundamental human needs, and the foundation for a well-functioning civil society. Such assumptions have been present in many romantic and nationalistic ideologies and also an essential part of the construction of particularly local and national media spaces or imagined communities (see Anderson 1983; Morley and Robins 1995) They have also been key to much phenomenological thinking on space, as well as for humanist geography, and the social theory of the Chicago school, regarding mobility mainly as a source of disorder, moral decay, and 'placelessness' (cf Relph 1976). Mobility here denotes an *absence* of enduring relationships within the family, local community, or the nation state.

The metaphysics of flow, or *nomadism*, on the contrary, celebrate mobility, fluidity, and the open-ended process of *becoming*. In academic writing such moral geographies are reflected in recent cultural anthropology, post-structuralism, and postcolonial theory, in which *routes* are regarded as more central for an understanding of modern societies than roots. Nomadism becomes a metaphor for all kinds of horizontal movement that resist and threaten the verticality of modern power structures. Interesting parallels can also be drawn to, for instance, Manuel Castells' (1992) before-mentioned writings on the global prominence of informationalized 'spaces of flows', making places of social proximity secondary to the ephemeral geographies of the digital interface, and to certain strands of architectural theory, such as Bernard Tschumi's (1994) assertion of the urban grid as an open-ended 'antinature'. From these viewpoints, which correspond to the megastructure architecture that influenced Expo 67 and 1960s Montreal, 'space is not there to tell you what to do, but to provide an opportunity to move and, in doing so, make things happen' (Cresswell 2006, 51).[1]

---

1 The pervasiveness of these ideological encodings is not unitary, however. As recent debates regarding 'globalization' show, for instance, a market driven globalism stands in sharp contrast to more sceptical, but possibly more nomadistic, multiculturalist positions advocating liberalized flows of people and information but increased regulations on for example money flows. To borrow Arjun Appadurai's (1996) terminology, this implies that interest driven political discourses ensure that the metaphysics of fixity and flow are expressed differently in relation to different *scapes* – such as *finance-scapes*, *mediascapes*, and *ethnoscapes* in the above example.

From Cresswell's discussion we can conclude that the metaphysics of flow have dominated Western societies from the late nineteenth century and onwards. This is also what Harold Innis (1951/1964) famously argued in his book *The Bias of Communication*, pointing to the growing prevalence of a spatial ideology sustaining the expansion of light, space-biased means of communication, and what several subsequent theorists have reasserted (e.g. Meyrowitz 1985; Carey 1989; Augé 1995). But it is no less important to consider the enduring social role of sedentarist moral geographies, which still today attain great explanatory power as to the global struggle for centrality and the authority to represent the world. This is a condition with broad historical and inter-sectorial relevance. For instance, while the clinical sciences during the late eighteenth century established abstract systems for producing value-free, systematized knowledge for the 'good of all people', the need to regulate the use of these systems and to control their expansion, also made them the object of more chauvinistic battles of national and regional prestige. Foucault (1963/1994, 42–3) describes the early hierarchization or duality between the 'ordinary', referring to medicalized regional centres in which diseases could not be set off from other social ills, and the 'extraordinary', referring to enclosed and exclusively medical spaces in which the clinical gaze isolated diseases within its strictly scientific knowledge structure. In a similar manner, as Mattelart (1996/2000) has shown in his history of network society (from 1794 and onwards), the structural understanding of global media flows as a sign of development has had a continuing tendency to coalesce with sedentarist ambitions to build national centres of control and reality maintenance.

A key conclusion of the above discussion is that the symbolic power of producing fixity must be understood in processual and relational terms. Fixity transforms into a desirable asset only when it is expressed and recognized in relation to, or rather within, dominant ideologies of communication, space and knowledge. Its confirmation necessitates a certain kind of gaze, which looks for regimes of spatial order and predictability in the midst of unbound circulation. Along this line of argument I will show how the IBC and the OCC at Expo 67 produced and celebrated an era of faster and more far-ranging information flows, while at the same time exposing the competitive logic of fixing these very flows. As historical cases they provide an extracted view of how the volatility of flows is always the precondition for the media's symbolic power to centralize. I will also provide a deeper account of how these pavilions – strangely realistic displays in the partly real, partly fictional, miniature world of Expo 67 – had the paradoxical (but fully logical) effect of further mystifying and epitomizing the workings of modern media.

The Flow-Space of Expo 67

Before turning to the strange spaces of the IBC and the OCC, we must consider the overall ideologies and visions that framed the production of Expo 67. The main point here is that Expo 67 worked as a test-ground for many of the most advanced architectural, infrastructural and media technological ambitions of the time. Above all, Expo 67 implied the climax of space-biased megastructure architecture, invoking a view of the future city as a society of flows, mediations, and open-ended mobility.

Under the universalizing theme *Man and His World*, Expo 67 became a huge success in terms of attention and attendance, attracting more than 35 million visitors during a period from April to October 1967. The exhibitions were part of the ongoing regeneration of Montreal, and took place mainly on two artificial islands, Île Sainte-Hélène and Île Notre Dame, created partly out of excavation debris from the ongoing Metro construction project (Lortie 2005). While integrated in the city's material infrastructure, Expo 67 was also a 'conceived space' (Lefebvre 1974/1991) transcending the boundaries of Montreal, articulating the city's ambition to become a Francophone world metropolis. City mayor Jean Drapeau even envisioned Montreal as the 'Paris of the New World', and in the *Official Guidebook* (1967, 7) Expo 67 was promoted as the utmost sign of this grand forecast: 'A city in full flow of expansion, where the contribution of every individual is appreciated and counts, Montreal is not merely a cosmopolitan metropolis: It is a city with a particularly universal concept – heightened by the advent of *Expo 67*.' The sedentarist vision of a prosperous heimat for the historically oppressed French majority population of Montreal and the province of Quebec was thus framed by the metaphysics of flow, through which the future space of Montreal was conceived of in terms of 'expansion', 'cosmopolitanism', 'metropolitanism' and 'universalism'.

Expo 67 represented an accentuated version of these intersecting values. Designed as a futuristic miniature world, fuelled by a thoroughly urban concept of the future, it worked as a medium for heightening Montreal's claims for becoming a world centre. Following the architectural *megastructure* movement, Expo 67 highlighted the integration of advanced urban transit networks and new media and communication technologies (Banham 1976). The most important influence came from the British Archigram group, whose biggest contribution to architectural history is perhaps Peter Cook's *Plug-In City* model from 1963–1964. Laid out as a landscape of inter-changeable tubes and capsules for monorail transit, covered walking, housing, work and leisure, the Plug-In City forecasted an open-ended city structure that would yield to spontaneous leisure and individual desires (Sadler 1998). Similarly, the Master Plan for Expo 67 stated that the site was to be marked by urbanism and visual continuity, foreshadowing 'the city of the future with mechanical traffic entirely separated from walkways' (Fiset 1968, 48; see also *Rapport*

Général, Tome III 1969, 1330). Consequently, the main transit systems, the *Expo Express*, the *Metro*, and the *Minirail*, were all above- or underground constructions, partly integrated with some of the main pavilions.

The Expo site was thus planned as a futuristic flow-space, based on thorough calculations of expected streams of people and goods, while yet upholding an aura of playfulness and fun. The encoding of integrated flows as a sign of futurity was further enhanced by new media, which were part of Expo's socio-material structure to an extent that by far surpassed the experiences of everyday life, utilized for information, surveillance, and as expositions in their own right. Like most event spaces and tourist resorts Expo 67 attained a heterotopian character, following an 'ecology of fantasy' (de Cauter 2004, Part II) separate from everyday life. What fascinated the visitors the most, according to surveys, were the exhibitions of multi-screen cinema. For instance, the most 'impressive' pavilion according to the visitors was the Telephone Pavilion (mentioned by 18 per cent), and the highest popularity ratings (ranging from 0–100) were achieved by the Telephone Pavilion (100), the Czechoslovak pavilion (95), and Labyrinth (90) (*Rapport Général*, Tome V 1969, 3058).

In the Telephone Pavilion, a joint venture between Canadian telephone companies, visitors could indulge in the Disney produced Circle-Vision 360° movie *Canada 67*, telling the story of *Canada and Communications*. The film was projected on nine screens mounted in a circle, with an audience of 1,500 wrapped in the middle 'in a cocoon of sight and sound' (Fulford 1968, 63). The Czechoslovak pavilion introduced the *Diapolyecran* multi-screen installation, in which a mural of 112 square cubes, each containing two slide projectors, generated unpredictable hallucinatory mosaics of forms and images. Entering the Labyrinth, which was produced by the National Film Board of Canada, people were drawn into a 45-minute adaptation of the Minotaur legend of ancient Greece. The film was presented on screens mounted on the floors and ceilings as well as on the walls of three different chambers, through which the audience was distributed (*Expo 67 Official Guide* 1967, 57).

While these media spectacles created new, or at least intensified, sensory experiences of space, there were also more subtle forms of mediation at work – forms that operated in-between the spectacular and the infrastructural dimensions of Expo 67. The best examples are the *Operations Control Centre* and the *International Broadcasting Centre*, which were both designed as technical displays, but whose nerves extended beyond the very pavilions, and whose design had more to say about modern communication ideology than the cinematic pavilions.

Flow Monitoring as Public Spectacle: The IBC and the OCC

Compared to other Expo pavilions, the IBC and the OCC were not merely experimental showcases of technological potentials, but also, and more significantly, administrative centres for observing and controlling the intense flows of people, goods, and information criss-crossing the Expo site, and connecting it to the surrounding world. At the same time they were designed as see-through laboratories with big panorama windows, exposing their inner spaces to a fascinated Expo audience. Rather than hiding media technologies and their ways of operation, the aim of the OCC and the IBC was to *reveal* how new media were used within two emerging cultural forms.

As I will show in this section, the combined effect of pavilion design and promotional discourses was a spectacle in which the mediatized monitoring of flows became the core attraction, and in which Expo 67 was reproduced as an integrated flow-space, linking up to the circuits of international communication. Then, in spite of their functional differences, the IBC and the OCC pavilions expressed a similar space-biased logic, which was on the one hand superficial and simulated, due to the temporary and exceptional nature and location of Expo itself, but on the other hand perfectly 'real', since it was thoroughly interlaced with the ideologies and developments marking the emerging global modernity of the 1960s. The pavilions testified that the media saturated world was not only something to *believe in*, but something to *live in*. They not only conceptualized flow-space, but produced it as a miniaturized *lived space*, no less real than the world outside. Within a lived space, as conceived by Henri Lefebvre (1974/1991), the distinctions between reality and myth, the perceived and the imaginary, are no longer legible.

The IBC was built and governed by the *Canadian Broadcasting Corporation* (CBC), and began operating already in early 1967 (Figure 12.1). It became well-known for colour-broadcasting the grand opening ceremony of April 27 to an estimated more than one billion people audience around the world via the Earlybird satellite – a historical moment of Canadian pride. As presented in guidebooks and other documents, the IBC constituted an epicenter of international media diffusion with the central control room as its main feature (Figure 12.2) – 'designed to beam television and radio programs anywhere in the world, wherever they originate on the Expo grounds' (*Expo 67 Memorial Album* 1968, 308). The notion of 'beaming' information underscores the scripting of the IBC as a privileged site of diffusion and control – a fixture in relation to both incoming and outgoing media flows. Simultaneously, the pavilion held a typical observational function, gathering incoming reports from three mobile radio units, and five mobile TV units 'often seen on the Expo site' (ibid.). The IBC and its additional units thus *performed the modern ideal of broadcasting* as a process of *observation and diffusion.* The localized production of television and radio flows became a matter of expressing the

**Figure 12.1   Exterior of International Broadcasting Centre, from the Expo 67
Memorial Album (de Lorimier 1968), private collection**

social centrality of broadcasting, while simultaneously monitoring the gazes
of a dispersed, global media audience.

Located close to the Métro station, many Expo visitors encountered the
OCC early on their visits. It was not open to visitors, but people could watch
the inside operations through huge glass windows. In contrast to the IBC, the
authority of the OCC was defined more exclusively in relation to the internal
flow-space of Expo:

> The hub of day-to-day activity is […] the Operations Control Centre on Île
> Sainte-Hélène. Dominated by a huge illuminated chart of the site, the control
> room can check any happening, from a lost child to a fire alert. Closed-
> circuit television cameras allow controllers to inspect likely bottlenecks and
> trouble spots. Roving reporters check on crowds and queues, reporting back
> by radio. Computers keep records of admission, car park capacities and
> traffic densities. It is possible to communicate with any pavilion by teletype
> machine, or with the public by means of eleven giant electronic boards
> spaced around the grounds […] Through windows at ground level the public
> can see the space-age equipment, the electronic wands that combine with
> humans to weave the expo spell. (Gladstone 1968, 20)

Figure 12.2   IBC control room and hostess, frontpage of Montréal '67 (March 1967), private collection

The above description gives an outright view of the metaphysics of flow, stating that 'bottlenecks', 'crowds', 'queues', and 'traffic densities' must be governed, and ultimately eliminated. The means for handling unwanted restrictions of flows was to apply new means for information gathering, processing and dissemination – radio, teletype, telephone, loudspeakers, and electronic information boards – that is, an integrated communication system with its own logic of flow. All information was ultimately gathered and processed by the OCC, including weather reports, airport information, etc, and passed on to visitors and staff on the grounds. The OCC was crucial for the encapsulation of Expo 67 as a regulated heterotopian flow-space. Its crew monitored an infrastructure that was presented as one of the most mediatized and sophisticated ones in modern history, logically associated with the 'space-age' (cf Colomina 2002). As stated in the *Rapport Général* (1969, 2051), 'for the first time in the history of international exhibitions, electronic information boards [the most complex and subtle of all existing systems (ibid.)] have served the purpose of communicating information and guiding the visitors.' Due to the panoptic view of abstract visitors and events, the OCC crew could work as directors of flow, or so it was told. They could even steer people in certain directions by means of information about happenings, special offers, and queues to avoid, as well as more direct interventions:

> In addition, this facility of crowd control has, in the event, also turned into something like party-giving on a huge scale. Should the situations room see a bottleneck on the site, the gallant Expo Band is ordered into action to siphon off some of the crowd; or if an area of the site looks dull, then a mobile pop group can be driven over there to liven up the proceedings. (Baker 1968)

Both the IBC and the OCC can be understood as small-scale models of how the ability of fixing flows (of information, crowds, events, and so on) in a mediatized society becomes sacralized and turns into a symbolic asset, even a form of art, of a kind similar to the clinical gaze, which leading medical thinkers of the late eighteenth century associated with 'the art of describing facts' (Foucault 1963/1994, 114). The metaphysics of flow, thus, cannot be uncoupled from the modern fascination with nodes or fixtures, attaching the means and skills of mobility maintenance to certain local settings, such as early-modern railway and telegraph stations, and more recently broadcasting centres and television towers. The power of these centres to produce flows stands in direct proportion to their power to generate fixity – which is obviously not a matter of bringing an end to flows, but creating spatial representations of emplaced technical mastery. As we will see, however, the exceptional public display of this kind of mastery also invokes a paradoxical state of displacement when viewed from the outside.

Strangeness as Serialized Displacement

What makes the IBC and the OCC particularly interesting in the context of this book is their ambition to expose media governance in 'reality', behind the scenes, and to do so by means of an architecture of transparency. The pavilions represented the art of monitoring flows not only in their capacity of materialized media emplacements, or fixtures, and through public discourse, such as guidebooks, but also through their direct exposure of the sacred high-tech spaces of professionalized communicative practice. As we have seen, the OCC revealed controllers and 'space-age equipment' in operation behind panorama windows. Similarly, the IBC was designed as a partly transparent pavilion:

> The International Broadcasting Centre is situated in Cité du Havre [the Downtown waterfront area] and visitors will be taken on guided tours. Through large slanted window panels they may watch artists and technicians at work. An observation gallery has been provided for one of two large television studios, the largest the CBC has ever built. Visitors will also be able to sit-in for all kinds of programs, news to opera, in the studio. (*Expo 67 Official Guide* 1967, 168)

The viewing positions sustained by the transparency of the IBC and the OCC were anything but clear-cut, however. Nor were they entirely bewildering or confusing. Rather, they were marked by a strange *ambivalence*, due to the supreme circumstance that glass architecture tends to mystify the distinctions between concealment and revelation, simulation and reality, spectacle and observation. To further explain this ambiguity I will deconstruct it, and represent it as a series of four displacements.

The first displacement stems from the condition that the Expo visitors, through the mediation of the exhibition heterotopia, were transported from their ordinary positions as citizens and media audience. As Nick Couldry (2000, 2003) found in his fieldwork among people encountering real sites of media production, such as the full-size but strangely artificial stage set of Coronation Street, the sudden transition from being in front of the television screen to being behind the scene, or 'on the actual place', attains a liminal quality. While finding oneself at the site of production is an extraordinary experience in itself, due to its temporary and highly regulated nature, it is also an experience that puts all previous media experiences into a new light. It has the potential of transforming the visitor's subjectivity – the future composition of his or her consuming gaze. This was very much the case at the IBC, where particular tour guides showed visitors around, exposing them to the inner holiness of modern broadcasting, and where people could also participate as a live studio audience (cf Livingstone and Lunt 1994).

While these efforts were framed by a discourse of enlightenment, or what we today might call 'media literacy education', with their see-through panoramas and pedagogic hostesses-as-instructors, on a more structural level they also worked as spectacles reproducing the sacralized status of mediation. While Jonathan Crary (1999) in his history of modern public perception shows that most media, and especially twentieth century television, have operated as systems for the *management of attention* – as amalgamations of spectacle and control – the IBC constituted an extension of this condition. The pavilion re-articulated, through carefully planned tours and realistic displays, the pre-existing role of film and broadcasting as technologies of attraction (see also Gunning 1989), while at the same time preventing the audience from entering its most secret spaces – now relocated as the centre of attention.

Secondly, visitors were displaced in relation to their ordinary positions as spectators. When gazing at the central control rooms of the two pavilions, what people saw was something quite strange given the context of Expo 67 as an exhibition of colorful and extravagant performances. It was not so much the pedagogic exposure of technologies that was odd, but the lack of any obvious acts of performance – the exposure of seemingly routinized practices of professional observation. Designed according to the scientific ideal of systematized observation, as a kind of 'communication clinics', the control rooms celebrated the modern ideology of objectivity and professionalism among journalists, media technicians and controllers.

Although the late 1960s was a period in which such an ideology was contested, for example through the emergence of cross-over genres and 'new journalism' (see Wolfe 1973), the dominant view of journalists and media practitioners was reminiscent of the older view of physicians examining bodily symptoms, seeing through to underlying realities in a purely objective manner. A common depiction of the journalist was still the critical 'watchdog', working to reveal the truth about society.

A similar mythology framed the panoptic techniques that were in the hands of the surveillance personnel of Expo, and told to be 'the most advanced in the entire civil sector of Canada' (*Rapport Général*, Tome IV 1969, 2023, my translation). In the OCC control room 10 scientifically educated controllers (all men), assisted by 11 skilled telephone operators (all women), strived to attain an abstract, panoptic view of the flows of people at Expo 67 (Figure 12.3). There was allegedly something ghostly about these people and their silent and gestureless use of monitors, audio systems, switch-boards and telephones, all converging around a cartographic overhead panel representing the Expo area. Was it a scientific obsession with remote control? Or simply voyeurism hidden behind a façade of rationality? Probably, the silent sight evoked different albeit systematically structured connotations adhering to a mediatized, scripted gaze (Lagerkvist 2004; Jansson and Lagerkvist 2009) formed in symbiosis with popular media phantasmagoria such as science fiction movies with mad scientists or eccentric technological experts.

Figure 12.3   Operations and Control. © Library and Archives Canada

While chronologically subsequent, a thematically striking parallel here is the ambiguous hero Harry Caul (Gene Hackman) in Francis Ford Coppola's movie *The Conversation* from 1974, a wiretapping expert obsessed with his own privacy and the art of fixing other people's talk on tape. Acquainted to the male engineers working at the OCC, mastering new media as extensions of their bodies (cf. McLuhan 1964), Harry Caul represents a more or less detached, machine-like person, whose own construction and use of sound recording equipment attains a corporeal presence in his life. Caul is hired to spy upon a loving couple, and even manages to mix a soundtrack of their distant conversation as they walk around Union Square in San Francisco. Eventually, however, he is made the object of his own surveillance techniques. While *The Conversation* was produced shortly after the Watergate affair, in a much more ambiguous era and context than Expo 67, and provides a complex critique of the commercial spectacles and dubious ethics saturating the surveillance industry, the two events are part of the same representational circuits. They operate within the same mythologized lived space (Lefebvre 1974/1991), through which media control unfolds as an alien yet socially pervasive force of modern society and everyday life.

The third displacement came with the strange viewing position that the IBC and the OCC seemed to call upon. According to Foucault, the secret of observation is the capability to attain knowledge from a distance. And that is what the Expo goers were encouraged to do. The promotional imagery, as we have seen, invited the audience to examine the truth 'behind the scenes', that is, to parallel the observing gaze appropriated by the controllers inside the pavilions (especially the OCC crew). Viewers were encouraged to discover and examine the material and technological basis, even the 'truth', of the expanding visual culture of broadcasting and surveillance. Glass windows created an X-ray sight of what modern information hubs, normally hidden behind the machineries and administrative walls of everyday life, *looked like* in reality. In addition, visitors were not just observers. They were themselves immersed in mediation, either as the anticipated audience of popular media, or as the objects of crowd control. Gazing at the IBC and the OCC was to observe the systems observing oneself, getting an abstract view of the textures one was oneself part of producing. The utmost sign of this condition was the cartographic overhead flow-panel in the OCC, epitomized as the interface of man-machine interaction. Due to its large-scale, iconographic design it was immediately understandable and could thus work as 'evidence' of the odd clinical gaze – that spectators were transformed into observers, and that the object of their observation was also real processes of observation and monitoring.

The fourth and final displacement has to do with the significance of glass walls, and was perhaps the most crucial one, since it made the other three displacements seem unreliable. Both the IBC and the OCC were designed to provide sweeping vistas, enabling people *to gaze in without being seen* by the people working inside. While visitors were able to *see* what was normally hidden, they could thus not take part in, or interfere with, the processes inside the pavilions. Nor could they hear, or in any other way sense, what was going on (unless they participated as a live studio audience). In this context, as opposed to the 'purity' of Foucault's clinical gaze, the elimination of all other senses left the spectators partly knowing, partly bewildered.

This raises important questions as to the experiences of the visitors. Was it really a media centre they were seeing? Where the people inside the communication clinics actually observing or disseminating anything, and if so, what? And was it really an observing gaze the pavilions evoked? These are not necessarily questions that the visitors asked themselves – but they are crucial to our understanding of the spatial ambiguity these pavilions, more than any others, created.

Evidently, at the IBC and the OCC not so much was actually revealed. What the visitors could *not see*, for instance, was that behind the glass walls the OCC had much less practical use and was less efficient than the spectacle gave the impression of. In the evaluation from the Operations Control Unit a long list of technical problems were reported, underscoring that the cultural form of new media was to some extent only hypothetical, or superficial (see *Rapport Général*, Tome IV 1969, 2040–54). Some examples:

- the 32 cameras of the CCTV system were far from enough to cover as much of the site as intended, especially since the flow projections partly failed;
- the cameras were useless after night break;
- most of the cameras were fixed, and filled merely an 'ornamental function' (ibid., 2048). The evaluation asked for a system with fewer but more flexible cameras;
- the TV screens in the control room were positioned in an angle that made it difficult for the staff to discern the picture;
- the teletype machine at the command table was experienced as too noisy and was therefore mostly used after closing-time, at 10 pm, for assembling weather reports;
- the amount of technical equipment in the control room left too little space for conducting ordinary paper work.

Considering shortcomings like these, the evaluation concluded that 'the OCC constituted an interesting element of the exposition, which captivated the visitors, although they were just able to view it from outside' (ibid., 2054, my translation from French).

Glass walls were thus the ideal means for putting up an illusion, an artificial world that appeared to be accessible, but yet, as Thomas Mical (2001, 95) has argued, remained as alien as a cinematic space: 'Glass architecture announces hygienic modernity and the dream of a purely transparent society. But in practice glass walls hide what they propose to expose, projecting distorted and phantom reflections from within their intended transparency.' This means that while the glass walls of the IBC and the OCC celebrated a modern ideology of the observing, or perhaps 'communication clinical', gaze, they produced sights that rather called for a cinematic gaze. Standing outside was a means of being encapsulated and carried away. As Crary (1999, 72–74) argues, spectacles and surveillance do not necessarily stand in opposition to one another, but are historically aligned. While the direction of the gaze is inverted, spectacles work as pervasive means of controlling people's mode of attention – so long as spectatorship is not turned into penetrating observation. Like the lens of the cinematic projector, the panorama windows of the IBC and the OCC enhanced the spectacle's social role as a technology, not only of attraction, but also of separation (see also Debord 1967/1994).

Accordingly, it was clearly not an observing gaze that the Expo visitors enacted, but just an imitation of it. As opposed to the medical doctor, or the journalist or surveillance expert, the ordinary Expo visitor standing outside any of the pavilions had no code for the translation of visible facts. As Foucault (1963/1994, 107) notes, while concerned with the immediately visible, or symptoms, the authority of the observing gaze lies in its capacity to 'see', or 'listen to', what is *behind the obvious*: 'The observing gaze refrains from intervening: it is silent and gestureless. Observation leaves things as they are; there is nothing hidden to it in what is given.' In the case of the OCC and the IBC, however, the registration of visual keys opened not as much a realm of objective truth, as an ambiguous realm of modern media ideology and spatial metaphysics.

Conclusion

In this chapter I have attempted to show how the *International Broadcasting Centre* and the *Operations Control Centre* at Montreal's Expo 67 reproduced, by means of a replication of the observing gaze, what I understand as an overarching spatial ideology of modern communication. I have argued that the aura of media emplacements, whether 'permanent' or as in this case temporary and partly simulated, emanates from, and legitimizes, an ideological pattern that articulates flow and circulation as taken for granted assets, but at the same time, precisely because of the ideal of a fluid flow-space must elevate the capacity of producing *fixity* as a keystone to social power. This observation suggests that within the ideology of modern communication the *metaphysics of fixity and flow* are always complementary, even mutually reinforcing, rather than competing (Cresswell 2006). The social, political and economic logic of producing *fixtures*, understood as strong points of circulation and symbolic power, is a matter of localizing the power to monitor and fix flows.

However, fixing flows, as we have seen, is in fact the very opposite of reducing them. In the process of controlling and disseminating primary flows (e.g. through surveillance and broadcasting), fixtures generate additional flows of incoming and outgoing information, as well as various forms of social and material mobility. What emerges is a multilayered flow-space, whose social significance extends beyond the flows themselves and into the multisensory fabric of social life – in this case represented by the mediatized Expo site and its related contexts in Montreal and elsewhere. Flow-space and its fixtures thus constitute what Lefebvre (1974/1991) calls a *lived space*, stressing that the imaginary realm is always interlaced with social praxis and power.

The identification of these structural dynamics finds resonance in my deconstruction of the IBC and the OCC as sights of *serialized displacement*. While I can show no evidence of audience experiences, nor any explicit accounts of the extent to which these pavilions were understood as 'strange', my ambition has been to pinpoint the more abstract *spatial ambiguity of anticipated viewing positions*, and how these were produced in interplay with architecture, media discourses, and the ideological structures of modernity.

As Nick Couldry (2003) argues, the symbolic power of modern media stems from their ritualized monopoly to decide and express what is *worth mediating*, marginalizing things that are not mediated, while hiding their own logics and machineries of operation. Exposing the processes of media production and surveillance through huge windowpanes, the IBC and the OCC altered this state, while still departing from, and pointing to, the heterotopian character of mediation. Inviting visitors to watch communication professionals at work 'behind the scenes', the pavilions epitomized a version of the modern observing gaze (Foucault 1963/1994), which in turn reproduced the spatial ideology of fixtures as privileged sites of control. While the experience of observing the processes normally concealed was strange in itself, glass architecture also separated the spectators from what they observed. Altogether, the IBC and the OCC represented and realized the modern art of monitoring flows through the medium of an incontestable, yet incomprehensible, cinematic spectacle – socially immersive, while at the same time beyond reach.

References

Anderson, B. 1983. *Imagined Communities. Reflections on the Origins and Spread of Nationalism*. London: Verso.

Appadurai, A. 1996. *Modernity at Large: Cultural Dimensions of Globalization*. Minneapolis: University of Minnesota Press.

Augé M. 1995. *Non-Places: Introduction to an Anthropology of Supermodernity*. London: Verso.

Baker, J. 1967. 'Expo and the Future City', *Architectural Review*, 846 (Special Issue on Expo 67).

Banham, R. 1976. *Megastructure: Urban Futures of the Recent Past*. London: Thames and Hudson.

Baudrillard, J. 1983. *Simulations*. New York: Semiotext(e).

Baudrillard, J. 1991/1995. *The Gulf War Did Not Take Place*. Bloomington: Indiana University Press.

Bauman, Z. 2000. *Liquid Modernity*. Cambridge: Polity Press.

Bourdieu, P. 1996/1999. *On Television*. New York: New Press.

Carey, J.W. 1989. *Communication as Culture: Essays on Media and Society*. New York: Routledge.

Castells, M. 1992. *The Informational City: Economic Restructuring and Urban Development*. Oxford: Blackwell.

Cauter, L. de 2004. *The Capsular Society: On the City in the Age of Fear*. Rotterdam: NAi Publishers.

Colomina, B. 2002. 'Enclosed by Images: Architecture in the Post-Sputnik Age', in Levin, T.Y., Frohne, U. and Weibel P. (eds) *Ctrl[Space]: Rhetorics of Surveillance from Bentham to Big Brother.* Karlsruhe: ZKM.

Couldry, N. 2000. *The Place of Media Power: Pilgrims and Witnesses of the Media Age.* London: Routledge.

Couldry, N. 2003. *Media Rituals: A Critical Approach.* London: Routledge.

Crary, J. 1999. *Suspensions of Perception: Attention, Spectacle, and Modern Culture.* Cambridge, MA: MIT Press.

Cresswell, T. 2006. *On the Move: Mobility in the Modern Western World.* London: Routledge.

Debord, G. 1967/1994. *The Society of the Spectacle.* Cambridge, MA: Zone Books/MIT Press.

Expo 67 Official Guide. 1967. Montreal: Maclean-Hunter Publishing.

Fiset, E. 1968. 'The Master Plan', in Lorimier, J-L. de (ed.) *Expo 67, Montréal, Canada: The Memorial Album.* Montreal: Thomas Nelson and Sons.

Foucault, M. 1963/1994. *The Birth of the Clinic: An Archeology of Medical Perception.* New York: Vintage Books.

Foucault, M. 1967/1998. 'Of Other Spaces', in Mirzoeff N. (ed.) *The Visual Culture Reader.* London: Routledge.

Fulford, R. 1968. *Remember Expo: A Pictorial Record.* Toronto: The Canadian Illustrated Library.

Gladstone, J. 1968. 'Magic Island', in Lorimier, J-L. de (ed.) *Expo 67, Montréal, Canada: The Memorial Album.* Montreal: Thomas Nelson and Sons.

Gunning, T. 1989. 'The Cinema of Attractions: Early Film, Its Spectator and the Avant-Garde', in Elsaesser, T. and Barker, A. (eds) *Early Film.* London: British Film Institute.

Innis, H.A. 1951/1964. *The Bias of Communication.* Toronto: Toronto University Press.

Jansson, A. and Lagerkvist, A. 2009. 'The Future Gaze: City Panoramas as Politico-emotive Geographies', *Journal of Visual Culture,* April.

Lagerkvist, A. 2004. '"We See America": Mediatized and Mobile Gazes in Swedish Post-War Travelogues', *International Journal of Cultural Studies* 7:3, 321–42.

Lefebvre, H. 1974/1991. *The Production of Space.* Oxford: Blackwell.

Livingstone, S. and Lunt, P. 1994. *Talk on Television: Audience Participation and Public Debate.* London: Routledge.

Lorimier, J-L. de (ed.) 1968. *Expo 67, Montréal, Canada: The Memorial Album.* Montreal: Thomas Nelson and Sons.

Lortie, A. 2005. 'Montreal 1960: The Singularities of a Metropolitan Archetype', in Lortie, A. (ed.) *The 60s: Montreal Thinks Big.* Montreal: Douglas and McIntyre/Canadian Centre of Architecture.

Mattelart, A. 1996/2000. *Networking the World 1794–2000.* Minneapolis: University of Minnesota Press.

McLuhan, M. 1964. *Understanding Media: The Extensions of Man*. New York: McGraw-Hill.

Meyrowitz, J. 1985. *No Sense of Place: The Impact of Electronic Media on Social Behaviour.* New York: Oxford University Press.

Mical, T. 2001. 'Berlin's Cinematic Spaces: Dialectical Analepses and Fugues', *Spectator – The University of Southern California Journal of Film and Television*, 21:1, 91–105.

Montréal '67. 1967. March issue. Montreal: City of Montreal.

Morley, D. and Robins, K. 1995. *Spaces of Identity: Global Media, Electronic Landscapes and Cultural Boundaries*. London: Routledge.

Rapport Général sur l'Exposition Universelle de 1967: Tome III–V (1969) Montréal: La Compagnie Canadienne de l'Exposition Universelle de 1967.

Relph, E. 1976. *Place and Placelessness.* London: Pion.

Robertson, R. 1992. *Globalization: Social Theory and Global Culture*. London: Sage.

Sadler, S. 1998. *The Situationist City.* Cambridge, MA: MIT Press.

Sturken, M. and Cartwright L. 2001. *Practices of Looking: An Introduction to Visual Culture.* Oxford: Oxford University Press.

The Conversation (dir. Francis Ford Coppola, 1974).

Tschumi, B. 1994. *Architecture and Disjunction*. Cambridge, MA: MIT Press.

Wolfe, T. 1973. *The New Journalism*. New York: Harper and Row.

Chapter 13

Modern Moon Rising: Imagining Aerospace in Early Picture Postcards

Johanne Sloan

A series of photographic postcards issued by the Reutlinger Photographic Studio in Paris, and mailed in the years 1906 to 1908, feature elegant women in the vicinity of bright crescent moons, surrounded by atmospheric, dark, star-studded skies (Figure 13.1). On some postcards these figures hover casually in the lunar region and on others they stand or sit directly on the moon-shape; draped in fabric or wearing diaphanous dresses, these women inhabit the cosmos without any indication of a rocket, biplane, airship or other mechanical conveyance that might have deposited them there. The Reutlinger moon postcards contributed to a burst of imagery, lasting from the opening years of the twentieth century until World War I, whereby aerospace (a term used here to encompass sky, atmosphere, and cosmos) was represented as an accessible and populated zone. New kinds of flying machines were making their way through the sky at this time, and to some extent this new imagery attests to how the space above the surface of the planet was rapidly becoming a technologically-conquered territory. The fanciful moon imagery in question suggests a kind of imaginative excess, however, which also coincided with this territorialization. The 'outer space' occupied by these fashionable Parisian space-travellers can indeed be regarded as the strangest part of this newly configured aerospace. And I want to suggest that the potential meaning of this alluringly remote and strange spatial zone would be re- negotiated every time one of these postcards was bought and sent.

The Reutlinger postcards represent one of the means by which the moon came down to earth at this historical moment, in the sense that the newly-invented picture postcard helped to transform this extraterrestrial body, and aerospace more generally, into intelligible visual representation, commodified cultural artifact, and mode of communication. The postcard moonscape is inevitably modern, therefore, even in the absence of that key iconographic element – the machine or invention which would convincingly suggest space travel – because the picture postcard itself attests to modern modes of consuming and disseminating visual images. The picture postcard craze of the first decade of the twentieth century offered the public an ephemeral array of little images, for the purpose of personal communication. And if this kaleidoscopic image-world is a characteristic of modern mass culture, what

Figure 13.1   Postcard issued by Reutlinger Photographic Studio, Paris, postmarked 1906

differentiates the nascent postcard is precisely that individual consumers were called upon to personalize and annotate these images as they sent them off to lovers, family members, friends, and acquaintances. We cannot learn very much about the Parisian woman named 'Paulette' who signed her name to multiple Reutlinger moon postcards, all sent to a 'Mademoiselle Sempé', but such concrete examples of postcard-sending give some indication of how this strange lunar imagery became enmeshed in networks of everyday, sentimental, and social transactions.

Taking Flight

The rapid development of aviation in France from around 1906 triggered a widespread fascination with flight, conveyed in poetry and prose, and in picture-form in newspapers and illustrated magazines.[1] The pioneers of French aviation and their new heavier-than-air flying machines also appeared on postcards: Louis Bleriot, first to fly across the English Channel in 1909, aviator/plane designers such as Gabriel Santos-Dumont and the Voisin brothers, for instance, are all pictured repeatedly. These postcards are generally derived from black-and-white photographs and are printed matter-of-factly on non-glossy paper; they show the aviators at ground level, posing next to their aircraft, or a dark line of spectators stretches across the horizon, while the small shape of an experimental aircraft is silhouetted against a pale and featureless sky (Figure 13.2). To humanize these views and because the first aviators quickly became celebrities, an insert of the pilot's face was often added to the picture; still, these are rather stark images. But as the first boxy biplanes got off the ground, this kind of 'straight' photographic document was very rapidly complemented by postcards featuring fanciful, colourful versions of air-travel. Those Reutlinger women aloft in a lunar zone were hardly the only fantastically airborne people to appear in popular imagery, in other words.

French postcard makers issued many photographic images showing ordinary people (many women, but also men and children) 'flying' through skies that are tinted various shades of blue and embellished with picturesque clouds, aboard facsimile aeroplanes adorned with garlands of multi-coloured flowers. These studio productions produce the illusion of seeing airborne bodies and faces up-close – a visual proximity which the more realistic aviation imagery was unable to provide. Realism is not the goal here, though: the occupants of these planes are not wearing the goggles, helmets and warm clothing appropriate to piloting a machine through the air, but rather, they wear regular street clothes, even as they look perfectly at ease. In fact, the 'pilots' in question are invariably

---

1   Robert Wohl (1994, 38) writes that, 'the capital of aviation before the First World War was indisputably Paris', while celebrity status was conferred on the first generation of flyers.

20 BÉTHENY (25 Août 1909) — La foule acclame PAULHAN qui bat le record du monde

Figure 13.2   Postcard (no publisher) from 1909 documenting a flight by the French aviator Louis Paulhan

exuberant and smiling, and this celebratory mood is enhanced by the floral arrangements and by the heightened colour itself. All this helped to convey the impression that aerospace could be a veritable pleasure-dome, available in some sense (imaginatively if not physically) to one and all.[2]

The multi-categoried world of early postcards included many other variants on flying machines or flight, as well. The *Bon Marché* department store in Paris issued comical postcard-advertisements showing a futuristic aerial delivery service to be someday offered by the store; there are postcards of people holding umbrellas above their heads as they somehow glide above various European cities; on another card postmarked Philadelphia 1899, a gleeful woman rides a bicycle through the air above the city of Hamburg, balanced on a cord, one end of which is tethered to a crescent moon. I am suggesting a contiguity between this expanded category of fantasy-flight imagery, and the moon postcards, even if the Reutlinger space-travellers appear to have ventured somewhat further afield in this aerial landscape. The dream of reaching out to the moon is of course quite ancient, but this lunar

---

2 The fantasy-aeroplane genre also became a standard feature of French fairground photographers, who provided a range of props and backdrops as an enhancement of the portrait genre; see Chéroux (2005). The 'paper moon' was a comparable studio prop across the United States and in the UK, but does not seem to have taken hold in France.

desire fuses to some extent with the modern promise of flight in the early years of the twentieth century.

Communicating with Pictures

Postcards provided a new way of communicating with pictures at the beginning of the twentieth century. These early postcards must be regarded as complex pictorial/textual constructions, as the images themselves are very often hybrid amalgams of photography, graphic design, painting or drawing, while the postcard will normally include a combination of printed, stamped, and handwritten texts. Postcards are striking examples of how mass-produced objects are appropriated and personalized, as each scribbled or beautifully cursive mark on the surface of the purchased item is the inscription of subjectivity, and a way of anchoring that object to a distinct time and place. The images on postcards cannot be understood without considering this communicative dimension, while the postal system which enables the (spatial) distribution of postcards is also key, as part of a modern network whereby information and goods can be exchanged efficiently and rapidly. Postcards are literally set in motion: addressed and mailed, they describe trajectories between people and through social space; as 'moving images' of a kind they were therefore part of the dynamic visual culture of the period.

While some types of mailing cards were introduced earlier, picture postcards were officially introduced in the 1890s: in this initial 'undivided back' era, one side of the postcard was exclusively reserved for a name and address, so that the message – if any – had to be squeezed onto any available space on the picture side. The form of the postcard which is familiar today, with a picture on one side and a so-called 'divided back' on the other, was gradually introduced in England in 1902, in Canada in 1903, in France in 1904, and the US in 1907 (Staff 1967, 66). In the years leading up to World War I, large parts of the Western world seemed to go postcard-mad. It has been recorded that approximately 300 million postcards were produced in a single year in France, c. 1905 (Cheroux 2007, 196), although even this might be a conservative estimate.[3] At any rate, we might well imagine that the air was filled with postcards at this time. The cheap postage – and in many places, multiple daily deliveries – meant that people could use postcards to communicate casually and rapidly, and it is tempting to compare this flurry of exchanges to the contemporary phenomena of email, text-messaging and instant image transmission. What I want to emphasize is that the cultural

---

3   It is very difficult to ascertain statistics about postcards and, indeed, serious research on postcards is hindered because 'most of the information on this novelty item was not collected, catalogued, or housed like other historical materials' (Woody 1998, 22).

identity of the early postcard is quite different from its later incarnation as the exemplary touristic souvenir, sent when someone takes a holiday. Instead, these first postcards are very often embedded in the immediacy of urban experience, and are an extension of everyday urban transactions.

Postcards are not the obvious object of study for any one scholarly discipline, and indeed these diminutive, mostly commercial, anonymous and ephemeral artifacts have often been overlooked.[4] *Delivering Views*, an important book edited by Christraud M. Geary and Virginia-Lee Webb in 1998, examines the genre of ethnographic postcard from a contemporary postcolonial perspective, and this continues to be an influential methodological model (Geary and Webb 1998). Other scholars have invented distinctive interdisciplinary nodes for their study of postcards.[5] And then, most accounts of modern art and most art-oriented histories of photography have relegated postcards to a marginal position.[6] Even the 2007 exhibition *The Stamp of Fantasy: The Visual Inventiveness of Photographic Postcards*, organized by Clément Chéroux (curator of photography at the Centre Georges Pompidou), while vaunting the 'inventiveness' of these half-forgotten images, still tends to privilege the taste and judgment of the artist/collector, who is able to recognize in these images the 'form of poetry present in raw material state in popular culture' (Cheroux 2007, 203). This notion of 'raw material' reinforces the rather old-fashioned idea that the kitsch properties of pop culture must be elevated and redeemed by art.

Still, this exhibition held at prestigious European venues including the *Jeu de Paume* in Paris, signals an important shift in consciousness with regard to the cultural worth of this kind of image. The photographic historian and theorist Geoffrey Batchen has argued for the category of 'vernacular photographies' as a way to address such heterogeneous and everyday examples of photographic activity. This approach is part of a discursive momentum, over the past 25 years or so, away from a singular, monolithic, modernist history of photography, towards an acknowledgment of what John Tagg famously characterized as photography's 'complex and discontinuous history', embedded in a wide range of social and institutional practices (Tagg 1989, 11). Batchen's discussion

---

4   The term 'commercial' is too broad, really, to describe the diversity of postcard production, which could range from large companies boasting millions of cards in their inventory, to small-town entrepreneurs who provided local views and enabled local citizens to put their own homes and faces onto postcards.

5   For instance, Cary Nelson has brought his expertise in the areas of twentieth-century poetry, politics, and communications to the study of military-themed postcards that include some kind of poetry; see Nelson (2004).

6   Books about Dada and Surrealism, or those dealing with collage and photomontage, often include a few fantasy-genre postcards but only to demonstrate that crucial moment when serious artists took notice of these mass-produced items, and went on to refine the aesthetic manipulation and ironic juxtaposition of photographic imagery.

of the ordinary familial snapshot – an exemplary instance of vernacular photography – raises important questions about the imponderably large, multi-sited 'archives' of such images. Batchen emphasizes the conventionality and iterability of these pictures, but also wants to account for the fact that any given snapshot, while appearing banal and unoriginal to most viewers, might well be the most valued, moving, affective picture in an individual's possession. These insights suggest an interpretive framework that does more than merely isolate a few unusual specimens (whether snapshots or postcards), to then treat these as singular, aesthetically autonomous objects. Certainly the mass production, multiplicity and mobility of postcards cannot be ignored. And if postcards are meant to circulate through social space, this is to say that these images-on-the-move acquired meaning as they passed through the imaginative orbit of successive individuals. In the first years of the picture-postcard craze, the postcard was enthusiastically written on, mailed, tacked up, pasted into albums, exchanged and collected; these social habits and material circumstances are integral to the identity of these cultural objects.

The moon postcards in question were produced by the Reutlinger Photographic Studio in Paris, a well-established family firm that had been producing *cartes-de-visites* and other forms of photographic portraiture since the mid nineteenth century. Under the direction of Leopold Reutlinger, the nephew of the original founder, the studio in its twentieth century incarnation became one of the main suppliers of photographs for the *belle époque* entertainment scene, running the gamut from respected dramatic actors and opera stars to popular dancers and cabaret showgirls.[7] Many such photographs became postcards, while the studio also branched out into the production of novelty or fantasy postcards. Graphic inventiveness abounds throughout this studio's production, with fanciful art nouveau borders and shapes, plant-like letters of the alphabet, boldly outlined photographic details, etc. Female figures are a constant feature of the Reutlinger postcards, whether these are recognizable and named entertainers such as the dramatic actress Jeanne Granier, the opera star Lina Cavalieri, and the dancer Otero, to cite only a few of these personalities; or, as was the case with the lunar ladies, when anonymous models played a role in the studio's allegorical tableaux. These postcards are hybrid images: parts of the original black-and-white or sepia-coloured photograph are left untouched, but delicately-coloured washes or dabs of vibrant pigment are judiciously applied to the image, sometimes to the point of almost obliterating the photographic element. The same iconographic motif or pictorial effect would also be repeated across different categories of postcard, with rather different results: thus an expanse of photographic darkness serves as the smoky depths of a stage behind Jeanne Granier, while a similarly shadowy expanse becomes the nocturnal garden through which an

---

7 See Berlanstein (2001) for a discussion of the shifting social status of these entertainers.

attractively somnambulic woman wanders. In the case of the Reutlinger moon postcards, the creation of an otherworldly quality involved the deployment of artificial lights and studio props, but the pictorial effects of glowing moon and alluringly star-filled skies are primarily achieved through the manipulation of that same, foundational photographic opacity – and yet again, the strategic introduction of a few painterly touches makes all the difference.

The Moon as Destination and Metaphor

The affinity between fantasy-flight and imaginary moon voyages has been noted, but of course the dream of understanding, perceiving, and ultimately reaching the moon has a long history in the scientific and literary texts of the Western tradition. Marjorie Hope Nicholson's book *Voyages to the Moon* looks at notable examples from the 15th to eighteenth centuries, and her study is structured according to the different modes of accomplishing this journey: through supernatural means, with bird-like wings, with flying machines, etc. (Nicholson 1948). Cyrano de Bergerac's seventeenth-century space-traveller, for instance, has multiple dew-filled vials attached to his body, so that he could be air-lifted off the surface of the earth as the dew evaporated. A charming illustration of this very contrivance shows a moon that is somehow almost within reach of the human body. Although the technologies and means of representation change after the period covered by Nicholson, the moon certainly persists as an object of speculation and phantasmatic projection, and some fundamental desires remain constant, regarding how humans might become less earthbound.

By the early twentieth-century lunar imagery was apparent in high and low cultural spheres, through representations ranging from scientific projects to various forms of entertainment.[8] On the scientific front, the Paris Observatory published its *Atlas photographique de la lune*, under the direction of the astronomers Moritz Loewy and Pierre Puiseux. Published in 12 volumes between 1894 and 1909, its 80 photogravures provided detailed views of sections of the moon.[9] The combination of new telescopes and improved photographic techniques offered a visual contact with the moon never before possible, resulting in new cosmological knowledge. This scientific imagery had

---

8   I do not consider Pierrot imagery (derived from the *Commedia dell'arte*) in this essay, but as an example of contemporaneous avant-garde culture it is worth mentioning the composer Arnold Schoenberg's *Pierrot Lunaire*, a so-called 'melodrama', consisting of poetry spoken against an instrumental background, which premiered in Berlin in 1912. The work used twenty-one poems by Albert Giraud, originally published in French, and the 'moonsick' theme is explored throughout; see Green and Swan (1993, 199–204).

9   For a description of the Loewy and Puiseux Atlas, see Whitaker (1999, 151).

its direct pop-culture counterpart, as well, in the 'moon-voyage' available at the 1900 *Exposition Universelle* in Paris. This was one of an array of thrilling proto-cinematic entertainments whereby still images were endowed with a semblance of movement, often through elaborately-designed sets, sounds, and light effects. For example, one entertainment/mechanism simulated the Trans-Siberian train journey, by having visitors seated in a railway car, while painted scenery went whizzing past; another device recreated the visual and spatial effects of a balloon trip.[10] The moon-voyage in question initially used a sequence of scientifically-precise lunar photographs to illusionistically suggest to viewers that they were seeing the moon from up close. But then, as Rosalind Williams describes, this episode of realism dramatically segued into a fantastical voyage through space to a distant star. 'So convincing was the illusion that the spectator was hardly aware of crossing the line between reality and fantasy, of moving from a painstaking reproduction of the moon's surface to a wholly imaginary simulation of a journey beyond the galaxy' (Williams 1982, 75). Whether or not this imaginary voyage utilized the Paris Observatory's photographs, what this means is that the same genre of realistic astronomical imagery could be re-directed to very different ends – legitimate scientific inquiry or ephemeral pop culture diversion.

The films of George Méliès are of course essential to any discussion of imaginary representations of the moon in the first decades of the twentieth century. If Méliès's moon also becomes accessible through a combination of science and fantasy, the 'science' part of this equation is only loosely associated with astronomical inquiry, and it is instead the emergent technology of cinema which becomes the true object of admiration. In narrative terms, there is a mock-scientific dimension to the lunar-themed stories of *Voyage to the Moon*, 1902, and *The Eclipse of the Sun by the Full Moon*, 1907; the latter, less well-known film is of particular interest to this discussion. It can be noted that the space travellers of the earlier film are not exactly heroic figures, while in *The Eclipse* a group of pompous, old-fashioned astronomers appear in the opening scenes, but become quite irrelevant once the moon itself appears as a protagonist, interacting with other cosmic characters. The cinematic apparatus allows these absurd scientists with their clumsy machines to be bypassed, providing the viewer with immediate access to an anthropomorphized celestial zone. The eclipse itself is played out as an eroticized encounter between the sun and the moon, and subsequently the camera moves to another part of the extra-terrestrial atmosphere inhabited by humanized planets, moons, comets, shooting stars, etc. Méliès's version of outer space comes across as an activated, populated, libidinous and litigious social space, including a traffic-accident-like collision, with the cosmic citizens yelling at each other just as they would on the streets of Paris (Figure 13.3). The film closes with a more

---

10   Rosalind H. Williams (1982, 75) notes that 'at the 1900 exposition 21 of the 33 major attractions involved a dynamic illusion of voyage.'

Figure 13.3   Still-image from ***The Eclipse of the Sun by the Full Moon*, 1907,
a film by Georges Méliès**

serene, feminized meteor shower, but the lewdness and hilarity of the earlier sequences are most memorable.

In Méliès's universe, feminine personifications or articulations of astronomical phenomena abound: shooting-star girls, moon-ladies, various winged creatures, etc. Elizabeth Ezra (2000) has argued that while these airborne figures are everywhere in Méliès's films, women are not empowered through the image of flight or cosmic travel, as the narratives are likely to include some form of patriarchal slap-down, just when they are on the verge of achieving social agency or technological prowess.[11] The Reutlinger moon-voyagers certainly have much in common with Méliès's celestial creatures, not least because outer space is apparently accessed through the realistic visual technologies of photography or cinematography. While the filmmaker hired performers from the world of popular theatre (acrobats from the *Folies Bergères* play the lively lunar inhabitants in *Voyage to the Moon*, for instance (see Frazer 1979, 97)), so too do the models in the moon postcards resemble the many Parisian opera, theatre, dance and music-hall stars who appeared on other Reutlinger postcards. These are not timeless deities but rather historically-

---

11   See the chapter, 'The Amazing Flying Woman', in Ezra (2000, 89–116).

specific urban types, who evoke a lively nocturnal world of theatricality and feminine glamour. Ezra's argument about Méliès's disempowered flying women cannot so easily be transferred over to the moon-postcards in question, however. In contrast to the satirical and slapstick storylines which characterize the Méliès films, the postcard image – by definition – is not embedded in a single narrative sequence. Inserted into the postcard-sending circuits of the time, these particular lunar images would have been bought, collected, and sent in idiosyncratic sequences, and would be endowed with additional layers of meaning through inscriptions and annotations. This is an important point to which I will return.

The Reutlinger moon-postcards should be regarded, though, as a significant contribution to the science-fiction-like takeover of aerospace during this period. We have seen that the lunar and aerospatial environment was construed as a socialized, inhabited territory in turn-of-the-century representations: at the Exposition moon-voyage attraction one bought a ticket to become a space traveler along with hundreds of other visitors, and in the films of Méliès, outer space is crowded with people, is indeed the very site of human comedy. Many of the fantasy-flight postcards reinforce this impression of a populated cosmos. The Reutlinger women-on-the-moon seem to occupy the outer limits of this newly available aerospace, though, as if they had managed to escape the crowded conditions to be found elsewhere in the *belle époque* sky.

The Reutlinger postcards suggest a more extreme form of escapism relative to other aerospace imagery, by launching Parisian women beyond the domesticated extra-planetary *arrondissements*, to a more remote and even lonely zone where people and objects are scarce. These women fly solo, unencumbered by balloons, zeppelins, aeroplanes, or any of those machinic inventions which would imply that a voyage into space is necessarily a collective project, the result of shared knowledge and expertise. Out here, the stars distributed throughout the darkened sky create an illusion of great distances, and each solitary Reutlinger woman is oriented towards a panoramically-wide space. Neither the bodily alignment of the women nor the positioning of the camera which is apparently aimed at them suggest a downward or earthbound gaze; the points of view in these images are directed upwards and outwards. If this expansive spatial construct could, ironically, be contained within the reduced dimensions of the postcard, perhaps this is because such imagery remained intelligible as a variant of the landscape genre.

The Lunar Landscape

The sequence of Reutlinger postcards, together with the broader category of fantasy aerospace imagery, can be regarded as (strange) permutations or extensions of the landscape genre. Certainly huge numbers of the picture postcards produced during this era were landscapes of some description: local scenery, gardens, seaside or mountainous views, famous historic sites, exotic locales, colonial territories, etc. The French postcard industry would zealously set out to document place: every region, every town, every neighbourhood of Paris. Naomi Schor (1992, 216) commented about the methodical 'postcarding of Paris', that this visual project represents 'the positive taxonomic imperative, the bourgeois apprehension of the real as composed of interchangeable microunits'. Schor's comments about Paris can be extended to the postcard representation of France more generally: despite the heterogeneity of the locales, the majority of postcard landscapes can indeed be considered as 'microunits' of the academic European tradition, repeating the same pictorial conventions to achieve spatial coherence and ensure visual pleasure. The majority of postcard views become intelligible through the structural device of a centrally-positioned, slightly elevated, stable viewer: this point-of-view indeed becomes the benchmark of postcard banality. The ideological result of all this legibility and stability, spread out over innumerable 'microunits', was the promise of visual control over the world, and with regard to the representation of nature, the promise of a state of equilibrium between human agency and the natural world. Of course the genre of landscape as it developed within the European tradition included many possibilities for introducing elements of fantasy and exoticism into an otherwise conservative genre. Artists learnt to beautify or allegorize an ordinary stretch of terrain through various pictorial enhancements, while built into the landscape genre is a promise of imaginative dislocation, as a glimpse of scenery will transport the viewer to other sites (non-European locations) and other times (arcadian or antique pasts.)

It must be acknowledged that most of the postcards produced during this period were largely unaffected by the modernist experimentation with landscape, which had been on-going since the mid-nineteenth century. Landscape was central to the development of modern art, as several generations of artists became fascinated by the aesthetic and ideological flexibility of this genre. By the turn of the twentieth century, the translation of natural sites and forms into landscape image had become more abstracted, while the illusionistic representation of space had itself become an object of artistic inquiry. With Impressionist art, for instance, the air itself – or the space between people and objects – is rendered palpable and visible; with an accumulation of small, multi-coloured brushstrokes painters evoked an airspace full of movement and vitality, so that the most evanescent aspects of nature seemed imbued with a modern sensibility. Following on these experiments, the pictorial space of landscape pictures would be further flattened and compressed (Post-impressionism to Fauvism), and ultimately fractured and fragmented (Cubism, Futurism, etc.) In avant-garde terms, this was where experimentation with the landscape genre had led, by the time picture postcards emerged. Postcards are part of a commercialized, mass-produced image-world, which is to say they arise out of a very different milieu than the artworks mentioned above. Nonetheless, the lunar environments depicted in the Reutlinger postcards can be thought of as inventive engagements with the landscape genre, providing a parallel challenge to the spatial conservatism of traditional landscape imagery.

The aero/lunar landscapes in question share a range of basic pictorial characteristics with the academic landscape tradition, by recreating an enveloping natural environment for human action, by differentiating spatially between foreground interest and background context, by establishing strong figure/ground relationships, and so on. The moon postcards are different from conventional landscape views in various ways, though. Due to the very low horizon or complete absence of a horizon line, the human figure is no longer firmly situated on terra firma, and can no longer be measured against the horizontal span and solidity of the earth. The 'nature' which is featured in these images, meanwhile, is not materially-dense ground, rocks, water or trees, but rather the more de-materialized natural entities of air, sky, or atmosphere. At the same time these postcards summon up a natural realm that has retained something strange, something 'wild'. Here it is worth remembering that some of the most impressive examples of landscape art in the Western tradition manage to suggest a dialectical tension: on the one hand, the artwork will evoke the wildness, strangeness, sublimity and unintelligibility of natural sites and phenomena, but at the same time the very genre of landscape provides an aesthetic system with which to frame, control, and contain that natural world. To provide only one example of this art-historical legacy: Joseph Leo Koerner describes how the German Romantic painter C.D. Friedrich began his landscapes with a strictly organized pictorial system, but then, in order to truly capture the elemental force of nature, he proceeded 'to "disorganize" the achieved continuity of traditional pictorial space ... by invoking at the horizon an infinity that no system can arrange' (Koerner 1990, 113). What is at stake with these few sequences of postcards is not the brash deconstruction of a genre that was characteristic of avant-garde projects, but these ephemeral little images nonetheless manage to valorize spatial ambiguity, to persuasively detach the human figure from the illusion of a stable ground, and also to problematize the very concept of nature – thus contributing to an on-going debate about the modern meaning and appearance of landscape.

An Urban Moonscape

The postcard moonscapes should not, however, be considered solely as representations of a natural realm. Despite its alluring remoteness and otherness, the Reutlinger lunar environment still resonates with the social space that lies somewhere out of sight, below these ascendant women. I have already mentioned that these 'space travellers' are kin to a tribe of Parisian entertainers, actors, dancers, and so on, but the nocturnal urban world inhabited by such figures is itself relevant in certain ways. If the venues of Parisian nightlife were by this time flooded with artificial light, it can be argued that these star-studded lunar landscapes are as electrified and energized as the modern city at night. Electric light was still an object of fascination in the early years of the century: the 1900 *Exposition Universelle* which hosted the moon-voyage attraction also featured a Palace of Electricity lighting up the night-time sky with a vengeance. This was not the first such apparition in Paris, as earlier exhibitions had introduced electricity to the public in a comparably dramatic fashion, but by the turn of the century the surrounding cityscape was subject to ever-increasing illumination, and in the latter years of the nineteenth century there is much evidence of artists and writers dwelling on the blurred boundaries between interior and exterior spaces, between day and night, between natural and artificial forms of illumination. The metaphoric qualities of artificial light are at times explicitly compared to the moon as a natural light source. Two well-known paintings can be mentioned in this respect: Edgar Degas's *La Chanson du Chien* (1877), and Edouard Manet's *Bar au Folies Bergère* (1882) both feature several round white lamps within their respective scenes of night-time entertainment. In both instances the 'lamps' are mysterious, non-naturalistic, moon-like orbs, and the light they cast onto people and things is like nothing in the previous history of painting. In fiction, Emile Zola's novel about a department store, entitled *The Ladies' Paradise*, describes an interior space enlivened by electric lights that resemble 'illuminating moons', while the dazzle of light transforms the shopping emporium into a kind of 'dreamy firmament' (Zola, quoted in Asendorf 1993, 100).[12] Walter Benjamin would analyze the glitter-effect of the commodity as it first becomes evident under such conditions, and indeed the experience of bedazzlement and enchantment triggered within brightly-lit department stores, exhibitions, and other public spaces becomes a key modern trope for his Arcades project.[13]

The discourse around the modern spectacle of artifical light, as compared to the wattage and aesthetic impact of the real moon, achieves a kind of rhetorical climax with the Futurists. While living in Paris, in 1909, the Futurist leader F.T. Marinetti published a text entitled 'Kill the Moonlight', while the following year a manifesto against Venice proclaimed, 'Let the reign of holy Electric Light finally come, to liberate Venice from its venal moonshine of furnished rooms' (Marinetti 1971, 55). For these avant-garde artists and poets,

---

12   Also, from Guy de Maupassant, writing in 1909: 'And the electric globes – like shimmering, pale moons, like moon eggs fallen from the sky …', quoted in Benjamin (1999, 570).

13   Benjamin suggests, however, that the glare of electricity could be too intense: 'For someone entering the Passage des Panoramas in 1817, the sirens of gaslight would be singing to him on one side, while oil-lamp odalisques offered enticements from the other. With the kindling of electric lights, the irreproachable glow was extinguished in these galleries… All at once they were the hollow mold from which the image of 'modernity' was cast' (Benjamin 1999, 874).

the moon was a ready-made signifier of old-fashioned values and aesthetics, bespeaking a nostalgic attachment to the objects, images, ideas and cities of the past. In contrast, any kind of synthetic moon-effect could be embraced as a necessary gesture toward the future.

Given this contemporaneous discourse, we might then ask whether the Reutlinger moon-women appear as if trapped in a moonlit past, or poised on the threshold of a future world; whether they inhabit a naturally-occurring celestial space, or alternately, occupy the resplendent realm of the enthroned commodity. I would suggest that such tensions remain unresolved, that these postcards create a spatial interval for a subject who is energized by an electrified modern moon, while still partly under the melancholy spell of that old-fashioned moon. It is possible that the moon acquires a new imaginative potency when it is associated with technologically advanced forms of power, light, and transportation, and with a world of rapidly moving images, goods, and information, and with urban life. Thus the hybridized space of these postcards stages a productive encounter between the sublimity of nature and the promises of a modern, technologized world, between interior and exterior realms, between collective life and subjective experience.

Image Transactions

I have suggested that the early twentieth-century picture postcard (and more specifically, the strange spatiality of one postcard series) acquires meaning once it is inscribed and sent, and once its communicative potential is unleashed within the terms of a modern visual culture. Reutlinger's lunar landscapes took their place amidst the ever-expanding opportunities for visual stimulation and consumption in late nineteenth and early twentieth century Paris. As has often been described, this included panoramas and proto-cinematic entertainments, alongside the dazzling department stores and *expositions universelles*.[14] As well, several photographically-illustrated magazines became available to the French public at this time – *La vie au grand air, Fémina, Je sais tout, L'illustration*[15] – joining an already-existing production of caricatures, advertisements, posters and illustrations. This is the saturated visual context in which postcards were bought, sent, exchanged, and collected in such great numbers. The picture-postcard phenomenon invited people to make choices from amongst this changing and moving image-world – to create and transmit to others their own meaningful sequences of images. It is crucial to this discussion that

---

14   Along with Williams and Asenforf, some key discussions of late nineteenth-century visual culture are provided by Jonathan Crary (1999) and Vanessa Schwartz (1998).

15   The impact of these illustrated magazines is discussed in Moore (2004, 72).

the cosmos-in-miniature pictured on postcards would become part of a communicative cultural gesture.

As a contribution to the debate about visual culture, W.J.T. Mitchell provides the methodologically valuable suggestion that we should consider not only 'vernacular images' (a concept developed by Batchen, as we have seen), but 'a larger field of ... vernacular visuality or everyday seeing' (Mitchell 2002, 248). Mitchell advocates treating 'visual culture and visual images as 'go-betweens' in social transactions, as a repertoire of screen images or templates that structure our encounters with other human beings' (ibid., 243). This notion is very persuasive in relation to postcards, as image/objects which are continually on the move between people; the postcard is the epitome of a 'go-between.' And so for 'Paulette', that woman who affectionately signed her name to more than one such Reutlinger moon postcard in 1906, and for her friend 'Mademoiselle Sempé' who was the recipient of these postcards sent in sequence, these image-transactions became an opportunity to become active participants in the visual culture of the period. Paulette and other correspondents made choices from amongst a seemingly endless array of subject-matter, pictorial styles, and sentimental attitudes. Out of this streaming multiplicity of images choices were made, idiolectic semiotic sequences were created (Barthes uses the word 'idiolect' to designate the distinctive speech of an individual[16]), and messages, or sometimes only a name, were inscribed. It is indeed the case that 'Paulette' on one occasion did no more than sign her name – albeit with dramatic flourish and flowing ink – to a moon postcard. The selection, sending, and sharing of picture postcards would become a kind of associative game of images and signification. Images were set in motion, and crucially, directed towards others. The strange lunar space of the Reutlinger postcards was therefore part of this constellation of moving images, and it was up to pairs or clusters of correspondents to negotiate their access to these imaginative and communicative spatial networks.

Down to earth

The significance of this aero-spatial imagery, as it circulated and proliferated in the early years of the twentieth century, comes into sharp focus because this cultural episode was cut short by the First World War. Ten years after some Reutlinger moon postcards were exchanged by Parisian friends, it was possible to purchase an entirely new kind of postcard, showing the ground-level devastation caused by aerial bombardment. The aircraft which could only barely lift off the ground such a short time before had been transformed

---

16   'There is a plurality and a co-existence of lexicons in one and the same person, the number and identity of these lexicons forming in some sort a person's idiolect' (Barthes 2003, 121).

into war machines used for surveillance, mapping, communication, the surreptitious crossing of borders, and the strategic dropping of bombs. With the war, that apparently open and accessible zone above the surface of the planet was militarized, and the metaphoric consequences of this occupation were immediately registered in the visual universe of postcards. Alongside those images showing the grim results of aerially-delivered bombs, patriotic postcards issued by France and other countries would eulogize a generation of aviators and the flying machines involved in the war effort. With this new iconography, the modern aerospace fantasies which had circulated in the pre-war era were re-directed, and firmly anchored to serviceable war-time sentiments (see Nelson 2004).

The militarization of aerospace meant that every possible sightline – from one aeroplane to another, from one target to another, from the ground to the sky and back again – had to be precisely articulated and charted. The spatial ambiguities of the moon postcards stand out against the relentlessness of this visual regime. The lunar landscapes were constructed, as we have seen, from the perspective of a lightly-poised woman occupying the threshold of a panoramically-open space, while it could also be said that the imaginative potential of this space was extended by the mobility of the postcards themselves. For a short time, the picture postcards which inaugurated the new century provided glimpses of a strange new space.

References

Asendorf, C. 1993. *Batteries of Life: On the History of Things and their Perception in Modernity*. Berkeley, CA: University of California Press.

Barthes, R. 2003. 'The Rhetoric of the Image', in Wells, L. (ed.) *The Photography Reader*. New York: Routledge.

Batchen, G. (2001). *Each Wild Idea: Writing, Photography, History*. Cambridge, MA: MIT Press.

Benjamin, W. 1999. *The Arcades Project*. Cambridge, MA and London: Harvard University Press.

Berlanstein, L.R. 2001. *Daughters of Eve: A Cultural History of French Theater Women from the Old Regime to the Fin de Siècle*. Cambridge, MA: Harvard University Press.

Chéroux, C. 2005. 'Portraits en pied ... de nez: l'introduction du modèle récréatif dans la photographie foraine', *Études photographiques*, 16, May, 89–107.

Cheroux, C. 2007. 'The Small Change of Art', *The Stamp of Fantasy: The Visual Inventiveness of Photographic Postcards* (exhibition catalogue for Fotomusuem Winterthurm, Jeu de Paume). Gottingen: Steidl.

Crary, J. 1999. *Suspensions of Perception: Attention, Spectacle, and Modern Culture*. Cambridge, MA: MIT Press.

Ezra, E. 2000. *Georges Méliès: The Birth of the Auteur*. Manchester: Manchester University Press.

Frazer, J. 1979. *Artificially Arranged Scenes: The Films of Georges Méliès*. Boston: G.K. Hall and Co.

Geary, C.M. and Webb, V.-L. (eds) 1998. *Delivering Views: Distant Cultures in Early Postcards*. Washington, DC: Smithsonian Institution Press.

Green, M. and Swan, J. 1993. *The Triumph of Pierrot*. University Park, PA: Pennsylvania State University Press.

Koerner, J.L. 1990. *Caspar David Friedrich*. London: Reaktion Books.

Marinetti, F.T. 1971. 'Against Past-Loving Venice', in Flint, R.W. (ed.) *Marinetti: Selected Writings*. New York: Farrar, Strauss and Giroux.

Mitchell, W.J.T. 2002. 'Showing Seeing: A Critique of Visual Culture', in Holly, M.A. and Moxey, K. (eds) *Art History, Aesthetics, Visual Studies*. New Haven, CT: Clark Art Institute and Yale University Press.

Moore, K. 2004. *Jacques Henri Lartigue: The Invention of an Artist*. Princeton, NJ and Oxford: Princeton University Press.

Nelson, C. 2004. 'Martial Lyrics: The Vexed History of the Wartime Poem Card', *American Literary History*, 16:2, 263–89.

Nicholson, M.H. 1948. *Voyages to the Moon*. New York: The MacMillan Company.

Schor, N. 1992. 'Cartes Postales: Representing Paris 1900', *Critical Inquiry*, 18, Winter, 216.

Schwartz, V. 1998. *Spectacular Realities: Early Mass Culture in Fin-de-Siècle Paris*. Berkeley, CA: University of California Press.

Staff, F. 1967. *The Picture Postcard and its Origins*. New York: Praeger.

Tagg, J. 1989. *The Burden of Representation: Essays on Photographies and Histories*. Amhurst, MA: University of Massachusetts Press.

Whitaker, E.A. 1999. *Mapping and Naming the Moon: A History of Lunar Cartography and Nomenclature*. Cambridge and New York: Cambridge University Press.

Williams, R.H. 1982. *Dream Worlds: Mass Consumption in Late Nineteenth-Century France*. Berkeley, CA: University of California Press.

Wohl, R. 1994. *A Passion for Wings: Aviation and the Western Imagination 1908–1918*. New Haven, CT and London: Yale University Press.

Woody, H. 1998. 'International Postcards: Their History, Production, and Distribution (Circa 1895 to 1915)', in Geary, C.M. and Webb, V.-L. (eds) *Delivering Views: Distant Cultures in Early Postcards*. Washington, DC: Smithsonian Institution Press.

Chapter 14

Strange Exhibitions: Museums and Art Galleries in Film

Steven Jacobs

Beyond Artists' Biopics

Housing valuables, images, meanings and symbols, art museums and art galleries are attractive places for filmmakers.[1] Moreover, museums and galleries are not only physical spaces, they are also institutions that embody specific economic, social and cultural values. Associated with high culture and education, museums emanate a gravity that lends itself easily to both glamour and mockery. In addition, museums are often presented as cultural shrines that celebrate and construct national identities and collective memories. Often, they are seen as treasure chambers displaying objects of exotic and extinct cultures. Consequently, both lucid temples of enlightenment and dark cabinets of curiosities, cinematic museums are often presented as realms outside the ordinary world. They are a kind of heterotopias for their visitors, who, to a large extent, can be categorized into specific types. Apart from the evident category of artists and connoisseurs, rather tourists, snobs, dandies, iconoclasts, thieves, secret lovers, spies and haunted or cursed characters are the eminent museum visitors in cinema. In films, apparently, museums provide a kind of harbour to people who are haunted, hiding or in transit: tourists searching for the commodified strangeness of the exotic; snobs, dandies and iconoclasts in-between high culture and mass culture; thieves and spies transgressing laws; secret lovers between moral codes and hedonistic pleasure; and characters haunted by mummies, wax figures or mesmerizing painted portraits between life and death. In films, therefore, the otherness or strangeness of museums often involves a state of transgression and a certain *in-betweenness*.

In films, museums are even turned into places of conflict and estrangement for artists, who are nonetheless their most obvious category of visitors. Museums, galleries, and exhibitions abound in modern artists' biopics (see Walker 1993; Guieu 2007, 253–66). Van Gogh visits an exhibition of

---

1   The topic of museums in films has been addressed previously, see Louagie (1996), Fisher (2002), and also Grasskamp (2006). Museum locations are frequently mentioned in Reeves (2006).

Impressionist painting in *Lust for Life* (Vincente Minnelli, 1956) whereas Toulouse-Lautrec takes a girl to the Louvre to admire the Venus de Milo in *Moulin Rouge* (John Huston, 1952). For modern artists, the gallery is the place where they interact with the public. Cinema, however, often sees the exhibition as an ordeal or apotheosis. Galleries and museums are turned into places of conflict. A drunken Toulouse-Lautrec gives offence to a woman at the *vernissage* of one of his shows, *Basquiat* rides the sneers of the Soho demimonde in Julian Schnabel's 1986 film, Warhol nearly dies in his studio-cum-gallery in *I Shot Andy Warhol* (Mary Harron, 1996), and Frida Kahlo can barely make it to her show in *Frida* (Julie Taymor, 2002).

Museums and galleries feature in films about fictitious artists too, such as in William Dieterle's *Portrait of Jennie* (1948) and Fritz Lang's *Scarlet Street* (1945). The final scene of *Portrait of Jennie* shows the painting of the title on display in the Metropolitan Museum as a kind of apotheosis – something that is emphasized by the fact that the last shot of this black-and-white film is in colour. Christopher Cross (Edward G. Robinson), the clerk and amateur painter in *Scarlet Street*, also exhibits his primitivist paintings in a prominent art gallery (Gunning 2000, 307–39). However, his works are presented as creations by his mistress. The gallery thus becomes part of the game of mistaken and hidden identities that is crucial to the narrative. Lang's film ends with an insane and derelict Cross walking past the gallery and staring at the portrait of the murdered woman which once embodied his erotic fantasies. *Scarlet Street* is emblematic of the ways museums and art galleries are depicted in film. They are often connected to suppressed passions, criminal activities, fatal encounters, a sense of doom, and death.

The Tourist Gaze

Feature films often present museums as strange places in the sense that they are situated in foreign countries. They are part of the lure and dangers of the exotic so cherished by blockbuster movies. As famous tourist attractions, museums are often part of establishing shots as well as montage sequences, which situate the story in a particular city. Museums therefore contribute to the construction of a cinematic space answering to what John Urry (1990, 138–40) has called the tourist gaze, which reduces the city to a series of postcard images. Focusing on stereotypes, many Hollywood films present themselves as moving postcard collections. In light of the close connection between the James Bond series and the development of global tourism, it should be noted that the opening teaser of *The World is Not Enough* (Michael Apted, 1999) is set against the backdrop of Frank Gehry's Bilbao Guggenheim Museum – a landmark building that stimulated the tight relationships between museum architecture, gentrification, global tourism, and city branding.

This type of tourist gaze also appears when a museum is presented as part of a larger collection of urban monuments. In films such as *Three Coins in the Fountain* (Jean Negulesco, 1954), *Born Yesterday* (George Cukor, 1951), *Funny Face* (Stanley Donen, 1957), or *The Talented Mr Ripley* (Anthony Minghella, 1999), main characters visit museums as part of a sightseeing tour including the monuments of Rome, Paris, or Washington DC. In such a cinematic narrative, museums contribute to the construction of a visually glorious but topographically nonsensical sequence (see Bass 1997, 84–99). This is particularly striking when the sites are juxtaposed one right after the other as in *On the Town* (Gene Kelly and Stanley Donen, 1949), in which three sailors have a 24-hour leave to visit almost all the major New York sights. A musical and thus part of a studio-bound tradition par excellence, the film was innovating for its location shooting. Its museum interiors, however, are unmistakably sets. First, the protagonists visit the Museum of Anthropological History, where a display of non-western art is the stage for a jazz dance number. Next, after stating 'let's try all other museums', a series of institutions succeed one another in a montage sequence: the Metropolitan Museum of Art, the Museum of the City of New York, the Museum of Surrealist Art and the Museum of Modern Art.

This postcard-like succession of museums, which mimics the superficiality of the tourist visit, is precisely what modernist directors of the 1950s and 1960s have ridiculed or criticized. Jean-Luc Godard, who prefigured the age of mass tourism and the appropriation of the world through industrially produced images in the famous postcard sequence in *Les carabiniers* (1963), created the ultimate cinematic museum visit in his *Bande à part* (1964) (see de Baecque 2004, 118–25). In this film, three protagonists kill some time by breaking the record, as a voice-over indicates, of tourist Jimmy Johnson from San Francisco who visited the entire Louvre in nine minutes and 45 seconds. The scene opens with a panning shot of the museum's exterior followed by shots showing the trio running through the museum rooms. Lampooning cultural consumerism and prefiguring the present-day art museum as a machine designed for the quick and efficient circulation of large numbers of visitors, this famous speeded-up dash through the Louvre as well as the postcard sequence in *Les carabiniers* laid the foundation of some of Godard's later video works that consider the institution of the museum. In *The Old Place*, his 1999 video essay commissioned by the New York Museum of Modern Art, for instance, Godard equates the logics of museums (bringing pictures together, generating meaning through their juxtapositions) with the editing processes inherent to the medium of video.

Cinematic Self-reflection

Museum scenes enable Godard to reflect on the nature of images and the medium of cinema. This has also been the case in the works by other directors. Perhaps the two most fascinated museum scenes ever created on film, the ones in Roberto Rossellini's *Viaggio in Italia* (1954) and Alfred Hitchcock's *Vertigo* (1958), deal with a wide range of topics (some of them discussed further on in this essay) but they also investigate the mechanisms of the look and the ways these can be evoked in cinema (see Figures 14.1 and 14.2). In *Viaggio in Italia*, Rossellini unmistakably refers to the construction of the postcard city cherished by Hollywood. The film, which explicitly deals with the perception of Italy by a British married couple, comprises a scene in which the woman protagonist (Ingrid Bergman) visits the archaeological museum in Naples (see Mulvey 2000, 95–111). She is accompanied by a guide, who kills her with all sorts of obligatory pieces of information on the masterpieces of the Farnese collection. The museum visit is a pivotal scene in the film, which in its entirety is structured on a series of visits to tourist attractions that seamlessly follow contemplative car drives. Facing the powerful and raw sensuality of the classical sculptures, the Bergman character starts to realize that her (tourist, Romantic, and spiritualist) preconception of Italy and classical culture does not tally with reality. Furthermore, the contemplation of art works and the museum experience enable Rossellini to investigate the cinematic representation of the act of looking by means of virtuoso camera movements. Often, camera movements start at an art work and end up, without intercutting, at Bergman's subdued facial expressions taking in both objective and subjective viewpoints into one single sliding whole. In so doing, Rossellini employs the museum, an environment dedicated to the sophisticated look, as a context for analyzing the cinematic gaze.

The museum as a motif of cinematic self-reflection has also been used by Hitchcock in *Vertigo* (see Peucker 1999, 141–56; Jacobs 2006 and 2007, 55–63). Scottie (James Stewart) shadows Madeleine (Kim Novak) into the museum where he watches her staring at the painted portrait of the deceased Carlotta Valdez in peace and quiet. In this scene, Hitchcock emphatically follows the conventions of the Hollywood editing logic of shot/reaction shot. Evoking Scottie's gaze, the camera vertiginously switches from the real to the realm of representation: between the bouquet on the museum bench and the identical flower piece in the painting, between Madeleine's bun and the identical coil on the painted portrait. As in the museum scene in *Viaggio in Italia*, the theme of the erotic desire of the gaze is elaborated by means of the motifs of tourist attractions and the contemplation of artworks. Both films comprise elaborate scenes without dialogue that are marked by a slow, contemplative rhythm, which matches perfectly the solemn silence and circumspection of classical museum spaces. The museum setting in particular enables both directors to investigate the theme of looking by means of emphatic close-ups of faces,

Figure 14.1   *Viaggio in Italia* (Roberto Rossellini, 1954)

Figure 14.2   *Vertigo* (Alfred Hitchcock, 1958)

highly unusual juxtapositions of action and reaction shots and bravura camera movements. In short, the museum is presented as the perfect place for the concentrated and contemplative gaze – a gaze that also, as we shall see, can turn museums into uncanny places of mystery.

Snobs, Dandies and Iconoclasts

In most feature films, however, museums are not visited to scrutinize or enjoy art works. Instead, they are showcases for snobs and dandies, who visit museums because they are strange, incomprehensible, or hostile places to others. The already discussed museum montage sequence in *On the Town* ends in the Museum of Modern Art, which enables the filmmakers to scoff at abstract art – a topic not uncommon in Hollywood films of the 1940s in particular (Waldman 1982, 52–65). A painting is turned upside down (a common trope to ridicule abstraction) and one of the sailors starts speaking French to one of the onlookers – in Hollywood cinema, a popular token of un-American 'strangeness' implying decadence, snobbery and pretentiousness. In *Born to be Bad* (Nicholas Ray, 1950), for instance, one of the visitors of an art gallery has a heavy French accent. Furthermore, if museum visitors are not speaking French, they often give grandiloquent speeches on aesthetics such as Diane Keaton's character in *Manhattan* (Woody Allen, 1979). Strikingly, the first encounter between the film's protagonists takes place in one of the rooms of the Whitney Museum. Their conversation comprises references to the work of Diane Arbus and minimalist pieces such as a certain 'Plexiglas sculpture' (Judd) and a 'steel cube' (Smith), which, according to Keaton, 'was perfectly integrated and it had a marvellous kind of negative capability.' Later in the film, we see the main characters strolling in the Museum of Modern Art. Although Allen's film is marked by his characteristic self-irony, *Manhattan* presents museums as stages for snotty and neurotic characters. Cinematic museums, clearly, are not built for the enjoyment of art but are sterile environments meant for dry education. 'This place is a joint for a bunch of sexless women who have not any love in their life', one of the characters of *Shadows* (John Cassavetes, 1959) states while visiting the sculpture garden of the Museum of Modern Art. Apparently, museums are the perfect places to strike a sophisticated pose. In *Born Yesterday* (George Cukor, 1951), a film about the mistress of a corrupt business man who is given a crash course by a critical journalist, the main characters visit a series of cultural institutions in Washington DC including the National Gallery. Strikingly, the museum's exterior forms the background of the closing scene of the film, in which the girl and the reporter run away from a life of inauthentic feelings and experiences. Importantly, the building itself, which was designed by John Russell Pope in the 1930s, is impressive. 'It's more gorgeous than the Radio City Music Hall', the girl says, 'and it smells nice'. The prestige of a museum visit is clearly determined by its architectural context.

A similar architectural grandeur characterizes the Capitoline Museum in Rome in *Three Coins in the Fountain* (Jean Negulesco, 1954). Designed by Michelangelo, the building came to play an important role in the history of museums since the papal palace already opened to the public in 1734 as an art collection. In the film, the characters visit the painting collection. First, two dandies find themselves in the ancient art section. However, since 'we can't live in the past forever', the scene ends in a fictitious section of modern art containing works by Calder, Picasso, Tanguy and Masson among others, made by the Twentieth Century Fox art department. A girl wants to impress the art lovers by simulating a contemplative gaze and by juggling with 'strange' and 'difficult' words such as 'neo-impressionist.' The museum as a space of artificial poses and idle talk also occurs in *LA Story* (Mick Jackson, 1991), in which Steve Martin describes a painting to his friends as sexy: 'Look how he's painted the blouse sort of translucent. You can just make out her breasts underneath and it's sort of touching him about here. It's really ... pretty torrid, don't you think?' However, when the painting is revealed, it's only a large red rectangle.

In short, cinema seems to endorse Bourdieu's famous critique of the modern art gallery as an instrument of social distinction (Bourdieu and Darbel 1991). Since art and museums seem to be created as means of social prestige in the first place, cinema loves to present museums as privileged environments where cultured Victorians meet. *Room with a View* (1986) and *Maurice* (1987), both costume dramas by James Ivory, comprise scenes situated in the National Gallery and the room of the Assyrian statues in the British Museum respectively. Similarly, in *The Age of Innocence* (Martin Scorsese, 1993), the main character contemplates Rubens paintings in the Louvre. In the museum scene of *Dr Jekyll and Mr Hyde* (Victor Fleming, 1941), the protagonist's fiancée tells about her initiation into 'art and culture' and, facing a small replica of the Winged Victory of Samothrace, about her confrontation with the original one in the Louvre in Paris.

Dominating a grand staircase, the original version of the statue forms the appropriate background for a glossy fashion photo report in *Funny Face* (Stanley Donen, 1957), in which Audrey Hepburn's fluttering scarf echoes the wings of the ancient sculpture. Precisely because of their associations with high culture and upper-class values, museums are highly in demand for all kinds of elegant society events. The Metropolitan Museum and its vast glass-walled gallery housing of the Egyptian temple of Dendur are the locale for dressy cocktail parties in films such as *A Perfect Murder* (Andrew Davis, 1998) and *Six Degrees of Separation* (Fred Schepisi, 1993) whereas the Los Angeles County Museum of Art is the venue of a star-studded charity ball in *The Player* (Robert Altman, 1992). Demonstrating the museum's function as a tool of social distinction, these scenes also remind us of the role that emerging museums played in the formation of the bourgeois public sphere (see Bennett 1995, 25–33).

Both temples of high culture and posh party venues, museums, consequently, are also perfect victims of mayhem and blasphemy. This is the case, for instance, in many popular comedies: in *LA Story* (Mick Jackson, 1991), Steve Martin roller-skates through the LA County Museum of Art; in *The Return of the Pink Panther* (Blake Edwards, 1975), clumsy Inspector Clouseau wreaks havoc on the interior of the National Museum of Lugash; and in *Bean* (1997), the title character is sent to a Los Angeles museum, which purchased a Whistler painting – a valuable masterpiece, of course, to be wrecked by Bean. Art, apparently, entices destructive and childishly iconoclastic behaviour. In Tim Burton's *Batman* (1989), the Joker and his thugs break into a museum and enjoy daubing paintings. 'I've always been a little bit daunted by art because, rather than embracing, it's always been a very distancing, pretentious thing to me – and it should be the opposite, it should be inviting. So the museum scene is just a perverse fantasy, really, of not having any boundaries and recreating. It's classic anarchy', Burton stated (Smith n.d.).

Thieves

Presenting museums as elitist temples or habitats for snobs and dandies, Hollywood also emphasizes that museums are treasure chambers stuffed with strange, rare, and precious objects. As a result, cinematic museums are not only favourite attractions for tourist, snobs, and dandies but also for thieves. Sceptical of the ethical and edifying ambitions of museums, philosopher Nelson Goodman once stated that 'the only moral effect a museum has on (him) is a temptation to rob the place' (Goodman 1985, 53–62). In films too, museums represent big money and one of their main functions is to serve as targets for burglary. Pearls (*The Pearl of Death*, Roy William Neil, 1944), diamonds (*The Hot Rock*, Peter Yates, 1972), Amazon figurines (*L'homme de Rio*, Philippe de Brocca, 1964), Rembrandt portraits (*Stealing Rembrandt*, Jannik Johansen, 2003) and El Greco triptychs (*L'incorrigible*, Philippe De Brocca, 1975) have been stolen from cinematic museums. The sensational 1911 theft of the Mona Lisa from the Louvre immediately unleashed a swarm of comic silent films such as *Gribouille a volé la Joconde* (Albert Capellani, 1911) and it also inspired filmmakers later to produce *Der Raub der Mona Lisa* (Geza von Bolvary, 1931) and *On a volé la Joconde* (Michel Deville, 1966). Self-evidently, in films, museum robberies are invariably spectacular, such as in the classic caper movie *Topkapi* (Jules Dassin, 1964) with Maximilian Shell stealing a dagger set with jewels from the Topkapi Palace Museum in Istanbul – a set construction that bears no resemblance to the real thing. Ever since, the theft of a museum treasure has been a recurring trope in caper films. A famous recent example can be found in *The Thomas Crown Affair* (John McTiernan,

1999), in which Pierce Brosnan ingeniously steals, with the help of Magritte-inspired diversion tactics, a priceless painting by Monet. The exterior of the gallery is the Metropolitan Museum, though it is never identified as such in the movie, whereas the entrance concourse is the New York Public Library and the interior galleries are studio sets. The paintings themselves are props supplied by 'Troubetzkoy Paintings' – a firm specialized in copying masterpieces working regularly for Hollywood production designers.

The motif of intruding museums as thieves in the night proved a very powerful cinematic formula that transcends the genre of caper films. Sneaking into the strange territory of an uncannily empty, silent, and unlit museum rooms can be found in a wide variety of films ranging from a French director's film such as *Les amants du Pont Neuf* (Leon Carax, 1991), a Hollywood blockbuster as *The Da Vinci Code* (Ron Howard, 2006), a poetic documentary as *La ville Louvre* (Nicolas Philibert, 1990) – all set in the Louvre – and an arthouse experiment as *Elegy of a Voyage* (Alexander Sokurov, 2002), which is situated in the Museum Boymans van Beuningen in Rotterdam. In all these titles, the filmmakers were unmistakably fascinated by the play of flashlights in nocturnal museum spaces. Restless shadows create the illusion that some of the characters on the paintings come to life. As a result, the museum is presented as a strange and mysterious place filled with secrets and uncanny powers. Public institutions, museums are turned into places of privacy and intimacy.

Secret Lovers

Given this perspective, museums become convenient meeting places for secret lovers and spies. In *The Kiss* (Jacques Feyder, 1929), the Greta Garbo character and her lover meet in a Hollywood interpretation of the museum of Lyon. Its rooms are in a remarkable Art Deco style, which were designed by the MGM art department. Some of the rooms are rendered by means of impressive tracking shots evoking the hectic rhythm of the guide dragging along a bunch of visitors. The lovers' encounter is interrupted by voyeuristic women – one of them stating that the room accommodates 'intimate pieces.' The situation is almost literally repeated in the museum scene in *Dr Jekyll and Mr Hyde* (Victor Fleming, 1941), in which the kissing Jekyll and his fiancée are caught by old women. 'Just another work of art, madam. It's not in your catalogue', the father of the bride-to-be assures. Secret lovers also choose a museum as the site of their rendezvous in *The Age of Innocence* (Martin Scorsese, 1993), a story about suppressed passions among members of the New York high society of the 1880s. 'I need to see you. Somewhere we can be alone ... the Art Museum in the Park', Daniel Day Lewis whispers to Michelle Pfeiffer. The following shot shows a gigantic skylighted museum room evoking the interior of the former building of the Metropolitan Museum, which opened in 1880.

Apparently, many filmmakers have been attracted by the telling contrast between the burning passion of the lovers and the solemn silence of museum spaces. Consequently, art galleries are perfect places at which to flirt. Their restrained coolness and the repressed desire needed to perceive the desire expressed in art works give the mind opportunities to open up to extraordinary encounters. In *Le plaisir* (Max Ophuls, 1952), which tells the story of the love between an artist and his model, the museum even serves as a place where the sketching artist picks up a girl. Master of the long take, Ophuls' sensuous camera movements rewardingly employs the impressive galleries, staircases, and vaults of the Louvre. Lovers also get acquainted in museums in the aforementioned scenes in Negulesco's *Three Coins in the Fountain* and Woody Allen's *Manhattan* (1979). Swanker of New York bohemia, Allen's characters often visit art world venues. In *Play It Again, Sam* (1972), Allen's character attempts to pick up a girl in front of a Jackson Pollock in an art gallery. ('What are you doing Saturday night?' 'Committing suicide.' 'What about Friday night?') Undoubtedly, because of other connotations attached to museums discussed in the following paragraphs, the museum visits of lovers often involve a sense of doom. In Vincente Minnelli's *The Clock* (1945), an army corporal meets a girl during his New York visit on a two-day pass. They spend an afternoon together and they end up at the Metropolitan Museum of Art. They have a long chat in the Egyptian Gallery (constructed on the MGM sound stage) but their encounter is overshadowed by his imminent departure for war.

In short, films such as *The Kiss, The Clock, The Age of Innocence* as well as *Far from Heaven* (Todd Haynes, 2002) and *Dressed to Kill* (Brian De Palma, 1980) present the museum as a place of impossible love, sneaking desires, or repressed sexuality. *Far from Heaven* tells the story of the impossible relationship between a perfect 1950s white housewife and her African-American gardener. A key scene is set in a small-town museum. Under the watchful eye of gossiping high-society matrons, both protagonists connect over a Joan Miró painting. 'Modern art, abstract art … it's just picking up where religious art left off, somehow trying to show you divinity', the man says. While the presence of an African American in a museum in the 1950s is seen as an intrusion by the small-town community, for the protagonists the museum becomes a place where they can act civil with one another. Passions, however, are subtly hidden under the veneer of love for high culture.

In *Dressed to Kill*, the lovers' museum encounter is also overshadowed by a sense of doom and repressed sexuality. In this museum scene, however, anxiety and erotic tension determine the atmosphere. A middle-aged, sexually frustrated housewife meets a strange man at a museum – the Metropolitan Museum stands for its exterior whereas the elaborately shown interior rooms are those of the Philadelphia Museum of Art. The entire scene is built up slowly and then culminates into a series of restless tracking shots evoking the woman's point of view. The man draws her attention while she is sitting on a bench in the wing of modern painting. Then he leaves and she pursues him dropping a glove (in cinema nearly always a token of crime). The museum scene turns into a game of disappearances and sudden reappearances.

The Art of Espionage

As secret meeting places, museums are not only popular among lovers. Criminals on the run, spies and secret agents as well are fond of museum excursions. As dandies and snobs use museums as a décor for social prestige, lovers, fugitives and spies use museums as fronts. *Absolute Power* (Clint Eastwood, 1997) tells the story of a thief who gets entangled in a political intrigue. For him, the museum serves as a perfect cover. The film opens with Eastwood as an amateur artist sketching from the old masters in the Corcoran Gallery of Art in Washington DC. Since they operate in secret, spies are attracted by the contemplative silence associated with museum spaces. However, in spy thrillers, the silence of museum galleries becomes strange, mysterious and sinister. In Hitchcock's *Torn Curtain* (1966), Paul Newman walks through the uncanny empty and silent rooms of the Alte Nationalgallerie in Berlin, the classicist architecture of which is mirrored by neoclassical sculptures. The museum functions as a labyrinth in which Newman tries to get rid of his invisible pursuer. The menacing silence at the museum is emphasized by the sound of Newman's and his pursuer's echoing footsteps.

A similar uncanny silence in a context of political intrigues characterizes the museum sequence in Francesco Rosi's *Cadaveri eccellenti* (1976), set in the Archaeological Museum in Naples. In this climactic scene, we follow the main character during his long walk through the classical sculpture collection. We see him walking past busts and heads of ancient emperors, rhetoricians, philosophers, and statesmen – an appropriate focus since the film's narrative presents the museum as a place where the future of the corrupt state is at stake. The museum, after all, is the place where the protagonist meets a politician who has information on the mafia connections of some of his opponents. Both characters, however, are shot by an off-screen sniper. Strikingly, Rosi employs evocatively the solemn silence of the museum. On the soundtrack, nothing can be heard from the dialogue between the characters. Both in *Torn Curtain* and *Cadeveri eccellenti*, the silence and emptiness of the galleries turn the museum – one of the public spaces of urban modernity – into a site of paranoia. A realm designed for the cultivation of sophisticated sensory perceptions, the museum becomes an environment of excessive anxiety. Temples of visual culture and training areas for the look, museums become suffocating labyrinths where characters are chased by invisible adversaries.

Wax Figures and Mummies

In *Torn Curtain* and *Cadaveri eccellenti*, museums are turned into sinister places shrouded in a sepulchral hush. Unmistakably, these spy thriller museum scenes interfere with the popular association of museums with mystery and danger. Accommodating valuable, exotic, and strange objects, museums are often produced as treasure chambers dominated by spiritual and atavistic powers. As Andrea Witcomb noted, in the popular imagination, museums are 'dark and musty places, full of strange objects. This is the representation, for example, of museums in [...] the *Indiana Jones* series. It is not the narrative of progress that is remembered but the exotic, the strange – museums as houses of mystery. This association suggests that despite their role within hegemonic discourses, museums are also associated with danger, the irrational, the uncontrollable' (Witcomb 2003, 24). Consequently, museums are often presented as places of witchcraft. They accommodate magical artefacts, such as in *It!* (Herbert J. Leder, 1966), in which a cursed statue withstands every attempt to destroy it, or in *Night Life of the Gods* (Lowell Sherman, 1935), in which a scientist turns the Metropolitan Museum's statues of Greek Gods into people. These are only a few of the infinite number of examples in which museums become realms of occult reincarnations – an item crucial to horror films featuring waxworks museums, such as *Das Wachsfigurenkabinett* (Paul Leni, 1924), *Mystery of the Wax Museum* (Michael Curtiz, 1933), or *House of Wax* (André De Toth, 1953). Another sub-genre of the horror film featuring museums is the mummy film. The most significant contribution to this genre, no doubt, was *The Mummy* (Karl Freund, 1932) starring Boris Karloff as Imhotep, an Ancient-Egyptian priest brought back to life by an archaeologist. In the nocturnal Cairo museum of antiquities, which is turned into a threatening environment by Freund's characteristic *chiaroscuro* lighting and mobile camera, Imhotep conjures up the spirits embodied in the museum's exhibits.

Museums also serve as settings in the numerous remakes and sequels of Freund's film. In addition, French popular cinema gave rein to archaeological museum horror in a similar cycle that started with *Belphégor* (Henri Desfontaines, 1927) about a phantom appearing in the middle of the night in the Louvre. In *Belphégor: Le fantôme du Louvre* (Jean-Paul Salomé, 2001), a recent contribution to the cycle, ghostly powers interfere with the up-to-date museum security and surveillance technology since the spirit escapes and makes its way into the museum's electrical system while experts are using a laser scanning device to determine the age of a sarcophagus.

Films featuring mummies and wax figures foster the popular association of museums with death. Archaeological museums, after all, display artefacts of strange and extinct cultures, which are represented by objects that relate to complex death rituals and life in the hereafter: tombs, mummies, death masks, funeral monuments, sarcophagi, sacrificial objects, etc. Furthermore, in the popular imagination, museums themselves are presented as tombs or sterile environments characterized by a sepulchral silence and solemnity. Films, consequently, present museums as fatal places or as environments where characters are confined in a mysterious past. In *Blackmail* (1929), Hitchcock chose the British Museum as one of the first employments of his typical climax at a famous location that is entertaining itself. Unmistakably, Hitchcock played on the associations between museums and death. The blackmailer tries to evade the police by sneaking into the rooms with ancient Egyptian statues. In a striking shot, the fugitive climbs a rope next to the huge stone face of an Egyptian colossus. Eventually, the museum also becomes literally a tomb since the man falls through the skylight of the museum's dome.

These associations also play an important part in other Hitchcock museum sequences, such as in the aforementioned scene in *Vertigo* (1958). In this film, after all, the museum visited by the protagonists is the California Palace of the Legion of Honour, which was conceived as a shrine for soldiers who fell in the First World War. Furthermore, the building forms the shrine for the portrait of a deceased woman who keeps the living in her grasp. Given this perspective, Hitchcock tails an entire tradition of films made throughout the 1940s in which haunted painted portraits play a significant role. The mysterious presence of an important absentee through a painting was a key motif, for instance, in prominent examples of film noir and also in some films directed by Hitchcock himself such as *Rebecca* (1940), *Suspicion* (1941), and *The Paradine Case* (1947) (see Felleman 2006, 25–55; Peucker 1999, 130–37; Walker 2005, 319–34). In each case, cinematic elements such as framing, editing, lighting, camera positions, and camera movements are used to bring into being the person portrayed. Haunting the characters, the deceased figure seems to look back at the beholders. These connotations also mark Dietrele's *Portrait of Jennie* (1948), which has its final scene situated in the Metropolitan Museum where schoolgirls admire the portrait. The film tells the story of the love between a painter and a girl, who turns out to be the ghost of a woman who died a long time ago. Apparently, when cinema deals with paintings, such as in museum scenes, it almost always shows instances of what Ernst Kris and Otto Kurz (1979, 73–9) have called 'effigy magic', that is the primitive belief that a person's soul resides in their image or effigy.

The mesmerizing effect of a painting and the idea that a portrayed person is looking back is a recurring trope in cinema, which certainly marks many scenes situated in museums. In line with this, filmmakers love to create a spatial and narrative continuum between the characters and the figures depicted in artworks: Madeleine seem to belong to the same distant realm as Carlotta in Hitchcock's *Vertigo*, Angie Dickinson tries to establish contact with a portrait in De Palma's *Dressed to Kill*, Audrey Hepburn mimics the statue in Donen's *Funny Face*, Ingrid Bergman's character finds her own body amidst the sensual nudes in Rossellini's *Viaggio in Italia*, a character recognizes himself in the mask-like face of a statue in Cassavetes' *Shadows*, teenagers feebly imitate the figures on the paintings of the Chicago Art Institute in

Ferris Bueller's Day Off (John Hughes, 1986), and the cleaning personnel and technicians become secondary characters to the ones depicted on renaissance and baroque masterpieces in Philibert's *La ville Louvre*. As in Thomas Struth's famous pictures of interiors of famous museums, through colours, costume, and gestures, a complex dramaturgy is established that unites the museum visitors with depicted figures (Belting 1998). In some instances, boundaries are completely abolished. In *Red Dragon* (Brett Ratner, 2002), a character visits the graphic collections of the Baltimore Museum of Art that stands for the Brooklyn Museum. He attempts to silence the voice in his head by eating William Blake's *Red Dragon* drawing and thus literally incorporating the piece of art. The mysterious nocturnal intruder of the Rotterdam museum in Sokurov's *Elegy of a Voyage* (2001) turns out to be a man that was present at the making of a Saenredam painting that was created hundreds of years ago.

Mausoleum

The association between the museum and death is not only established through the presence of depicted deceased people or through the stasis of the characters but also because the museum often becomes a site where connections with the past and with death are established. In films, museums are tombs. Popular culture and films illustrate philosopher Adorno's famous statement that 'museum and mausoleum are connected by more than phonetic association.' 'Museums', Adorno wrote, 'are like the family sepulchers of works of art' (Adorno 1997, 173–86). Rossellini called Beaubourg 'the tomb of a civilization' (Gallagher 1998, 679). Furthermore, museums not only contain images of the dead but are also monuments to entire nations, political constellations, and their histories. Storing national treasures and embodying collective memories, museums encompass entire histories. According to Louvre curator Germain Bazin, an art museum is 'a temple where Time seems suspended' (Bazin 1967, 7). This is clearly the subject of *Russian Ark* (Alexander Sokurov, 2002), which was entirely shot in the Hermitage in Saint Petersburg. The Hermitage was built in the eighteenth and early nineteenth century as a Winter Palace but serves as a museum since 1946. In several ways, Sokurov's film presents the museum as a sediment of history by means of his peculiar way of dealing with diegetic time. While the film is notoriously shot as a single take, history seems compressed since we are guided through the building by a timeless ghost-like character. Furthermore, the film starts in the eighteenth century but switches to other ages as if Sokurov touches upon the historical continuity between royal art collections in the age of Enlightenment and the modern art museum.[2]

---

2 On the continuity and discontinuity between royal collections and modern museums, see Bennett (1995, 33–40) and Duncan (1995, 21–47).

Almost unnoticeably within the confines of the single take, the palace is turned into a museum. When entering the Italian Gallery, we are confronted with strolling tourists and contemplating art lovers whereas in earlier scenes, the rooms were crowded with courtiers and servants of the tsars.

The funerary associations of the museum are enhanced by its architectural typology. National or civic monuments that commemorate and conserve the past, museum buildings are destined to last for eternity. In that sense, museum buildings express eternity and grandeur and both film directors and film characters are simply fascinated by their spectacular aspects. The monumental flight of steps up to the Greek Revival building of the Philadelphia Museum of Art, for instance, is a perfect stage for the triumphal run of *Rocky* (John Avildsen, 1976) whereas the spiral galleries of the New York Guggenheim Museum form the background for a bizarre spectacle in Matthew Barney's *Cremaster 3* (2002) as well as for a comical chase on an alien insectoid in *Men in Black* (Barry Sonnenfeld, 1997).

Monumentality and grandeur are values that were attached to the very first museum buildings. Its classical building type goes back to late eighteenth- and early nineteenth-century designs of architects such as Durand, Von Klenze, Schinkel and Soane and it remained the dominant model for a long time. It is characterized by long rows of rooms, the presence of overhead light and elements borrowed from classical temple architecture, such as a colonnade, a monumental staircase and a rotunda topped by a dome (see Searing 1986, 14–23; Pevsner 1976, 111–38). These elements refer to the presupposed sacral origins of art; they isolate the museum and its treasures from everyday life and they turn the museum visit into a ritual experience. From the beginning, this museum concept was fiercely criticized by architectural theorists and artists. As a matter of fact, the entire history of modern art, which paradoxically originated together with the notion of the museum, can be interpreted as a history of architectural and institutional alternatives for the museum. Of the Louvre, Paul Valéry famously said that 'neither a hedonistic nor a rationalistic civilization could have constructed a house of such disparities. Dead visions are entombed here' (Adorno 1997, 176–7). In the anti-museum discourse, it is repeatedly stated that museums freeze, suffocate, sterilize, kill, or bury art works. Also in this perspective, museums are interpreted as places of death.

This dimension is enhanced by the architectural references to sacral and sepulchral architecture, which are emphatically visualized in cinema. In popular culture, however, this association remained intact when classical forms have been exchanged for the modernist 'white cube', which combines the solemn character and sacral silence of the Greek temple with the smooth floors, white walls, and big glass surfaces of the clinic.[3] Popular culture embraces the clinical and sterile image of the museum as much as that of the museum

---

3   The notion of the gallery space as a 'white cube' refers to a famous text by Brian O'Doherty originally published in *Arforum* in 1976; see O'Doherty (1999).

as a tomb. Strikingly, in *Manhunter* (Michael Mann, 1986), Richard Meier's High Museum of Art in Atlanta stands for the mental ward, in which serial killer Hannibal Lecter is placed under restraint. Because of Lecter's haunting presence (or is it because of Meier's dazzling white architecture?), the detective visiting the killer feels anxious and he runs off the many ramps that turned the building into one of the paradigmatic museums of the postmodern 1980s – an era in which museums are no longer presented as temples to contemplate art but also as tourist destinations where visitors can stroll, talk, drink, eat and shop as well. An era, in short, in which frenetic attempts are made to deny that the museum is a place of death (cf. Davis 1990).

References

Adorno, T.W. 1997. 'Valéry Proust Museum', in *Prisms*. Cambridge, MA: MIT Press.
Baecque, A. de 2004. 'Godard in the Museum', in Temple, M., Williams, J. and Witt, M. (eds) *For Ever Godard.* London: Black Dog Publishing.
Bass, D. 1997. 'Insiders and Outsiders: Latent Urban Thinking in Movies of Modern Rome', in Penz, F. and Thomas, M. (eds) *Cinema and Architecture: Méliès, Mallet-Stevens, Multimedia*. London: BFI Publishing.
Bazin, G. 1967. *The Museum Age*. New York: Universe Books.
Belting, H. 1998. *Thomas Struth: Museum Photographs*. Munich: Schirmer/ Mosel.
Bennett, T. 1995. *The Birth of the Museum: History, Theory, Politics*. London: Routledge.
Bourdieu, P. and Darbel, A. 1991. *The Love of Art: European Art Museums and their Public*. Cambridge: Polity Press.
Davis, D. 1990. *The Museum Transformed: Design and Culture in the Post-Pompidou Age*. New York: Abbeville Press.
Duncan, C. 1995. *Civilizing Rituals: Inside Public Art Museums*. London: Routledge.
Felleman, S. 2006. *Art in the Cinematic Imagination*. Austin, TX: University of Texas Press.
Fisher, J. 2002. 'Museal Tropes in Popular Film', *Visual Communication*, 1, 197–201.
Gallagher, T. 1998. *The Adventures of Roberto Rossellini: His Life and his Films*. New York: Da Capo Press.
Grasskamp, W. 2006. 'Observing the Observer', *CIMAM Berlin Lecture*, accessed at http://www.cimam.org/arxius/recursos/Berlin_paper_Grasskamp.pdf.
Goodman, N. 1985. 'The End of the Museum?', *Journal of Aesthetic Education*, 19:2, Special Issue: Art Museums and Education, –62.
Guieu, J. 2007. 'Les biographies de peintres: un genre cinématographique', *Ligeia* 77–80, 253–66.

Gunning, T. 2000. *The Films of Fritz Lang: Allegories of Vision and Modernity.* London: BFI Publishing.

Jacobs, S. 2006. 'Sightseeing Fright: Alfred Hitchcock's Monuments and Museums', *Journal of Architecture*, 11:5, 595–602.

Jacobs, S. 2007. *The Wrong House: The Architecture of Alfred Hitchcock.* Rotterdam: 010 Publishers, 55–63.

Kris, E. and Kurz, O. 1979. *Legend, Myth, and Magic in the Image of the Artist.* New Haven, CT: Yale University Press.

Louagie, K. 1996. 'It Belongs in a Museum: The Image of Museums in American Film, 1985–1995', *The Journal of American Culture*, 19:4, 41–50.

Mulvey, L. 2000. 'Vesuvian Topographies: The Eruption of the Past in *Journey to Italy*', in Forgacs, D., Lutton, S. and Nowell-Smith, G. (eds) *Roberto Rossellini: Magician of the Real.* London: BFI Publishing.

O'Doherty, B. 1999. *Inside the White Cube: The Ideology of the Gallery Space (Expanded Edition).* Berkeley, CA: University of California Press.

Peucker, B. 1999. 'The Cut of Representation: Painting and Sculpture in Hitchcock', in Allen, R. and Gonzalès, S. I. (eds) *Alfred Hitchcock: Centenary Essays.* London: British Film Institute.

Pevsner, N. 1976. *A History of Building Types.* London: Thames and Hudson.

Reeves, T. 2006. *The Worldwide Guide to Movie Locations.* London: Titan Books.

Searing, H. 1986. 'The Development of a Museum Typology', in Stephens, S. (ed.) *Building the New Museum.* New York: The Architectural League.

Smith, G. n.d. Interview with Tim Burton, http://minadream.com/timburton/ EdWoodInterview.htm.

Urry, J. 1990. *The Tourist Gaze: Leisure and Travel in Contemporary Societies.* London: Sage.

Waldman, D. 1982. 'The Childish, the Insane, and the Ugly: The Representation of Modern Art in Popular Films and Fiction in the Forties', *Wide Angle*, 5.

Walker, J.A. 1993. *Art and Artists on Screen.* Manchester: Manchester University Press.

Walker, M. 2005. *Hitchcock's Motifs.* Amsterdam: Amsterdam University Press.

Witcomb, A. 2003. *Re-Imagining the Museum: Beyond the Mausoleum.* London: Routledge.

Films Cited

Absolute Power (dir. Clint Eastwood, 1997).

The Age of Innocence (dir. Martin Scorsese, 1993).

Les amants du Pont Neuf (dir. Leon Carax, 1991).

Bande à part (dir. Jean-Luc Godard, 1964).

Basquiat (dir. Julian Schnabel, 1986).

Batman (dir. Tim Burton, 1989).

Bean (dir. Mel Smith, 1997).
Belphégor (dir. Henri Desfontaines, 1927).
Belphégor: Le fantôme du Louvre (dir. Jean-Paul Salomé, 2001).
Blackmail (dir. Alfred Hitchcock, 1929).
Born to be Bad (dir. Nicholas Ray, 1950).
Born Yesterday (dir. George Cukor, 1951)
Cadaveri eccellenti (dir. Francesco Rosi 1976).
Les carabiniers (dir. Jean-Luc Godard, 1963).
The Clock (dir. Vincente Minnelli, 1945).
Cremaster 3 (dir. Matthew Barney, 2002).
The Da Vinci Code (dir. Ron Howard, 2006).
Dr Jekyll and Mr. Hyde (dir. Victor Fleming, 1941).
Dressed to Kill (dir. Brian De Palma, 1980).
Elegy of a Voyage (dir. Alexander Sokurov, 2002)
Far from Heaven (dir. Todd Haynes, 2002).
Ferris Bueller's Day Off (dir. John Hughes, 1986).
Frida (dir. Julie Taymor, 2002).
Funny Face (dir. Stanley Donen, 1957).
Gribouille a volé la Joconde (dir. Albert Capellani, 1911).
L'homme de Rio (dir. Philippe de Brocca, 1964).
The Hot Rock (dir. Peter Yates, 1972).
House of Wax (dir. André De Toth, 1953).
L'incorrigible (dir. Philippe De Brocca, 1975).
I Shot Andy Warhol (dir. Mary Harron, 1996).
It! (dir. Herbert J. Leder, 1966).
The Kiss (dir. Jacques Feyder, 1929).
LA Story (dir. Mick Jackson, 1991).
Lust for Life (dir. Vincente Minnelli, 1956).
Manhattan (dir. Woody Allen, 1979).
Manhunter (dir. Michael Mann, 1986).
Maurice (dir. James Ivory, 1987).
Men in Black (dir. Barry Sonnenfeld, 1997).
Moulin Rouge (dir. John Huston, 1952).
The Mummy (dir. Karl Freund, 1932).
Mystery of the Wax Museum (dir. Michael Curtiz, 1933).
Night Life of the Gods (dir. Lowell Sherman, 1935).
The Old Place (dir. Jean-Luc Godard, 1999).
On a volé la Joconde (dir. Michel Deville, 1966).
On the Town (dir. Gene Kelly and Stanley Donen, 1949).
The Paradine Case (dir. Alfred Hitchcock, 1947).
The Pearl of Death (dir. Roy William Neil, 1944).
A Perfect Murder (dir. Andrew Davis, 1998).
Le plaisir (dir. Max Ophuls, 1952).
The Player (dir. Robert Altman, 1992).

Play It Again, Sam (dir. Woody Allen, 1972).
Portrait of Jennie (dir. William Dieterle, 1948).
Der Raub der Mona Lisa (dir. Geza von Bolvary, 1931).
Rebecca (dir. Alfred Hitchcock, 1940).
Red Dragon (dir. Brett Ratner, 2002).
The Return of the Pink Panther (dir. Blake Edwards, 1975).
Rocky (dir. John Avildsen, 1976).
Room with a View (dir. James Ivory, 1986).
Russian Ark (dir. Alexander Sokurov, 2002).
Scarlet Street (dir. Fritz Lang, 1945).
Shadows (dir. John Cassavetes, 1959).
Six Degrees of Separation (dir. Fred Schepisi, 1993).
Stealing Rembrandt (dir. Jannik Johansen, 2003).
Suspicion (dir. Alfred Hitchcock, 1941).
The Talented Mr Ripley (dir. Anthony Minghella, 1999).
The Thomas Crown Affair (dir. John McTiernan, 1999).
Three Coins in the Fountain (dir. Jean Negulesco, 1954).
Topkapi (dir. Jules Dassin, 1964).
Torn Curtain (dir. Alfred Hitchcock 1966).
Vertigo (dir. Alfred Hitchcock, 1958).
Viaggio in Italia (dir. Roberto Rossellini, 1954).
La ville Louvre (dir. Nicolas Philibert, 1990).
Das Wachsfigurenkabinett (dir. Paul Leni, 1924).
The World is Not Enough (dir. Michael Apted, 1999).

Chapter 15

Hiding in Plain Sight:
Cinematic Undergrounds

David L. Pike

> Remember, Jesus was buried and resurrected in a cave ... Things happen in
> caves that don't happen anywhere else.
>
> Amateur spelunker in The Cavern

Subterranean Setting on Film

There are two long-delayed payoffs in Carol Reed's classic film noir, *The
Third Man* (1949): the bravura entrance of Orson Welles as the back-from-
the-dead, Mephisophelean charmer Harry Lime against the backdrop of
the ruins of the Prater Wheel in Vienna, and the even longer awaited descent
into the sewers of Vienna, scene of the final pursuit and extermination of
Lime. To be sure, the postwar Vienna in which the entire film takes place is
introduced and consistently filmed as a strange space, from Reed's trademark
tilted camera angles, to the crumbling ruins of the old imperial city, to the
sordid details of the black-marketeering plot. Graham Greene's screenplay
milks the setting for all of its disorientating possibilities, playing up the
cowboy naiveté of American abroad Holly Martins (Joseph Cotten) to create
an existential drama of a world radically destabilized by a cataclysmic world
war. Nevertheless, we are primed throughout the film to experience Lime and
the sewers as something even stranger, simultaneously the resolution and the
intensification of the film's thematics. In the first instance, it is the character
that makes the space strange. When Harry Lime boards the Prater Wheel with
his old friend Holly Martins, and they climb high above the streets of Vienna
to restage Jesus' Temptation on the Mount, we find a devil who transforms
whatever space he is in into an underworld, seducing us with the chilling yet
persuasive amorality of his logic. It is only once truly underground, however,
that Lime is stripped of the veneer of supernatural invincibility, revealed as a
cornered rat, dangerous but pitiful, trapped by the space he had long made his
own (Figure 15.1).

    Like the logic of myth from which popular cinema inherited much of its
subterranean tropes, the descent into the underworld is a privileged moment in
the narrative arc, a moment of danger and revelation, of death and return. But

Figure 15.1 **Deeper than the grave: Harry Lime's natural habitat turns against him. Frame grab from *The Third Man*. Dir. Carol Reed. Sc. Graham Greene. Perf. Orson Welles and Joseph Cotten. Rialto Pictures. 1949. DVD. Criterion Collection. 2007**

this mythic resonance is only part of the cinematic underworld, for one of the defining qualities of Reed and Greene's sewers, despite the satanic undertone in Lime's character, is their resolute materiality. They are an *other world* in the medieval sense of a space of adventure that follows different rules than the ordered world above, but they are also a physical world of brick and mortar that any member of the audience could visit, or could imagine visiting, should they so desire. This dialectic of familiarity and alienness, of the mythic and the material, began to dominate depictions of the underground in nineteenth-century spectacle; it was refined still further in the twentieth century through the verisimilar medium of the motion picture.

Subterranean settings have been so common in the cinema since its inception that we tend to take them for granted, assuming, first of all, that they all serve the same spatial function within the overall form of any film in which they appear; and, second, that the function they serve is unified and easily defined: either profoundly destabilizing or profoundly affirmative. Either, that is, we take the underground as the locus of subversion, opposition, and alternative culture, as, for example, in Luc Besson's anti-modernity fable *Subway* (1995), or we take it as an excuse for mindless spectacle, as, for example, the sewer lair of the eponymous reptilian quartet in *Teenage Mutant Ninja Turtles* (1990; three further feature films followed). I want to argue that neither of these assumptions is correct. Certainly, there is a general spatial identity to the underground, but within that general identity, we can distinguish myriad variations and a handful of key typological distinctions. And, certainly, underground settings possess a high potential for representing qualities subversive of or contrary to the 'normal' world above – heterotopias, to use Foucault's (1967/1998) influential formulation. But they also possess a high potential for mindless spectacle, playing their commercial role within the culture industry.

These disparate potentials are neither mutually exclusive nor reconcilable. Because, as is characteristic of what Lefebvre (1974/1991) terms lived or representational spaces, these settings are fragmentary rather than unified in their visualization (ibid., 41–2), the subversive and the spectacular tend to appear together and intertwined with one another. Moreover, we do not yet possess a theoretical framework for distinguishing between different uses of the underground in spatial representations. There is no easy way to disentangle the types of spaces or their primary meanings within the dominant conceptual space of mediated culture, the commercial cinema that gives them form and distributes them around the world. The first step in this process is to articulate the main types of subterranean setting, the primary narrative and thematic associations of each setting, and the different ways in which those associations have been mobilized in the commercial cinema. We need to ask how and why underground settings function differently than other cinematic settings; how and why different underground settings function differently themselves from one another; and how and why identical underground settings can possess different meanings from one film to the next. How can we determine when it is significant that a setting is subterranean and when it is mere novelty or accident, when, in other words, is it a *strange* space, and when is it simply a spectacular one?

We can begin by asking the following simple questions when analyzing a subterranean setting in a film:

1. Which subterranean space, or combination of spaces is being represented?
2. What degree of spatial and temporal specificity does the representation demonstrate?
3. How is the representation utilizing this space: generically; with contextual nuance; or as its thematic centre and in all of its contradictions?

In the following sections of this chapter, I will outline different ways in which answers to these questions help to focus an analysis of subterranean space in film. First, however, I need to detail my understanding of the third question, and, in particular, what I mean by the contradictions of underground space.

The Contradictory Associations of Underground Space

The degree to which a subterranean space estranges its viewers while entertaining them will depend upon the degree to which it activates the contradictory associations of the underground. These associations arise primarily from the coexistence within modern subterranean space of qualities attributed to modern infrastructure – controlled, ordered, quotidian, and banal – and qualities attributed to the archaic underground and the mythic underworld – shelter, riches, hidden knowledge, atavism, danger, death (Pike 2005, 2007b). We certainly experience these qualities simultaneously when we enter, say, the London Underground, where the colourful and orderly patterns of the Tube map and the authoritative numbers on the digital display of arriving trains reassure us of a perfectly functioning system, while the overcrowding, inevitable delays, and lack of ventilation remind us more of Dante – whence, for example, 'hell line', the nickname of the Northern Line. Nevertheless, we are not accustomed to articulating these different qualities simultaneously; indeed, the ways we categorize space militate against our doing so. Instead, we sense the contradictions of underground space as waves of reassurance (faith in the system) and uneasiness (mistrust of the system). When accidents do happen – the Kings Cross fire, the Underground bombings of 2006 – we experience them as ruptures in the fabric of normalcy rather than as formative of and integral to the identity of the space, the way we are using it, and the meanings we attribute to it.

The spectacle scene is predicated on the fact that what we are seeing does not happen *normally*; at the same time, the power and allure of such scenes lies in an underlying suspicion that what we are seeing *could* in fact happen in the normal course of events. The common assumption about underground spaces is that they work to bring this suspicion to conscious knowledge – in a traditional underground space (e.g., *Third Man*'s Vienna), we *expect* bad things to happen; in a traditional aboveground space we do not. The spaces of the modern underground play out this tension between the destabilizing associations of the mythic underground and the stabilizing work of the technological underground. But how far can we extend this insight into the realm of commercial cinema? Is it really significant that the climax of the blockbuster thriller, *Speed* (1994), involves a spectacular underground railway crash rather than, say, a bridge collapsing? After all, *Speed 2: Cruise Control* (1997) finishes in nearly identical fashion to the original film except that this time a luxury cruiser is jumping the dock rather than a subway train jumping the tracks. Any theorization of subterranean space must be able to distinguish between the incidental settings we find in *Speed* and innumerable other action movies; equally conventional settings where a fuller range of the meanings of subterranean space is mobilized, such as *The Taking of Pelham One Two Three* (1974) or *Alligator* (1980) or *The Matrix* (1999); and a much smaller number of less conventional films such as *Moëbius* (1996) or *Underground*

(1995) or *The Third Man* or *Kanal* (1957), where the contradictory identity of the subterranean setting is inseparable from the central meaning of the film as a whole.

In *Speed*, the spectacular spaces are important generically and abstractly, but not specifically. The subway is merely the last in a series of spectacular settings which begins in a skyscraper elevator shaft and spends about half the duration of the film aboard a speeding city bus. Each space provides a variation on the theme of a tightly confined but rapidly moving space from which the hero must devise a way to extricate himself before it crashes. Crashing through a construction barrier, and emerging, tobogganing upwards, into the open air of the streets of Los Angeles is little different from the morphologically identical but scaled-down sliding out from under the bus tightly embraced atop a flat plate of steel that concludes the previous set piece of action (Figure 15.2). Nor does the Los Angeles subway have the iconic recognition factor of the London Tube, the Paris metro, or the New York subway. And, indeed, the lion's share of subterranean settings, especially in action and suspense movies, are generic in this sense. Moreover, even when they do employ the monumental iconography of a specific city – the New York subway in *Die Hard with a Vengeance* (1995) or *Crocodile Dundee* (1986) or *Money Train* (1995) or *Men in Black 2* (2002) – they do so in a purely touristic or culinary fashion, citing an urban brand-name just as they would a make of car (say, the fleet of Mini Coopers in the storm drain tunnels of LA in the 2003 remake of *The Italian Job*) or a designer's clothes – as product placement and for simple variation rather than to make any specific formal or thematic argument.

Figure 15.2   *Speed*'s train jumps the tracks out of the underground and into the open air of Hollywood. Frame grab from *Speed*. Dir. Jan de Bont. Perf. Keanu Reeves and Sandra Bullock. 1996. DVD. 20th Century Fox. 2004

Nevertheless, the spectacular spaces in *Speed*, including the climactic scene's subterranean setting, do serve a function in the film's overall meaning. The film is quite self-conscious about the contrast it sets up between the quotidian, proletarian quality of its protagonists and the glamorous and unreal image of Los Angeles. The spectacular spaces are all working spaces – elevator, bus, subway – and the characters are patently *not* automobile owners in the automobile metropolis par excellence. Only working-class people take the bus in LA; the presence of a clueless tourist among the bus regulars is the exception that establishes this rule. Officer Jack Traven (Reeves) is an equally no-nonsense regular guy (compare him, for example, with Reeves's star turn as FBI rookie Johnny Utah in the earlier 1991 surfer-heist film *Point Break*). This is a cultural as well as a textual issue. With *Speed*, Bullock established her star persona as the peppy girl next door; Reeves honed his ability to straddle the line between mainstream idol and underground icon: both personae rely on the spatial associations of the action set pieces. The elevator sequence establishes his fearless heroism in the context of terrified business people; the central bus scene matches him with Bullock as they must negotiate postmodern spaces to which their old-fashioned mode of transport is unsuited – the *speed* of the title thus signals their dangerous willingness to push themselves beyond the set bounds of their situation. Paradoxically, the bus is able to maintain its velocity only in the spaces of present-day and future LA: a superhighway under construction, an airport. The subway sequence, finally, takes them into an LA space hidden in plain sight; it is public knowledge that LA does in fact possess a subway, but its existence is generally taken as some kind of spatial joke. Handcuffed to a metal pole in the non-existent space of everyday life in LA, Reeves and Bullock have no option but to jump the tracks, speeding the train out into the blazing sun and into the plain sight of 'real' LA: the picture-snapping tourists and gawkers who would define Bullock's and Reeve's careers for the next decade.[1] The subterranean setting is thus consistent with the film's spatial choices, but its potential strangeness is used against itself as something both the characters and the actors need to escape in order to succeed.

While not a significantly better film than *Speed*, *The Taking of Pelham One Two Three* is of more interest historically and from the perspective of the underground because the subterranean setting is integral to every aspect of the film. Nearly every minute of the film takes place either in the subway tunnels or in the underground MTA control room, and every detail of the heist's meticulous planning revolves around the habits and idiosyncrasies of the subway system and the criminals' intimacy with the protocol of the

---

1 It should be noted that Reeves seems to have possessed a more conflicted relationship with his *Speed* persona than Bullock (or to have had more options for negotiating it). Originally slated to co-star in *Speed 2*, he dropped out; nevertheless, he appears eventually to have capitulated to the pressure by appearing with Bullock a decade later in the chick-flick fantasy-romance-drama *The Lake House* (2006).

employees. Their goal is not, as in *Speed*, to escape at all costs from a situation that is important for the salvation of the world and for the demonstration of the hero's prowess but not for the specific qualities it may exhibit. Rather, their goal is to work within the rules of the space; in the event, they escape without a substantial crash or a major conflagration, although three members of the gang are killed in a shoot out for the simple reason that the system manager demonstrates a greater and more flexible knowledge of the space than all but the single surviving member, who is an ex-subway-motorman. A side effect of this duel of wits is the need to enumerate the rules of the space in which it takes place through a plethora of specific detail about the subway. Although not fully accurate in its depiction of the subway ecosystem, *Pelham* (unlike, for example, the recent television remake, which was shot in Toronto; the 2009 big-screen remake was released too late for evaluation here) does possess a docudrama realism perfectly consonant with its subway setting. This visual realism distinguishes *Pelham* in another way from the sensational quotation of similar spaces discussed above: the misanthropic humour of the dialogue ('what do they expect for their lousy 35 cents, to live forever?') and the fundamental absurdity of hijacking a subway train accurately reflect the desperate situation of a New York City that in the early 1970s was careening toward bankruptcy, with a crime-ridden and graffiti-filled subway system the graphic metonymy of the impending disaster. Now, this is not to say that *Pelham* is a meditation on the state of the city; like *Speed*, its primary function is to provide the visceral and momentary pleasure of a good genre picture. The context of the New York subway and the estranging qualities of the subway's symbolic meaning are mobilized as part of that pleasure, however, and, unlike in *Speed*, this means a pleasure grounded in the specific rather than the general. And, after the fact, *Pelham* did, like other such films, end up unwittingly engaging in arguments about underground New York. Its persuasive depiction of the subway as a microcosm of the city has had an enduring afterlife in the popular imagination. In many ways, cinematic New York still often resembles *Pelham*'s sardonic and gritty vision; the same certainly cannot be said for LA and the subway in *Speed* or the sewers in *The Italian Job*.

We do also find underground metaphorics as epitome of the film as a whole. In the Argentinian thriller *Moëbius*, a Buenos Aires subway train goes missing; it turns out that it has been trapped in an alternate space in the system, a vast moebius strip on which it circulates endlessly without ever reemerging. Rather than subordinating the contradictory associations of subterranean space to the demands of a popular genre film, *Moëbius* subordinates genre to the contradictory associations of subterranean space in order to express the contradictory qualities of historical trauma and collective memory. Like *Pelham*, *Moëbius* uses its setting for iconographic associations, and, like the train crash in *Speed*, the scenes of the 'disappeared' train passing the station in pixellated stop-action animation provide the intense visual pleasure of sensational motion. The *raison d'être* of this spectacle, however, is the figurative association introduced by that visual pleasure. As in *Pelham*, the quotidian banality of riding the subway grounds the pathos of the events and allows easy and immediate identification with the hapless victims. And, as in *Pelham*, the specific situation takes on historical significance as a metonymy for the city. But *Moëbius* exceeds the terms of this metonymy, weaving a visually haunting allegory of Argentina's dirty war and the way it has left its traces in the fabric of everyday life, more felt than seen, travelling the same rails as the present, overlapping but never intersecting, and with no simple way either to escape or to reintegrate into 'normal' life. The film's very title describes the estranging quality it attributes to its space, restoring a truly subterranean strangeness to the quotidian identity of the space while simultaneously insisting on the historical, everyday quality of that strangeness: the trauma of the dirty war is in no way alien to that space; rather, the film wants us to perceive *as alien* the erasure in the ordinary representation of the everyday, restoring the inherent uncanniness to a mechanized space generally depicted as rational and inorganic.

Toward a Typology of Cinematic Undergrounds

Thus far, the discussion has been restricted to the space of the subway, but this is only one – although certainly the most common – of the subterranean settings we find in the cinema; sewers, tunnels, cellars, underworlds, and the vertical city are the other broad categories I will discuss here. Each of these spaces articulates the contradictions of underground space in different ways, and so it is not surprising that each space tends to appear in different generic contexts and to be employed to different thematic ends. The subway, as is evident from the examples above, embodies the quotidian qualities of the underground; it is associated with work, daily routine, the average citizen, and the power of technology to rationalize and to give order to urban space. Its otherworld qualities address the problems faced by the quotidian when beset by powerful and unforeseen forces. In contrast, the sewer carries in its waters the atavistic and organic elements of the city – its waste, its filth, its repressions, its secrets. The subway is an extension of everyday life in its banality, its excitements, or its horrors; the sewer is a transformative experience, a descent to the underworld from which one returns, if at all, a changed individual. Victor Hugo's *Les Misérables* (1862) is the ur-text of the sewer; the same spatial qualities emerge in existential historical dramas such as *The Third Man* and *Kanal*. In popular cinema, the special province of the sewer is the monster movie. *Alligator* is seminal here, but the motif dates back to cinematic adaptations of *The Phantom of the Opera* and expands in scale to include the ocean depths from which the giant Godzilla and his mutant progeny were spawned by atomic experimentation in the 1950s (Pike 2007a).

The tunnel is a conduit, linking two separate spaces, and it mobilizes all of the dreams and anxieties endemic to such a mingling of differences. Like the arch with which it shares a morphology and a thematic function, a tunnel takes one to a different place; it also allows the qualities of that different place to follow one back home. Building a tunnel is the endeavour to return from a world in which one has been trapped, as in the genres of trench warfare, the prison escape, and the mining disaster; or the conquest of an unknown world, as in Maurice Elvey's 1935 drama of heroic engineering and conquered nature, *Transatlantic Tunnel*. Following a tunnel is an exploration of the unknown world, as in the various adaptations of Jules Verne's (1864/1977) novel, *Journey to the Centre of the Earth*, or the *Alien* series and many other explorations of the bowels of distant planets. Tunnels predominate in action and adventure movies, whether set in the past, the present, or the future; they also surface frequently as topographical markers in all manner of narrative, signalling a moment of transition or of vulnerability in a character's life and as the setting for a cataclysmic event, as, for example, the infamous rape scene in a dark pedestrian tunnel in Gaspar Noé's *Irréversible* (2002), or the steam tunnels beneath the mental institution where the climactic confrontation between Susanna (Wynona Ryder) and Lisa (Angelina Jolie) takes place in *Girl, Interrupted* (1999).

As opposed to the spaces I have just outlined, which are primarily public spaces, the cellar is an essentially private and individual space. Cellars and basements frequently allegorize the mind and the unconscious; they powerfully exhibit the tension between the underground as a shelter and as a trap. Their special province is the horror movie, since they are a space where we hide and into which we can be pursued and trapped, the site of our deepest fears and our most profound security. Few films spend all their time in the cellar, although David Fincher's *Panic Room* (2002) is constructed wholly around the spatial premise I have outlined here; rather, they reserve it for climactic revelations, such as the discovery of the preserved mother hidden in the basement in *Psycho* (1960) and the pods in *Invasion of the Body Snatchers* (1956), and confrontations, as in the final battle between Mel Gibson, Joaquin Phoenix and the invading aliens in *Signs* (2002). Certainly, the entire house can take on the spatial attributes of a cellar, as in *The Strangers* (2008), or either version of *Funny Games* (1997 and 2007). Even when the house is besieged as if a cellar or basement space, however, there will nearly always be a key scene, the underground of the underground, as it were, that reveals a key plot point or takes us beneath the surface, deeper into the characters, just as *The Third Man* takes us beneath the already putatively 'underworld' cityscape of Vienna.

Associated with the cellar as psychological spaces are the bomb shelter and the underground command centre, which expand the individual quality to the level of family and society, respectively, while retaining the same sense of isolation and secrecy. The familial sense of the space is evident in a film such as *Blast from the Past* (1999), where the fallout shelter functions as a comic metaphor for the excessively sheltered life of a naïve 35-year-old man who has been living underground since the 1950s. Since the 1970s, the private bunker has nearly always also come equipped with a social allegory of 1950s repression, usually quite self-consciously if not as outright camp, as in the song number 'Let's Do It For Our Country' in *Grease 2* (1982), set in a backyard fallout shelter as Louis tries to convince Sharon to sleep with him because nuclear war is imminent.

The social level is equally evident in scenes of secret and fortified government nerve centres and bunkers, inaccessibly omnipotent but also vulnerably paranoid – their security is always breached in the end. We find bunker paranoia and its exclusionary utopianism lampooned in Peter Sellers's caricature of Dr Strangelove in Stanley Kubrick's 1964 film of the same name. Faced with the *fait accompli* of mutual self-destruction, Strangelove proposes an emergency adaptation of deep mine shafts for a century's habitation by 'top government and military men' whose primary task would be the repopulation of the world, ten specially chosen brood women devoted to each red-blooded male. Reflecting on the difficult decision of who should be sheltered and who left behind, a nervous general is assured by Strangelove that a computer can do the job equitably, and will include all the top brass. Responding to the concern of the grief and guilt over the vision of the dead left behind, Strangelove asserts, 'When they go down into the mine everyone would still be alive. There would be no shocking memories, and the prevailing emotion will be one of nostalgia for those left behind, combined with a spirit of bold curiosity for the adventure ahead!' The satirical reversal of underground and aboveground proposed by Strangelove reveals the actual inseparability of spaces masked by the insistence on keeping them separate.

The underworld is the most metaphorical and the most ancient of the basic subterranean spaces, since its actual location underground is secondary to its status as a space that is *other* to the everyday world. The underworld accrues its subterranean status, of course, from the traditional location of the afterlife underground in many religions and mythologies including, most notably, the Christian hell. Like the sewer, the underworld implies a transformation, but the identity of that transformation is limited to an exploration of the boundary between life and death. Thematically, death can have a number of different functions; most frequently, however, it serves to compel a character to review and reconcile his or her life. It can also, as with Dante, incorporate a strong satirical charge, using the afterlife to critique the present day. Both qualities are present, for example, in Tim Burton's comedy *Beetle Juice* (1988), in which two novice ghosts unwisely recruit an antisocial demon from the afterlife to teach them how to haunt their suburban New England house effectively enough to frighten off the new owners. The comedy is gentler than *Strangelove* and the estrangement milder: following the gripping and rollicking central portion of the film where the mild-mannered homesteaders adjust to the uncanny fact that death has changed nothing but their ability to act on the space around them while the anarchic force of Beetlejuice runs roughshod over them, the film reconciles its polarities in a happy ending.

To this basic taxonomy, I want to introduce another form of space that is not so much underground as systemic, but which is perhaps the most specific to the medium of the motion picture. This is the vertical society, which combines attributes of the tunnel and the underworld, but which is primarily concerned with a global conceptualization of space rather than a specific setting, and which manifests itself in film most prominently through the motif of time travel. The origin of cinematic time travel is, of course, H.G. Wells's 1895 novel, *The Time Machine*, published, as has often been noted, in the same year as the first motion picture was exhibited. The time travel motif translates spatial change through a temporal grid, using a vertical stratification of society to explore issues of social inequality and historical change. In addition to the strict application of the time travel principle, as in Chris Marker's short film *La Jetée* (1962) or the two adaptations of Wells's novel (1960 and 2002), the time travel motif is often assumed as the premise of a science-fiction film, from *Aelita, Queen of Mars* (1924) and *Metropolis* (1926), through *Things to Come* (1936) and especially in the dystopic futuristic films inaugurated in the 1970s and codified by *Blade Runner* in 1982.[2] In these films, as in those employing other specific underground spaces, we find the vertical city sometimes as a sensational setting for action – as in so many recent dystopias, both animated and live action – sometimes as a thematic context for that action – as, for example, in Lang's *Metropolis* – and sometimes as the central vehicle of the film's thematics, as in *La Jetée*, where the richly paradoxical argument is expressed through the contradictory meanings of the underground (Pike 2005, 290-91).

We should regard the distinctions of this typology as heuristic rather than fixed divisions. In practice, there is plenty of overlap between different spaces; indeed, the overlap is often the most instructive aspect of the subterranean setting. The subway and sewer share attributes of the tunnel space which dictated their basic construction; when flooded, the subway and the tunnel assume the qualities of the sewer. When a hybrid form like the caving film focuses on goals, it uses tunnels (e.g., the innumerable variations on *Journey to the Centre of the Earth*); when it stresses catastrophe and loss, it uses claustrophobically closed-in spaces, as in *The Cavern* (2005) or *The Descent* (2005). Cast as an underground control centre, the cellar reveals collective, rather than individual neuroses, as in *Independence Day* (1996). Depicted in fantasy, a subway functions thematically as an otherworld, as in *Jacob's Ladder* (1990).

---

2 The substantial body of critical writing on time travel in film, especially *La Jetée*, has much to say regarding its meta-cinematic features but tends to ignore its relationship to space. See especially Coates (1987), Columpar (2006), and Del Rio (2001).

One of the functions of underground space in modern representation is to locate and visualize otherwise occulted individual and collective unconscious forces. The existence of such forces within any representation does not, however, guarantee a profound contribution to its meaning. But it seems likely that the underground settings of bare-bones women-in-danger films such as *P2* (2007) or *Creep* (2004) tell us more about contemporary cultural issues regarding gender and power than about the psyche of their protagonist-victims, just as the more subjective and surrealistic underground and underworld settings of films such as *Repulsion* (1965) or *What Dreams May Come* (1998) foreground the singularity of their narratives of individual breakdown. Focusing on the range and overlap of different spaces can help to determine the degree to which a space is amenable to either a psychoanalytic analysis such as Gaston Bachelard's (1969) *Poetics of Space* or a globally allegorical analysis on the lines of Fredric Jameson's (1981, 1986, 1987) theory of postmodern culture. Few films are limited to one extreme or the other; rather, the primary meaning will emerge from the varied interplay between individual – Bachelard's 'home' whose 'essence ... all really inhabited space bears' (5) – and collective – Jameson's assertion that 'the story of private individual destiny is always an allegory ... of the public ... culture and society' (1986, 69).[3]

Rather than fixed prescriptions of meaning, the heuristic qualities of this typology help to disentangle and clarify the interrelationship between the different qualities associated with underground space, as between the interplay of individual and collective. When overlapping, these spaces combine the different attributes rather than excluding one or the other. In Neil Gaiman and Dewi Humphrey's miniseries *Neverwhere* (1996), for example, the fantasy world beneath the streets of London retains the quotidian qualities of its sources in the subway map even as it asserts a transformative and otherworldly force to those qualities. Additionally, *Neverwhere* uses the London associations as a formative component in its otherworld; it is a spatially specific rather than a global otherworld, and this has important consequences to the meaning of the space.

---

3 My second ellipsis omits the crucial modifier 'third-world' in Jameson's argument; I do this not to deny important differences between first-world and third-world cultural production, but to stress an underlying commonality between them in the relationship of space, even as that commonality is inflected in quite different ways (as Jameson's own 1987 update of his essay's title as 'World Literature' indicates). On these issues and the controversial reception of Jameson's essay, see Szeman (2001, especially 807).

A Case Study: Subterranean Settings in Film Noir

In the space remaining, I want to unpack this taxonomy further by examining several examples in a single cinematic genre: the film noir. I find film noir particularly interesting in the context of underground space because its basic generic premise derives from the space of the underworld – the world of film noir is, famously and truistically, *dark* in every sense of the word – while individual films inflect that premise with the tropes associated with other aspects of the underground. My examples here will sketch the overall thematics of noir in Jules Dassin's *Night and the City* (1950); the subway in Fritz Lang's *Man Hunt* (1941), the sewer in Alfred Werker and Anthony Mann's *He Walked by Night* (1949), and the cellar in Alfred Hitchcock's *Notorious* (1946). Curiously, the tunnel as such does not figure in any noir or neo-noir film I have yet come across – I suspect this is because its primary qualities of heroic construction and escape are antithetical to the characteristic negativity of noir. When a tunnel does appear, as in the 2004 Québécois bank-heist-gone-wrong film *Le Dernier Tunnel*, it is identified more closely with the sewage system (or with the subway) than with the tunnel as such.

Although seldom analyzed in these terms, the conventions of film noir employ spatial topoi more, and more self-consciously, than do most cinematic genres. In addition to the dark city, mythically rendered in high-contrast night-for-night lighting and exploring the timeworn stations of the urban underworld – sex, crime, gambling, drugs, alcohol – film noir builds its meaning through specific iconic settings: the jazz-saturated nightclubs where the hero is lured to his doom by the femme fatale; the dark streets where he wanders and through which he flees; the seedy hotel rooms where he is run to ground; the diners, bus stations, and gas stations where his nocturnal life briefly intersects the lives of the 'normal' people who inhabit the daytime world. Rather than dwelling exclusively in these underworld settings, however, nearly every film noir uses them selectively to contrast with glimpses of the characters' lives aboveground. In *Night and the City*, for example, the dilemma of Harry Fabian (Richard Widmark) is sketched through the contrasting spaces of Nosseross's den of prostitution and vice the Silver Fox Club, the second-floor apartment of Fabian's girlfriend, and the ring where he dreams of making it big by controlling professional wrestling in all of London. Dassin uses these sites, and especially the London exteriors still riddled with the rubble of the Battle of Britain, to craft an allegory of the disappearance of an antiquated underworld of art and craftsmanship, represented by specialized artisan criminals out of a Dickens novel or the *Threepenny Opera*, and the rise of a modern underworld driven by greed and the bottom line. Never suited to the nine-to-five life of his girlfriend, Harry is also too ambitious and too sentimental for the new underworld. Cornered by the corporate underworld and rejected by the old artisan crooks, now interested only in the price on his head, Harry is finally run to ground along the Thames (Figure 15.3). He is last shown, back broken, his body dumped on the shore, with the crime boss Kristo (Herbert Lom) looking down from Hammersmith Bridge in the background and the sun rising, it seems, for the first time in the film. Dassin employs the vertical metaphor of his final shot to represent, not the victory of daytime over the night, but of the artless modern underworld over the old, an existential despair that even night has been conquered by commerce.

Figure 15.3   Harry Fabian on the run in the strange spaces of nocturnal London. Frame grab from *Night and the City*. Dir. Jules Dassin. Perf. Richard Widmark. 20th Century Fox. 1950. DVD. Criterion Collection. 2005

As in the majority of film noir, there are no strikingly subterranean spaces in *Night and the City*; its spatial vocabulary is limited to the metaphorical spaces of the underworld and the thresholds that allow a glimpse of the world above – daytime London. The meaning of these spaces is conventionally symbolic; consequently, Dassin can deploy them as visual icons in enunciating his own narrative through those symbols. In *Man Hunt*, Fritz Lang also uses the symbolic settings of film noir, but the literally underground scenes he inserts within those settings provide, paradoxically, a sense of the possibility of escape from the noir underworld unavailable in *Night and the City*. Although noirish in its portrait of a man trapped by his own foolishness, *Man Hunt* is also a conventional thriller in its clear sense of good and evil, involving the pursuit by German agents of Thorndike (Walter Pidgeon), an Englishman who has failed in his attempt to assassinate Hitler. Thorndike disposes of one of them on the tracks of a tunnel of the London Tube, but then is himself run to ground in a tiny cave in the English countryside, cornered by the master hunter, Quive-Smith (George Saunders). In the context of wartime, however, the cave turns out to be a shelter rather than a sign of animal atavism, just as the Tube tunnel had worked as Thorndike's ally against a better-armed enemy. When Lang returned to the subway setting in his late Hollywood feature, *While the City Sleeps* (1956), however, the subway tunnel no longer holds connotations of shelter and moral order; instead, it signifies the irredeemable deviance of a perverted serial killer who is chased to his death by a newspaper reporter through a New York subway tunnel. Rather than the conventional noir contrasts, however, Lang cynically plays out his dark plot in a daylight world in which the killer is impossible to identify. Like Peter Lorre's Beckert in Lang's *M* (1931) 25 years before, the 'lipstick killer' is quotidian in his banality and all the more deadly for his ability to move about in daylight; it is perfectly fitting he meets his death in the workaday underground of the subway.

He Walked by Night equally involves a killer, but Roy Morgan (Richard Basehart) is not an apparently everyday citizen in the world above; he is a nearly catatonic underground man who murders wordlessly and without remorse, devoting himself day and night to carrying out some unspecified, unmotivated, vaguely political 'plan'. Like Harry Lime in *The Third Man*, Morgan long eludes capture by using the sewers like a rat to move about the city unseen. Unlike the Mephistophelean Lime in the morally complex world of occupied postwar Vienna, and unlike the typical noir protagonist, however, there is nothing charismatic or ambivalent about Morgan's monstrousness. His use of the storm drain tunnels beneath Los Angeles identifies him as a creature rather than a man – the trademark shot shows him escaping the police by slithering sideways through a sidewalk drain opening – a creature so incomprehensible as only to be dealt with by hunting him down and shooting him dead. Rather than the twilit, seductive world of noir, and despite the promise of the film's title, Morgan moves primarily through the mundane cityscape of daylight LA. The strict division *He Walked by Night* makes between the underground and the sewers clearly marks Morgan as a cold-war monster, close kin of the giant irradiated ants that would make themselves equally at home in the same drains five years later in the horror classic *Them!* (1954).

Like *Man Hunt* and *He Walked by Night*, *Notorious* mixes elements of noir with the classically vertical dynamics of classical Hollywood cinema. The noir character is Alicia Huberman (Ingrid Bergman), persuaded by agent T.R. Devlin (Cary Grant) to marry an old flame (Claude Rains) who is spying for the Germans. But although she has a shady past, we are never really in doubt as to her allegiance; the plot plays out as a romantic thriller. Still,

Hitchcock films much of it as if it were noir, and creates a strong sense of Alicia trapped in a criminal netherworld that threatens to kill her. The fact that the evidence is hidden in the wine cellar rather than out in the open characterizes the conflict as a primarily individual love triangle rather than a social allegory, notwithstanding the anti-Nazi veneer. Alicia steals her husband's keys; she and Devlin sneak into the cellar during a party, find the evidence hidden in a wine bottle, and unsuccessfully erase the traces of their descent. Returning from this private underworld with the secret knowledge they need, the couple will have to brave the further private space of the upstairs bedroom where Alicia lies dying of poison, before escaping together to live happily ever after. In a final token of the film's libidinal core, the climactic moment finds the couple closing the car door on Rains's impotent lover, leaving him behind to suffer the consequences of his betrayal of his conspirators' plan through his own inability to resist Alicia's seductions.

In classic noir, as in *Moëbius*, the underworld setting is the indispensable vehicle of the film's theme; by contrast, the underground spaces in borderline examples of the genre such as *While the City Sleeps*, *He Walked by Night*, and *Notorious* denote their participation in the world we inhabit rather than their distance from it. Nearly every noir film has at least a moment where we glimpse the connection between its 'strange', mythic underworld and the 'ordinary' world we inhabit. Similarly, a spate of films at the recent turn of the century – *The Matrix*, *Dark City* (1998), and *The Truman Show* (1998), among others – made the vexed relationship between the 'real' world and the world of illusion the central conceit of their plot, arguing, like noir, that dirty and grimy matter was more enticing and more true than Hollywood fantasy. The truth, of course, is more complicated. Underground space is where we 'locate' or conceive what is 'strange' about our world: our lived spaces, our contradictions, our repressions, and our genuine dreams. But that very conception of strangeness – the division of spaces between above-, or licit, and underground, or illicit – belies the fact that in actual practice *every* space is composed simultaneously of perceived, conceived, and lived components, to borrow Lefebvre's (1974/1991, 38-41) useful terminology. Rather than presenting a pure expression of representational or oppositional practices or a pure force of estrangement – which is the assumption of many of these movies, and of many cultural theorists – underground space is a contradictory hodge-podge. In it, we can glimpse the archaic spatial practices that bestowed caves, mines, and tunnels with enduring associations of shelter, fear, and awe; the oppositional and alternative modes of living and thinking that we conventionally label as 'underground' or 'strange'; and the dominant spatial representation that since the late eighteenth century has made of the underground the conceptual receptacle for anything 'out of place' in modern society. To imagine the modern world vertically allows us to represent, to visualize, all of its contradictions and inequities in a powerfully immediate spatial model; however, it also locks us into a representational framework that accepts those contradictions and inequities as inevitable and eternal. Notwithstanding this important qualification, the underground remains an essential tool for exploring the range of strange spaces in the modern world and the variety of alternative spatial practices, for discovering different pleasures, and for renovating ideas discarded in the march of progress. For the past hundred years, the cinema has been perhaps the most influential medium for representations of subterranean space; consequently, any comprehensive analysis of these representations in the twentieth century must reckon with the cinematic underworld.

References

Bachelard, G. 1969. *The Poetics of Space*. Boston, MA: Beacon Press.

Coates, P. 1987. 'Chris Marker and the Cinema as Time Machine', *Science Fiction Studies*, 14, 307–15.

Columpar, C. 2006. 'Re-Membering the Time-Travel Film: from *La Jetée* to *Primer*', *Refractory*, 9 (4 July), <http://blogs.arts.unimelb.edu.au/refractory/2006/07/04/re-membering-the-time-travel-film-from-la-jetee-to-primer-corinn-columpar/>, accessed 30 May 2008.

Del Rio, E. 2001. 'The Remaking of *La Jetée's* Time-Travel Narrative: *Twelve Monkeys* and the Rhetoric of Absolute Visibility', *Science Fiction Studies*, 28, 383–98.

Foucault, M. 1967/1998. 'Of Other Spaces', in Mirzoeff. N. (ed.) *The Visual Culture Reader*. London: Routledge.

Hugo, V. 1862/1985. *Les Misérables*. Paris: Laffont.

Jameson, F. 1981. *The Political Unconscious: Narrative as a Socially Symbolic Act*. Ithaca, NJ: Cornell University Press.

Jameson, F. 1986. 'Third World Literature in the Era of Multinational Capitalism', *Social Text*, 15, 65–88.

Jameson, F. 1987. 'World Literature in an Age of Multinational Capitalism', in Koelb, C. and Lokke, V. (eds) *The Current in Criticism: Essays on the Present and Future in Literary Theory*. West Lafayette, IN: Purdue University Press.

Lefebvre, H. 1974/1991. *The Production of Space*. Oxford: Blackwell.

Pike, D.L. 2005. *Subterranean Cities: The World Beneath Paris and London, 1800–1945*. Ithaca, NY: Cornell.

Pike, D.L. 2007a. 'The Cinematic Sewer', in Campkin, B. and Cox, R. (eds) *Dirt: New Geographies of Cleanliness and Contamination*. London: I.B. Tauris.

Pike, D.L. 2007b. *Metropolis on the Styx: The Underworlds of Modern Culture, 1800–2001*. Ithaca, NY: Cornell.

Szeman, I. 2001. 'Who's Afraid of National Allegory? Jameson, Literary
    Criticism, Globalization', *The South Atlantic Quarterly*, 100: 3 (Summer),
    803–27.
Verne, J. 1864/1977. *Voyage au centre de la terre*. Paris: Garnier-Flammarion.
Wells, H.G. 1895/1993. *The Time Machine*. London: Everyman.

Films Cited

Aelita, Queen of Mars (dir. Yakov Protasanov, 1924).
Alien (dir. Ridley Scott, 1979).
Alligator (dir. Lewis Teague, 1980).
Beetle Juice (dir. Tim Burton, 1988).
Blade Runner (dir. Ridley Scott, 1982).
Blast from the Past (dir. Hugh Wilson, 1999).
The Cavern / WIthIN (dir. Olatunde Osunsanmi, 2005).
Creep (dir. Christopher Smith, 2004).
Crocodile Dundee (dir. Peter Faiman, 1986).
Dark City (dir. Alex Proyas, 1998).
Le Dernier Tunnel (dir. Erik Canuel, 2004).
The Descent (dir. Neil Marshall, 2005).
Die Hard with a Vengeance (dir. John McTiernan, 1995).
Dr. Strangelove; Or, How I Learned to Stop Worrying and Love the Bomb (dir.
    Stanley Kubrick, 1964).
Funny Games (dir. Michael Haneke, 1997).
Funny Games (dir. Michael Haneke, 2007).
Girl, Interrupted (dir. James Mangold, 1999).
Grease 2 (dir. Patricia Birch, 1982).
He Walked by Night (dir. Alfred Werker and (uncredited) Anthony Mann,
    1949).
Independence Day (dir. Roland Emmerich, 1996).
Invasion of the Body Snatchers (dir. Don Siegel, 1956).
Irréversible (dir. Gaspar Noé, 2002).
The Italian Job (dir. F. Gary Gray, 2003).
Jacob's Ladder (dir. Adrian Lyne, 1990).
La Jetée (dir. Chris Marker, 1962).
Journey to the Centre of the Earth (dir. Henry Levin, 1959).
Journey to the Centre of the Earth (dir. Eric Brevig, 2008).
Kanal (dir. Andrzej Wajda, 1957).
The Lake House (dir. Alejandro Agresti, 2006).
M (dir. Fritz Lang, 1931).
Man Hunt (dir. Fritz Lang, 1941).
The Matrix (dir. Andy and Larry Wachowski, 1999).
Men in Black II (dir. Barry Sonnenfeld 2002).
Metropolis (dir. Fritz Lang, 1926).

Moëbius (dir. Gustavo Mosquera, 1996).
Money Train (dir. Joseph Ruben, 1995).
Neverwhere (dir. Dewi Humphreys, 1996 – TV mini-series).
Night and the City (dir. Jules Dassin, 1950).
Notorious (dir. Alfred Hitchcock, 1946).
P2 (dir. Franck Khalfoun, 2007).
Panic Room (dir. David Fincher, 2002).
Point Break (dir. Kathryn Bigelow, 1991).
Psycho (dir. Alfred Hitchcock, 1960).
Repulsion (dir. Roman Polanski, 1965).
Signs (dir. M. Night Shyamalan, 2002).
Speed (dir. Jan de Bont, 1994).
Speed 2: Cruise Control (dir. Jan de Bont, 1997).
The Strangers (dir. Bryan Bertino, 2008).
Subway (dir. Luc Besson, 1995).
The Taking of Pelham One Two Three (dir. Joseph Sargent, 1974).
The Taking of Pelham One Two Three (dir. Félix Enríquez Alcalá, 1998).
The Taking of Pelham 123 (dir. Tony Scott, 2009).
Teenage Mutant Ninja Turtles (dir. Steve Barron, 1990).
Them! (dir. Gordon Douglas, 1954).
Things to Come (dir. William Cameron Menzies, 1936).
The Third Man (dir. Carol Reed, 1949).
The Time Machine (dir. George Pal, 1960).
The Time Machine (dir. Simon Wells, 2002).
Transatlantic Tunnel (dir. Maurice Elvey, 1935).
The Truman Show (dir. Peter Weir, 1998).
Underground (dir. Emir Kusturica, 1995).
What Dreams May Come (dir. Vincent Ward, 1998).
While the City Sleeps (dir. Fritz Lang, 1956).

Name Index

Subject Index